Stevenson & Reid Van Schaack

Annual Price Current (to the Trade.) Drugs,

Chemicals, Medicines, Dyes, Paints, Oils, Glass, Glassware, Toilet Articles, Chemical Products, etc.

Stevenson & Reid Van Schaack

Annual Price Current (to the Trade.) Drugs,
Chemicals, Medicines, Dyes, Paints, Oils, Glass, Glassware, Toilet Articles, Chemical Products, etc.

ISBN/EAN: 9783744762663

Printed in Europe, USA, Canada, Australia, Japan

Cover: Foto ©berggeist007 / pixelio.de

DRUG TRADE OF THE NORTHWEST.

We Dedicate

THIS, our FIFTH *Extended Trade List, which we have prepared at considerable labor and expense, with a desire to supply a* RELIABLE AND COMPLETE

Illustrated Catalogue,

Confident from the kind appreciation our past Volumes have received, that the present will be a welcome visitant to numerous friends in the Drug Trade.

VAN SCHAACK, STEVENSON & REID.

C. A. Sayre.

Rand, McNally & Co., Printers and Binders, Chicago.

1875.

Annual Prices Current,

(TO THE TRADE.)

Drugs, Chemicals

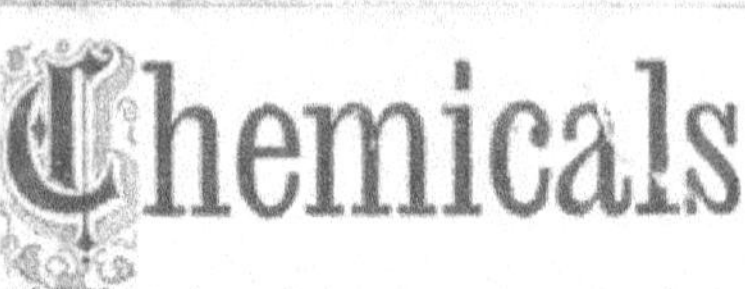

MEDICINES,

Dyes, Paints, Oils, Glass, Glassware, Proprietary Medicines, Perfumery, Toilet Articles, Chemical Products, etc.

Van Schaack, Stevenson & Reid,

WHOLESALE DRUGGISTS,

Paint and Glass Dealers,

92 and 94 Lake Street, Corner Dearborn

Opp. Tremont House, and Commercial Hotel, Chicago, *New York Office, 58 Cedar Street.*

TOILET
SOAP
TOILET
POWDER
TOOTH
BRUSHES
FRENCH
AND
ENGLISH
PERFUMERY
PHARMACY

To the Drug Trade:

The kind appreciation which has attended our previous endeavors to furnish the Trade with a COMPLETE and *reliable illustrated* list, has caused us to renew our efforts to maintain the *supremacy* in this our *fifth yearly issue.* While it is fully understood that prices are subject to market changes, still the list will be found exceedingly convenient in replenishing stocks, or for those embarking in the trade.

We offer a Drug Stock *unequaled* in the Northwest, in extent and variety, embracing almost every item connected with the *Drug, Chemical, Paint, Oil, Glass, Dye Stuff, Perfumery and Patent Medicine* trade.

We beg leave to assure our friends, that we are thankful for their favors of the *past*, and that no effort shall be wanting to deserve their confidence in the *future.*

When large quantities of goods are ordered, and there is an opportunity to do better than *quoted prices*, our customers may confidently rely upon having the advantage of it.

We aim to handle goods only of the *best obtainable* quality, and *most reliable manufacture*, neatly put up by experienced employees. We sell on *prompt* time, and the prices attached to orders will be *as low* as reliable goods can possibly be handled.

☞ Parties ordering for the first time will see the propriety of giving City (preferred) or other references as to their responsibility. *Our terms are invariably Cash* within 30 days. As we sell goods on close margins, we shall expect prompt pay.

We shall not follow the custom of sending out traveling agents on almost every road diverging from the city, being satisfied from experience that goods can be sold at a *lower margin* of profit by customers sending us their orders *direct* by mail, and we will rely on *low prices, prompt shipments*, and *superior quality* of the goods we handle. Experience fully justifies us in our judgment, that this principle is the *correct one*, and the Trade may rest assured of getting the benefit of any decline in value of goods, and the prices will be as low as if the purchaser were in market.

Very respectfully,

VAN SCHAACK, STEVENSON & REID.

MARCH 1, 1875.

Office and Salesroom of **VAN SCHAACK, STEVENSON & REID,** *92 and 94 Lake St., cor. Dearborn, Chicago.*

INDEX TO CONTENTS

— OF —

VAN SCHAACK, STEVENSON & REID'S

PRICES CURRENT, ETC.

FAMILY MEDICINE

CHEMICALS, DRUGS, ETC.

☞ Observe, the price of oz. vials is *included* in Chemical quotations.

ACID, Acetic, No. 8, carboys	per lb.	$ 14
" " " 8, in bottles	"	16
" " Glacial	"	1 15
" " "	per oz.	18
" Arsenious (Lump)	per lb.	18
" Benzoic, English	per oz.	35
" " German	"	25
" Boracic	per lb.	1 00
" Butyric	"	3 75
" Carbolic, Crystals, Calverts'	"	1 50
" " " "	per oz.	30
" " " Nichols'	per lb.	1 35
" " " "	per oz.	20
" " Solution, Calverts'	per lb.	80
" " " Nichols'	"	45
" " " Common Disinfectant	"	25
" Chromic, in oz. vials, P. & W.	per oz.	45
" " Nichols', in oz. vials	"	45
" Citric	per lb.	1 40
" " in 112 lb. kegs	"	1 35
" Fluoric, in 1 lb. bottles	"	2 00
" " in 1-2 lb. "	"	2 75
" " in 1-4 lb. "	"	3 50
" " in 1 oz. vials	per oz.	50
" Gallic, in oz. vials	"	22
" Hydrocyanic, in oz. vials, U. S. P.	"	18
" Hypophosphorous, in oz. vials	"	35
" Lactic, Conc., in 1 oz. vials	"	50
" " Dil., in 1 oz. vials	"	25
" Malic	"	3 25
" Muriatic, in bottles	per lb.	6
" " by carboy	"	4 to 5
" " C. P., in bottles	"	25
" Nitric, in bottles	"	15 to 18
" " by carboy	"	13 to 15
" " C. P., in bottles	"	30
" Oxalic	"	20
" Phœnic, in oz. vials	per oz.	20
" Phosphoric Glacial	per lb.	1 60
" " " oz. vials	per oz.	25
" " Dil.	per lb.	35

☞ For Addenda to Chemicals, etc., see page 48.

Acid, Picric	per lb.	$1 00
" Pyrogallic, in oz. vials	per oz.	60
" Prussic, (see Acid, Hydrocyanic.)		
" Pyroligneous	per lb.	10
" Succinic, in oz. vials	per oz.	60
" Sulphuric, in bottles	per lb.	5
" " by carboy	"	2¾
" " C. P.	"	35
" Sulphurous	"	25
" Sulphuric, (fuming, Nordhausen)	"	40
" " Aromatic	"	45
" Tartaric, crystals	"	57
" " powdered, P. & W.	"	58
" Tannic	"	1 75
" " in oz. vials	per oz.	22
" Valerianic, in oz. vials	per oz.	60
Aconitia, in 1-8 oz. vials	"	26 00
Acorn Coffee	per lb.	18
Agaric, (white)	"	85
Alcohol	per gall.	Market Rates.
" by bbl.	"	
" Cologne spirit	"	
" absolute "	"	4 00
Allspice	per lb.	18
" ground	"	25
Almonds, bitter	"	50
" sweet	"	40
Alum	"	5
" by bbl.	"	4
" ground	"	8
" " by bbl.	"	5
" burnt, or *Exsic.*	"	40
" powdered	"	12
" " by bbl.	"	9½
" Ammonio-Ferric	"	1 10
" " " in oz. vials	per oz.	16
" Roche	per lb.	20
Ambergris	per oz.	
Ammonia, Arseniate, in oz. vials	"	50
" Aqua, 3f.	per lb.	10
" " 3f. by carboy	"	6
" " 4f.	"	12
" " 4f. by carboy	"	9
" Liq. Concent.	"	22
" Acetate, *Spts. Minderus*	"	35
" Carbonate, in jars	"	25
" Citrate, in oz. vials	per oz.	20
" Hypophosphite, in oz. vials	"	35
" Muriate (Sal Ammoniac)	per lb.	16
" " " " Pow'd	"	25
" Nitrate, Cryst.	"	40

Ammonia, Nitrate, Fused	per lb	$ 40
" " Granular	"	40
" Oxalate	"	1 75
" Phosphate	"	1 85
" Spirits (Alcoholic)	"	40
" " (Aromatic)	"	45
" Sulphate	"	12
" Valerianate, in oz. vials	per oz.	65
Ammonium, Bromide	per lb.	1 10
" " in oz. vials	per oz.	15
" Hydrosulphuret	per lb.	75
" Iodide, in oz. vials	per oz.	70
Aniline, Black	per lb.	75
" Blue (water)	"	7 50
" Brown	"	2 75
" Green (silk)	per oz.	1 50
" " (wool)	"	75
" Orange	per lb.	2 00
" Purple	"	10 50
" Scarlet	"	2 50
" Red, Cryst	"	$2 to 2 25
" Violet	"	10 50
" Yellow	"	7 50
Annatto, English, in Rolls	"	45
" Para, Prime	"	60
Anodyne, Hoffman's	"	45
Antimony (Butter of)	"	35
" Black Sulphuret, Powd.	"	6
" (James' Powder)	"	1 00
" (Kermes Mineral)	"	2 15
" Oxide, white	"	1 10
" Sulphuret Precip.	"	70
" Potass. Tartrate, Cryst.	"	1 00
" " " Powd.	"	85
Apples, Orange	"	15
Argols, Red	"	18
Arrow Root, American	"	12
" Bermuda, Bulk	"	40
" " Taylor's, in 1-2 lb. and 1-4 lb.	"	45
" St. Vincent	"	25
Arsenic, White, Powd. Com.	"	8
" " Cryst. (See Acid, Arsenious.)		
" Yellow Sulphuret, (Orpiment)	"	35
" Donovan's Solution	"	35
" Fowler's "	"	18
" Iodide, in oz. vials	per oz	75
Ashes, Pearl	per lb.	15
" " by cask	"	13
" Pot, Crude	"	10
" " " by cask	"	7½
Ashphaltum	"	8

Atropia, Fleming's Solution, in 4 oz. bottles	per bot.	$ 50
" in 1 dr. vials	per dr.	3 50
" Sulphate, in 1 dr. vials	"	3 50
Axle Grease, Bidwell's, in wood	per doz.	
" " Frazer's, "	"	1 25
" " " quart, in tin	"	2 00
" " " 1-2 gall.	"	3 00
" " " 1 "	"	4 75
" " Van S., S. & R., in wood	"	
BALSAM Copaiva	per lb.	95
" " Solidified	"	1 35
" Fir	"	60
" Peru	"	3 00
" Sulphur	"	45
" Tolu	"	1 35
Bark, Angustura	"	45
" Bayberry	"	12
" " Powd.	"	16
" Canella Alba	"	16
" " " Powd.	"	25
" Cascarilla	"	18
" " Powd.	"	25
" Chiretta	"	75
" Cinchona Calisaya, Quill P. & W.	"	1 30
" " " " " Powd.	"	1 45
" " Red	"	50
" " " Powd.	"	75
" " " Quill P. & W.	"	1 25
" " " " " Powd.	"	1 35
" " Yellow	"	25
" " " Powd.	"	30
" Cinnamon, (See Cassia).		
" Elm Slab	"	18
" " Ground, in bulk	"	18
" " Powd., "	"	20
" " " in 2 oz. pkgs.	"	35
" Hemlock, (see Herbs).		
" " Powd.	"	20
" Lemon Peel	"	28
" Mezereon	"	30
" Oak, White, (see Herbs).		
" " " Powd.	"	20
" Orange Peel, Sweet	"	25
" " " " Ground	"	27
" " " " Powd.	"	35
" " " Bitter	"	28
" Pomegranate, (Bk. of Fruit)	"	25
" " (Bk. of Root)	"	60
" Poplar, (see Herbs).		

Bark, Poplar, Powd.	per lb.	$ 20
" Prickly Ash, (see Herbs).		
" " " Powd.	"	35
Bark, Quercitron	"	7
" Sassafras	"	18
" " Powd	"	30
" Soap Tree, (Quillaya)	"	30
" Wild Cherry, in bulk	"	12
" " " Ground	"	13
" " " Powd	"	25
☞ For other Barks see Shakers' list of Barks, etc., page 22.		
Barley, Pearl	per lb.	9
" " By keg, 100 lbs.	"	8
" Robinson's Patent	per doz.	1 50
Baryta, Carbonate	per lb.	1 20
" Muriate, (Chloride of Barium)	"	40
" Nitrate	"	30
" Sulphate (Crude Barytes)	"	4
Bath Brick	per 100,	5 00
" In 2 doz. boxes	per box,	1 50
Bay Rum, True Imported	per gall.	3 75
" American, Distilled	"	3 00
" For Barbers, Domestic	"	2 50
Beans, Calabar	per oz.	10
" St. Ignatius	per lb.	1 00
" Tonqua	"	1 15
" Vanilla, Long	"	32 00
" " Short	"	27 00
Bebeerine, Pure, 1 oz. vials	per oz.	3 50
" Sulphate, 1 oz. vials	"	2 50
Berries, Buckthorn	per lb.	40
" Cubebs	"	15
" " Powd	"	20
" Juniper, Italian	"	8
" " Powd	"	20
" Laurel	"	16
" Orange Apple	"	15
" Persian or French	"	75
" Poke	"	35
" Prickly Ash	"	35
" Sumac	"	18
Bismuth, Citrate, in 1 oz. vials	per oz.	55
" " and Ammonia, in 1 oz. vials	"	55
" Metallic	per lb.	2 75
" Oxychloride	"	2 75
" Sub. Carbonate	"	3 50
" " Nitrate	"	2 25
" Tannate	per oz.	75
" Valerianate	"	2 00
Black Drop	per lb.	3 50
Blue Mass, P. & W., in 1 lb. jars	"	1 00

Blue Mass, Powd	per lb. $	1 45
" English	"	1 45
Blue Vitriol	"	13
" By cask	"	11
" Powd	"	35
Blueing, Barlow's, large	per doz.	60
" " small	"	30
" Liquid	"	50
" Sawyer's, Pepper Box, large	"	60
" " " " small	"	35
" Van Deusen's, large	"	50
" " small	"	25
Bole, Armenia	per lb.	12
Borax	"	17
" Powd	"	25
Boxes, Lip Salve, Metal, Screw Cap	per doz.	38
" " Porcelain	"	1 50
" " Scotch Plaid	"	2 50
" Pill Chip, Nested	per paper	15
" " Oval, (Brandreth's)	per gross	30
" " Round, Long, (Jayne's)	"	65
" " Willow, Nested	per paper	25
" Pill Paper, American, Nested	"	15
" " English	"	35
" " Bronze, No. 6	per gross	85
" " " " 7	"	1 15
" " " " 8	"	1 45
" " " " 9	"	1 75
" " " " 29	"	1 50
" " " " 30	"	1 75
" " " " 31	"	2 00
" Powder, German, (Nested, 1 to 6)	per nest,	12
" " " (" 1 to 9)	"	30
" " Prescription, 1 white, (watered)	per gross	2 25
" " " 2 " "	"	2 50
" " " 3 " "	"	2 75
" " " 20 (Nested, 1, 2, 3,)	"	2 50
" " " 56 gilt	"	2 50
" " " 57 "	"	2 75
" " " 58 "	"	3 00
" " Seidlitz, Paper	"	2 25
" " " Tin, small	per 100	4 00
" " " " large	"	5 50
" Tin, Ointment, 1-4 oz	"	65
" " " 1-2 oz	"	75
" " " 1 oz	"	90
" " " 2 oz	"	1 25
" " " 3 oz	"	1 75
" " " 4 oz	"	2 25

TURNED-WOOD BOXES.—

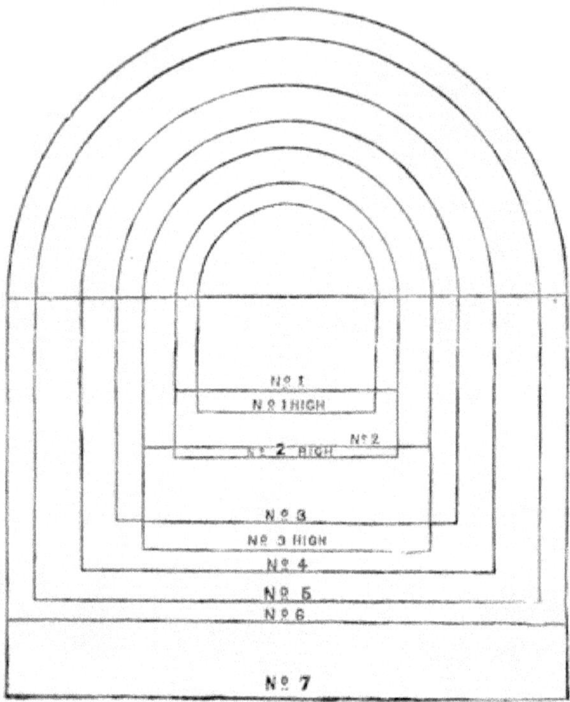

(The circular lines in the cut show the circumference, and the squares the height, outside measurement.)

No. 2, or 1-4 oz. (see p 151)	per gross	$	50
" 3, or 1-2 oz.	"		65
" 4, or 1 oz.	"		90
" 5, or 2 oz.	"	1	25
" 6, or 3 oz.	"	1	75
" 7, or 4 oz.	"	2	00
Boxes, Wood, Turned, (Nested, Nos. 1, 2 and 3)	"		60
" " " Tooth Powder, polished, 2 oz.	"	1	50
Brackets, (see page 153)			
Breast Tea	per lb.		50
Brimstone			5
Bromine, in 1 oz. vials	per oz.		30
" Chloride, in 1 oz. vials	"	1	25
Brucia, in 1-8 oz. vials	"	3	80
" Sulph., in 1-8 oz. vials	"	3	80
Burgundy Pitch, Com.	per lb.		10
" " True	"		12

Butter of Cocoa	per lb.	$	75
" Antimony	"		35
CADMIUM, Bromide, in 1 oz. vials	per oz.		45
" Iodide, " "	"		75
" Sulphate, " "	"		55
Caffeine, in 1-8 oz. vials	per drm.		65
Calomel, American	per lb.	2	10
" " in 1 lb. bottles	"	2	15
" Eng., Howard's, in 1 lb. bottles	"	2	85
Camphor	"		35
Candles, Paraffine, 6's and 4's	"		30
" Sperm, 6's and 4's	"		35
" Patent Wax, 6's and 4's	"		50
" " " colored, 6's and 4's	"		55
Cantharides	"	1	85
" Powd	"	2	00
Carbon, Bisulphuret	"		45
Carmine, No. 40,	per lb.	10	00
" " 40, in 1 oz. vials	per oz.		85
" " 12, " "	"		65
" " 6, " "	"		50
Cassia, in Mats	per lb.		33
" Buds	"		75
" Fistula, (Purging)	"		30
" Powdered	"		45
Castor	per oz.		25
Catechu, Black, True	per lb.		12
" Gambier	"		10
" " By bale	"		8½
Cayenne Pods	"		35
" Powdered	"		38
Cerate, Calamine, (Turner's)	"		60
" Cantharides	"	1	00
" Resin, (Basilicon)	"		35
" Simple	"		55
" Sub. Acet. Lead, (Goulard's)	"		75
" Savin	"		75
Cerium, Oxalate, 1 oz. vials	per oz.		45
Chalk, French, Lump	per lb.		18
" " " Powd	"		10
" " Tailor's, (Soapstone)	per box,		45
" Precip. Eng	per lb.		18
" Prepared Drops	"		7
" Red, Fingers	"		8
Chalk, White, Lump	"		3
" " " By bbl	"		2
" " Crayons, 1 gross boxes	per gross		18
Chamois Skins, (see Sundries, page 202)			
Charcoal, Animal	per lb.		8

Charcoal, Willow, Powd	per lb.	$	12
Cherry Laurel Water	"		50
China Clay	"		5
Chinoidine, in 1 oz. rolls	per oz.		14
Chloral Hydrate, in 1 oz. vials	"		25
" "	per lb.	2	25
Chlorine Water	"		50
Chloroform, Nichols'	"	1	25
" P. & W.'s	"	1	15
" Squibb's	"	2	25
Chocolate, Baker's No. 1	"		38
" " " Premium	"		42
" " Sweet, German	"		30
" " Vanilla, Double	"		60
Chromium Chloride, in 1 oz vials	per oz.		40
Cinchonia, in 1 oz. vials	"		45
" Sulphate, in 1 oz. vials	"		38
Cinnabar	per lb.	2	35
Civet	per oz.	7	50
Cloves	per lb.		55
" Ground	"		60
Cobalt	"		25
Coccul. Indicus	"		15
Cochineal, Hond	"		75
Cocoa, Baker's Cracked, bulk	"		44
" " " in 1 lb papers	"		46
" " No. 1	"		60
" " " Homœopathic	"		
" " Shells	"		18
" Butter	"		75
Codeine, in 1-8 oz. vials	per oz.	9	00
Collodion, in 1 oz. vials	per doz.	2	25
" in 1 lb. bottles	per lb.	1	15
" Cantharidal	per doz.	3	75
" Acetic Rubefacient	"	5	00
Colloyd, Nichols' Styptic	"	5	50
Colocynth Apple	per lb.		40
" " Powd	"		45
Composition Powder, in 1-8 lb. pkgs.	"		35
Condensed Milk	per doz.	3	00
Confection, Roses, in 1 lb. jars	per lb.		50
" Senna, in 1 lb. jars	"		50
Copper, Ammoniated	"	1	20
" Arseniate, in 1 oz. vials	per oz.		55
" Arsenite, in 1 oz. vials	"		45
" Cyanide, in oz. vials	"		40
" Iodide, in 1 oz. vials	"		80
" Nitrate	per lb.		80
" Oxide Blk. in 1 oz. vials	per oz.		20
" Sulphate, (see Blue Vitriol).			
" " C. P.	per lb.		40

Copperas	per lb.	$ 3
" by bbl.	"	1¾
Corks, } (see Van Schaack, Stevenson & Reid's price		
Corkwood, } list, page 82.)		
Cork Presses, Bronzed	each,	75
" " Iron	"	40
Corkscrews, (see Sundries, page 223)		
Corrosive Sublimate	per lb	2 00
Cowhage, in 1 oz. vials	per oz.	85
Crab Orchard Salts	per lb.	38
Cream Tartar, Crystals	"	44
" " Grocer's, 10 lb. cans	"	30
" " Powd., pure	"	47
" " " " by bbl.	"	44
Creasote	"	1 10
Crocus Martis	"	12
Crucibles, Hessian, Nested, 4 oz.	per nest	10
" " " 8 oz.	"	18
" " " 16 oz.	"	30
" " " Quart	"	50
" " " 1-2 gall.	"	90
" " " gall.	"	1 50
Cudbear, Opt.	"	25
Cuttle Fish Bone	"	25

DEXTRINE	per lb.	20
Dispensatories, U. S. P., Wood & Bache	each,	8 00
" Eclectic, King's	"	9 00
Distilled Water	per gall	25
Dover's Powder, P. & W.	per lb.	2 00
Dragon's Blood, in Reeds	"	1 00
" " in Mass	"	50
Drug Mills, Ames'	each,	9 00
" " Swift's, Gen.	"	12 00
" " " Imit.	"	10 50
" " Troemner's, large	"	30 00
" " " small	"	18 00
Dye Colors, large	per doz.	1 75
" " small	"	1 00

ELETERIUM	per drm.	1 00
Emetine, in 1-4 oz. vials	per oz.	3 75
Emery Flour	per lb.	7
" " by keg	"	6
" Grain, all numbers	"	10
" " " by keg	"	8½
Epsom Salts, P. & W.	"	4
" " by bbl., P. & W.	"	3
Ergot, Whole, Fresh	"	90

Item	Unit	Price
Ergot, Powd. Fresh	per lb.	$1 00
Ergotine, in 1-4 oz. jars	"	80
Essences, Flavoring, for Confectioners, Soda Water Manufacturers, etc See also Juices, page 29.		
Banana	per lb.	1 50
Peach	"	1 50
Pear	"	1 50
Pine Apple	"	1 50
Raspberry	"	1 50
Strawberry	"	1 50
Ether, Acetic	"	90
" Butyraceous	"	2 00
" Butyric	"	2 75
" Chloric	"	90
" " Concentrated	"	1 50
" Sulphuric	"	60
" " Concentrated	"	80
" " Fort Nichols'	"	90
" " Washed	"	70
Extracts, Eclectic and Active Principles—(see List, page 124)		
" Fluid—(see Tilden & Co.'s List, page 119)		
" Solid—(" " " " 123)		
Extracts, Flavoring, Burnett's (see p. 383)		
Almond	per doz.	2 50
Cinnamon	"	2 50
Cloves	"	2 50
Jam. Ginger	"	2 50
Lemon	"	2 50
Nectarine	"	2 50
Nutmeg	"	2 50
Orange	"	2 50
Peach	"	2 50
Rose	"	2 50
Vanilla	"	4 00
Extracts, Flavoring, Common—		
Lemon, 2 oz.	"	1 00
Vanilla, 2 oz.	"	1 50
All other Flavors from $1.00 to $1.50 per doz.		
Extract Logwood, in bulk, 12 and 24 lb. boxes	per lb.	10½
" " in 1's, 30 lbs. in box	"	13
" " in 1-2's, "	"	15
" " in 1-4's, "	"	17
" " assorted, "	"	16
Extracts, Solid, Allen's English—		
Extract Belladonna	"	3 25
" Cannabis Indica	per oz.	75
" Conium	per lb.	1 50
" Dandelion	"	1 25
" Gentian	"	1 20
" Hyoscyamus	"	3 25

Extracts, Solid, Miscellaneous—over

Extract Aloes	per oz.	$	30
" Bark Precip. P. & W.	per lb.	2	25
" " " " 1 oz. jars	per oz.		25
" Calabar Bean, 1-8 oz. jars	per 1-8 oz	1	00
" Colchicum Acet.	per oz.		35
" Dandelion, Shaker's	per lb.		50
" Ferri Pomatum	"	1	50
" Flesh, Nichols'	per oz.	1	25
" Ignatia Bean	"	1	25
" Jalap	per lb.	6	00
" " Powdered	"	6	50
" " Resin	per oz.	1	00
" Nux Vomica, P. & W.	"		45
" Opium, Aq., Ellis'	"	1	85
" Ox Galls	"		25
" Rhei, Ellis'	"		65
" Scullcap, Merrill's	"		45

FARINA, Hecker's, 1's	per lb.		15
" " 1-2's	"		16
Farine, Sea Moss	per doz.	2	25
Faucets—No. 0, Wood	"	1	00
" " 2, "	"	1	25
" " 4, "	"	1	50
" " 6 "	"	1	75
" " 8 "	"	2	00
" " 9, "	"	3	50
" " "	"		
Fish Sounds	per lb.	1	25
Flake White, Kremnitz	"		25
" Powdered	"		30
Flowers, Arnica	"		15
" Chamomile, German, new crop	"		55
" " Roman "	"		30
" Elder, in bulk	"		30
" Lavender	"		12
" Malva	"		50
" Roses, Red, Eng.	"	2	00
" Saffron, American	"		60
" " Spanish	"	10	50
Flowers, Tilia, or Linden	"		75
Foil, Tin—Thick	"		38
" Medium	"		45
" Thin	"		55
Fuller's Earth	"		5

GALLS, Aleppo	"		25
" " Ground	"		30
" " Powdered	"		35

Item	Unit	Price
Garlic	per lb.	$ 25
Gelatine, Cox's	per doz.	2 25
" French White	per lb.	1 00
" " Pink	"	1 50
(See also Isinglass, page 29.)		
Glauber Salts	"	3½
" " by bbl	"	2½
Glue, Carpenter's Ohio, "A"	"	25
" " " "B X X"	"	22
" " City, "C"	"	16
" White, Cooper's "A Ex."	"	50
" " " "No. 1 Ex."	"	45
" " " "No. 1 X"	"	35
Glycerine, (bottle included) Bower's	"	65
" " " P. & W., Concent.	"	40
" " " Van Schaack, S. & R.'s	"	45
" " " Vienna	"	40
" Bulk, Perfumers'	per gall	2 00
" " Van Schaack, S. & R.'s	per lb.	25
Grains, Paradise	"	25
Granville's Lotion	"	50
Guaiac Wood, Rasped	"	5
Gum Aloes, Barbadoes	"	45
" " " Powd	"	55
" " Cape	"	15
" " " Powd	"	22
" " Socotrine	"	70
" " " Powd	"	75
" Amber	"	60
" Ammoniac	"	75
" Arabic, 1st, Trieste	"	60
" " 2nd	"	50
" " 3d	"	35
" " sorts	"	25
" " Powd. 1st	"	65
" " " 2nd	"	60
" Assafœtida, Extra	"	30
" Benzoin	"	75
" " Powd	"	1 00
" Catechu, True	"	12
" " " Powd	"	30
" Copal	"	45
" Damar	"	40
" Elemi	"	60
" Euphorbium	"	30
" " Powd	"	45
" Galbanum	"	1 00
" Gamboge	"	90
" " Powd	"	1 00
" Guaiac	"	48
" " Powd	"	55

Gum, Hemlock	per lb.	$	75
" Kino	"		25
" " Powd	"		40
" Mastic	"	2	75
" Myrrh	"		50
" " Powd	"		58
" Olibanum	"		35
" Opium	"		
" " Powd	"		
" Sandrach	"		50
" Scammony, Aleppo, Powd	"	10	50
" Shellac, Campbell's	"		95
" " English	"		85
" " Native	"		75
" " White	"	1	15
" Spruce	"	1	25
" Storax, Liquid	"		60
" Tamarac	"	1	00
" Tragacanth, Flake	"	1	00
" " " Powd	"	1	25
" " Sorts	"		50
" Turpentine, or Thus	"		8

HERBS—Shaker's list of Herbs, Roots, Barks, and Flowers, Pressed in 1 lb., 1-2 lb., 1-4 lb., and 1 oz. Packages.

Six cents per lb. advance on this list for all Herbs in ounces. Barks and Roots come only in pound packages. Prices subject to market changes.

Common Names.	*Botanical Names.*	*Properties.*	*Per lb.*
Abcess root	Polemonium reptans	Ast. Sud. Feb.	$ 30
Aconite leaves	Aconitum napellus	Nar. Dia.	30
" root	" Radix	"	30
Agrimony	Agrimonia eupatoria	Ast. Ton. Dia.	28
Alder bark, black	Prinos verticillatus	Deo. Ast. Ton. A-sep.	20
" red or tag	Alnus serrulata	Ast. Deob. Ton.	20
Alum root	Heuchera pubescens	Ast. A-sep. Det.	30
Angelica leaves	Archangelica atropurpurea	Ton. Bal.	25
Angelica root	" "	Aro. Car. Emm.	30
Angelica seeds	" "	Aro. Sti. Car.	50
Anise seed	Pimpinella anisum	Aro. Car. Pec.	
Appletree bark	Pyrus malus	Ast. Ton.	25
Archangel	Lycopus sinuatus	Ast. Ton. Bal.	32
Arnica flowers	Arnica montana	Vul. Sti. Diu.	15
" root	" " Radix	Sti. Diu. Dia.	40
Ash bark, Mountain	Pyrus americana	Ast. Ton. Det.	25
" " Prickly	Xanthoxylum americanum	Sti. Aro. Ast.	25
" berries, "	" "	"	35
" bark, white	Fraxinus americana	Ast. Ton.	30
Asparagus root	Asparagus officinalis	Ape. Diu.	40
Avens root	Geum rivale	Ton. Ast. Sto.	30
Backache brake	Asplenium filix-fæmina	Pec. Dem.	35
Balm, Lemon	Melissa officinalis	Sto. Dia. Sud.	25
" Sweet	Dracocephalum canariense	Aro. Sto. Dia.	25

Common Names.	*Botanical Names.*	*Properties.*	*Per lb.*
Balm of Gilead buds	Populus balsamifera	Bal. Sto.	$ 65
Balsam, Sweet	Gnaphalium polycephalum	Sto. Sud. Sed.	25
Barberry bark	Berberis vulgaris	Ton. Ref. Est.	30
Basil, Sweet	Ocymum basilicum	Ara. Sti.	30
Basswood bark	Tilia americana	Emo. Dis.	25
Bayberry bark	Merica cerifera	Aro. Sti. Ast.	12
Beech bark	Fagus ferruginea	Ast. Ton.	25
" leaves	" " Folium	Ast. Alt.	30
Belladonna leaves	Atropa belladonna	Nar. Ano.	38
Bellwort	Uvularia perfoliata	Her. Dem.	38
Benne leaves	Sesamum indicum	Dem. Lax.	35
Beth root	Trillium purpureum	Ast. Ton. Pec.	30
Betony weed	Pedicularis canadensis	Ner. Ton. Dis.	90
Birch bark, black	Betula lenta	Ton. Ast.	20
Bird peppers	Capsicum baccatum	Sti. Car. Rub.	35
Bitter root	Apocynum androsæmifolium	Ton. Cath. Sud.	35
Bittersweet false, bark of root	Celastrus scandens	A-bil. Dis. Diu.	40
Bittersweet herb	Solanum dulcamara	Herb. Deo. Alt.	30
Blackberry root	Rubus villosus	Ast. Ton.	25
Blazing Star root	Aletris farinosa	Ton. Sto. Nor.	35
Blood root	Sanguinaria canadensis	Dia. Deo. Exp.	12
Blue flag	Iris versicolor	Diu. Eme. Cath.	25
Boneset	Eupatorium perfoliatum	Sud. Ton. Em.	20
Borage	Borago officinalis	Sto. Dia.	26
Boxwood bark	Cornus florida	Ton. Ast. Sti.	25
" flowers	"	"	35
Brooklime	Veronica beccabunga	Dis. Alt. Diu.	44
Buck bean	Menyanthes trifoliata	Ton. Deo.	50
Buckhorn brake	Osmunda spectabilis	Muc. Dem.	44
Buckthorn berries	Rhamnus catharticus, Bacca	Cath.	40
Bugle	Lycopus virginicus	Pec. Deo. Ton. Sty.	25
Burdock leaves	Lappa major, Folium	Feb. Sud.	22
" root	" Radix	Her. A-scor.	25
" seeds	" Semen	Diu. Car. Ton.	25
Butternut bark	Juglans cinerea	Cath. Ton.	25
Canada Thistle root	Cirsium arvense	Diu.	32
Cancer Root plant	Epiphegus virginiana	Ast. Ton.	30
Canker weed	Nabalus albus	Ast. Ton.	30
Cardinal Flowers, blue	Lobelia syphilitica	Diu. Sud. Her. Deo.	28
" " red	Lobelia cardinalis	Ver.	38
Carduus, spotted	Cnicus benedictus	Dia. Diu.	30
Carrot leaves, wild	Daucus carota	Diu. Emm.	
" seed, wild	"	Diu. Deo.	30
Catnip	Nepeta cataria	Sto. Car. Sud.	20
Celandine, garden	Chelidonium majus	Diu. Her.	28
" wild	Impatiens pallida	A-bil Sto.	25
Centaury, low	Hypericum mutilum	Ast. Ton. Sto.	30
Cherry bark, wild	Prunus virginiana	Ast. Ton. Sed.	12
Chickweed	Cerastium vulgatum	Emo. Ref.	32
Cicely, sweet	Osmorrhiza brevistylis	Car. Aro.	60
Cicuta leaves	Conium maculatum	Nar. Deo.	20
Clary	Salvia sclara	Sti. Sto.	28
Cleavers	Galium aparine	Diu. Sud.	25
Clover heads, red	Trifolium pratense	Pec. Acr.	32
Cocash root	Aster puniceus	Sud. Sti. Sto.	31
Cohosh, black	Cimicifuga racemosa	Deo. Nar.	20
" blue	Leontice thalictroides	A-spas. Sti.	25
" red	Actæa rubra	Deo. Ner. Emm.	30
" white	Actæa alba	Car. Deo. Nar.	25
Columbo root	Frasera caroliniensis	Ton. Cath. Eme.	25
Coltsfoot	Tussilago farfara	Exp. Pec. Dem.	30
Comfrey root	Symphytum officinale	Pec. Dem. Bal.	25
Consumption brake	Botrychium lunarioides	Sti. Ton.	35

Common Names.	*Botanical Names.*	*Properties.*	*Per lb.*
Coolwort	Tiarella cordifolia	Diu. Ton.	$ 30
Cowparsnip leaves	Heracleum lanatum, Fol	Ner. Car. Diu.	31
" root	" " Rad	"	31
" seeds	" " Sem	Car. Aro.	1 00
Cramp bark	Viburnum tantanoides	A-spas.	30
Cranesbill	Geranium maculatum	Sty. Ast. Ton.	25
Crawley	Corallorhiza odontorhiza	Bal. Sto.	1 50
Cuckold	Bidens frondosa	Ast. Diu. Car.	30
Culver's root	Leptandra virginica	Cath. Diu. Ton.	30
Daisy flowers	Leucanthemum vulgare	Diu. Vul.	30
Dandelion herb	Taraxacum dens-leonis	A-bil. Ast. Ton.	25
" root	" "	Deo. Diu. Ast. Ton.	30
Dittany	Cunila mariana	Ner. Sud.	25
Dock Yellow	Rumex crispus	Ton. Deo. Her.	25
" Water	R. hydrolapapathum	Ast. Dia. Deo. Her.	31
Elecampane	Inula helenium	Exp. Ast. Sto.	25
Elder bark	Sambucus canadensis	Diu. Deo. Sud. Her.	28
" flowers	" "	Alt. Sud. Her.	30
" dwarf	Aralia hispida	Dia. Ton. Diu. Dem.	30
Euphorbia	Euphorbia ipecacuanha	Eme. Cath. Ton.	
Fern, Male	Aspidium filix mas	Ver. Ton. Ast.	31
" Sweet	Comptonia asplenifolia	Sto. Ast.	20
Fever bush	Benzoin odoriferum	Aro. Sto. Sti.	30
Feverfew	Pyrethrum parthenium	Ner. Sto.	40
Fever root	Triosteum perfoliatum	Cath. Ton.	31
Fire weed	Erechthites hieracifolia	Ton. Ast. Alt.	30
Fit root	Monotropa uniflora	Ner. A-spas.	1 65
Five finger	Potentilla canadensis	Ast. Emm. Em.	31
Fleabane	Erigeron canadensis	Ton. Ast. Diu.	25
Flower-de-luce	Iris sambucina	Diu. Deo.	31
Fox glove	Digitalis purpurea	Diu. Nar. Dia.	20
Frostwort	Helianthemum canadense	Ast. Ton.	30
Fumitory	Fumaria officinalis	Deo. Dia. Diu.	31
Garget	Phytolacca decandra	Deo. Cath. Alt.	25
Garlick	Allium sativum	Sti. Exp. Ton.	25
Gentian	Gentiana lutea	Ton. Sto. Act.	
Ginseng	Panax quinquefolia	Sial. Ner. Sti.	
Gold thread	Coptis trifolia	Sto. Ton. Ast.	75
Golden rod	Solidago odora	Car. Aro. Diu.	31
Golden seal	Hydrastis canadensis	Ton. A-bil. Sto.	30
Gravel plant	Epigæa repens	Dia. Dem.	30
Hardhack leaves	Spiræa tomentosa	Ast. Ton.	25
Heal-all	Prunella vulgaris	Ast.	25
Hearts-ease	Polygonum persicaria	Sud.	20
Hellebore, black	Helleborus niger	Diu. Cath.	32
" white	Veratrum viride	Eme. Cath. Nar. Ner.	30
Hemlock bark	Abies canadensis	Ast. Ton.	20
Hemlock leaves	"	Dia. Alt. Sud.	26
Henbane, black	Hyoscyamus niger	Ner. Nar.	35
Hoarhound	Marrubium vulgare	Sto. Pec. Deo.	24
Hollyhock flowers	Althæa roseam	Ast. Dem.	80
Horsemint	Monarda punctata	Diu. Ton.	28
Horseradish leaves	Cochlearia armoracia	Acr. Sti.	26
" root	" " Rad	"	28
Hyssop	Hyssopus officinalis	Exp. Ceph.	20
Indian hemp, black	Apocynum cannabinum	Eme. Cath. Ton. Diu.	35
" white	Asclepias incarnata	Ape. Diu. Alt.	30
Indian turnip	Arum triphyllum	Sti. Exp. Aro. Nar.	30
Indigo, wild	Baptisia tinctoria	A-sep. Ton. Dia.	25

Common Names.	*Botanical Names.*	*Properties.*	*Per lb.*
Ipecac American, or Indian physic	Gillenia stipullaca	Eme. Cath. Ton. Exp.	$ 50
Ivy, ground	Glechoma hederacea	Dem. Sto. Ton.	28
Jacob's ladder	Smilax herbacea	Diu. Sial.	30
Jassamine, yellow			35
Job's tears, per doz.	Coix lachryma	Diu.	
Johnswort	Hypericum perforatum	Ast. Bal. Diu.	25
King's clover	Melilotus leucantha	Emo. Dis.	31
Knot grass	Polygenum aviculare	Diu. Ner. Car.	31
Labrador tea	Ledum latifolium	Diu. Bal.	31
Ladies' sorrel	Oxalis stricta	Ref. A-sep. Diu.	38
Larkspur herb	Delphinium consolida	Nar. Acr.	31
" seed	" "	Diu. Nar. Acr.	2 75
Lavender	Lavendula spica	Car. Pec. Ner.	30
Lettuce Garden	Lactuca sativa	Diu. Ano.	28
" Wild	Lactuca elongata	Diu. A-scor. Nar.	28
Life root	Senecio aureus	Diu. Feb. Aro. Stom.	30
Lily, White Pond	Nymphæa odorata	Pec. Emo. Ast. Ton.	25
" Yellow Pond	Nuphar advena	" "	25
Linden flowers	Tilia americana	Sud. A-spas. Ner.	
Liverwort	Hepatica triloba	Ast. Dem. Pec. Deo.	35
Lobelia herb	Lobelia inflata	Eme. Diu. Exp.	26
Lovage leaves	Ligusticum levisticum	Car. Sto. Emm.	30
" root	" "	Dia. Car. Stom.	35
" seeds	" "	"	45
Lungwort	Pertusaria faginea	Pec. Sto. Dem.	40
Maidenhair	Adiantum pedatum	Exp. Car. Sto.	25
Mallow, low	Malva rotundifola	Dem. Pec.	22
" marsh, leaves	Althæa officinalis	Dem. Ast.	25
" " root	"	Emo. Dem. Ast.	25
Man root	Convolvulus panduratus	Cath. Diu. Pec.	31
Mandrake root	Podophyllum peltatum	Deo. Em. Cath. Nar.	20
Maple, red or soft	Acer rubrum	Ton. Ast.	30
" striped	Acer striatum	Ver. Ton.	28
Marigold flowers	Calendula officinalis	Sto. Aro.	70
Marjoram, sweet	Marjorana hortensis	Ton. Sto. Aro.	45
Marsh rosemary	Statice limonium	Ast. A-sep.	30
Masterwort leaves	Heracleum lanatum	Ner. Car. Diu.	31
" root	" "	"	31
" seeds	" "	"	1 00
Mayweed	Maruta cotula	Sud. Dia. Sto.	25
Meadowsweet	Spiræa Salicifolia	Ton. A-sep. Feb.	31
Mezereon, Nett	Daphne mezereum	Dia. Sti.	
Milkweed root	Asclepias cornuti	Diu. Ano. Sud.	30
Monarda	Monarda punctata	Diu. Sto. Ton.	31
Moosewood bark	Dirca palustris	Dia. Sti.	25
Motherwort	Leonorus cardiaca	Ner. Sto. Emm.	25
Mountain dittany	Cunila mariana	Sti. Ton. Ner. Sud.	30
Mountain mint	Origanum vulgare	Sud. Sto. Aro.	28
Mouse ear	Gnaphalium uliginosum	Sud. Stom.	25
Mugwort	Artemisia vulgaris	Deo. A-bil. Ner.	25
Mullein herb	Verbascum thapsus	Ano. Dem. Ast.	22
Nanny-bush, bark	Viburnum lentago	Ton.	31
Nettle flowers	Urtica dioica	"	30
" root	"	Ast. "	28
Nightshade, deadly	Atropa belladona	Nar. Ano.	38
Oak bark, black	Quercus tinctoria	Ast. Ton.	20
" red	" rubra	Ast. Sti. Ton.	20
" white	" alba	Ast. Ton. A-sep.	20

Common Names.	*Botanical Names.*	*Properties.*	*Per lb.*
Oak of Jerusalem	Chenopodium botrys	Ver. Sto. Emm.	$ 30
Orange peel, Nett	Citrus aurantium	Ton. Sto.	
Osier bark, green	Cornus circinati.	Ast. Deo. Ton.	30
Oswego Tea	Monarda didyma	Sti. Sto Ton. Feb.	31
Pappoose root	Leontice thalictroides	Deo. Nar. Emm.	25
Parilla, yellow	Menispermum canadense	Ton. Lax. Alt.	25
Parsley leaves	Apium petroselinum	Dem. Diu.	25
" root	" "	Ape. Dem. Diu.	30
" seeds	" "	Diu. Dem.	1 00
Peach bark	Amygdalus persica	Cath. Ton. Sto.	31
" leaves	" "	Ver. Ton. Lax.	31
" pits	" "	Ton. Sto.	60
Pennyroyal	Hedeoma pulegiodes	Car. Sti. Sto. Aro.	22
Peony flowers	Pæonia officinalis	Ner.	75
" roots	"	Ner. Ver.	60
Peppermint	Mentha piperita	Sto. Sti. Sud.	25
Pilewort	Amaranthus hypochondriacus	Ast. Her.	28
Pine bark, white	Pinus strobus	Sti. Diu. Lax. Pec.	25
Plantain leaves	Plantago major	Ref. Vul. A-sep. Det.	28
Plantain, bitter, or Gall of the earth	Gentiana quinqueflora	Det.	90
" spotted	Goodyera pubescens	Det.	45
Pleurisy root	Asclepias tuberosa	Dia. Sud. Aro.	30
Poke berries	Phytolacca decandra	Deo. Alt.	35
" leaves	" " Folium	"	25
" root	" " Radix	Deo. Cath. Alt.	25
Polypody	Polypodium vulgare	Pec. Dem. Cath.	38
Pomegranate	Punica granatum	Ast. Ton.	
Poplar bark	Populus tremuloides	Ton. Alt.	18
Poppy capsules	Papaver somniferum	Nar. Ano.	40
" leaves	" Folium	"	31
" flowers	" Petals	"	75
Princes pine	Chimaphila umbellata	Diu. Sti.	25
Queen of the meadow herb	Eupatorium purpureum	Diu. Aro.	25
" root	" " Radix	"	25
Queen's delight	Stillingia sylvatica	Cath. Alt. A-Sep.	30
Raspberry leaves	Rubus strigosus	Ast. Ton.	22
Rock brake	Pteris atropurpurea	Sec. Ast. Ton.	45
Roman wormwood or Rag weed	Ambrosia artemisæfolia	Sti. Ton. Ver.	28
Rose willow	Cornus sericea	Ast. Ton.	30
Rosemary flowers	Rosmarinus officinalis	"	31
" leaves	" " Folium	"	25
Rue	Ruta graveolens	Ton. Diu. Sto.	28
Saffron	Carthamus tinctorius	Diu. Sto. Aro.	
Sage	Salvia officinalis	Sud. Sto. Bal.	25
Sanicle root, black	Sanicula marilandica	Ton. Ast. Diu.	45
" " white	Eupatorium ageratoides	Deo. Sto. Ton.	45
Sarsaparilla, Am.	Aralia nudicaulis	Alt. Dem. Deo.	30
Sassafras	Sassafras officinale	Sti. Ape. Ton. Alt.	
" pith	"	Dem. Muc.	
Savin	Juniperus communis	Sti. Emm. Deo. Acr.	28
Scabious	Erigeron philadelphicum	Diu. Sud. Her.	25
" sweet	" annuum	Diu. Sed.	25
Scabish	Œnothera biennis	Dem. Sto.	75
Scrofula plant	Scrophularia nodosa	Deo. Ton.	31
Scullcap	Scutellaria laterifolia	Ton. Sud. Ner.	55
Scurvy grass	Cochlearia officinalis	Sti. Ape. Diu	20
Senna, Am.	Cassia marilandica	Cath. Diu. Deo	75
Side-saddle plant	Sarracenia purpurea	Cath. Deo	30

Common Names.	*Botanical Names.*	*Properties.*	*Per lb.*
Skunk cabbage root	Symplocarpus fœtidus	A-spas. Ver. Acr.	$ 30
Snakehead	Chelone glabra	A-bil. Ton. Ver.	25
Snakeroot, Button	Liatris spicata	Sti. Diu. Bal.	25
" Canada	Asarum canadense	Her. Aro. Ner.	30
" Rattle	Goodyera pubescens	Det.	
Soapwort	Saponaria officinalis	Alt. Ton.	30
Solomon's seal	Polygonatum giganteum	Ast. Dem. Bal.	30
" small	Smilacina racemosa	Dem. Ast.	30
Sorrel, sheep	Rumex acetosella	Ref. A-scor. Ton.	31
" wood	Oxalis acetosella		31
Southernwood	Artemisia abrotanum	Sti. Ton. Ner. Det.	28
Spearmint	Mentha viridis	Feb. Diu. A-eme.	20
Speedwell, Virginia	Veronica officinalis	Diu. Dia. Exp.	60
Spikenard	Aralia racemosa	Pec. Bal. Sto.	25
Spleenwort	Asplenium angustifolium	Diu.	45
Spurred rye.	Acinula	Nar.	
Squaw vine	Mitchella repens	Diu. Emm.	28
" weed	Senecio obovatus	Her. Acr.	31
Star flower	Aster novæ-angliæ	Her. Sto. Ner.	35
Stone root	Collinsonia canadensis	Diu. Sto.	30
Strawberry leaves	Fragaria virginiana	Ast. Feb. Ref.	30
" vines	" "	" "	30
Succory	Cichorium intybus	Ton. Ape. Deo.	25
Sumach bark	Rhus glabra, Cort	Ast. Ton.	25
" berries	" Bacca	Ast. Ref. Diu.	25
" leaves	" Folium	Ast. Ton. Diu.	25
Summer Savory	Satureja hortensis	Sto. Aro.	25
Sunflower, wild	Helianthus divaricatus	Car. A-spas. Lax.	
Sweet or King's clover	Melilotus leucantha	Emo. Dis.	30
Sweet flag	Acorus calamus	Aro. Sto.	25
Sweet-gale burrs	Myrica gale	Ast. Aro. Her.	75
" herbs	" Herba	Pec. Ast. Aro.	30
Tamarack bark	Larix americana	Ape. Exp. Bal.	25
Tanzy, double	Tanacetum vulgare	Sud. Emm. Ver. Sto.	20
Thimble weed	Rudbeckia laciniata	Diu. Ton.	30
Thornapple leaves	Datura stramonium	Nar. Acr.	25
" seeds	" Semen	"	25
Thyme	Thymus serpullum	Ton. Sto. Aro.	35
Tilia flowers	Tilia americana	Sud. A-spas. Ner.	
Turkey pea	(Corydalis) dicentra canadensis	Ton. Diu. Alt.	35
Unicorn root	Chamælirium luteum	Ton. Sto.	40
Vervian	Verbena hastata	Sud. Ton.	25
Violet, blue	Viola cucullata	Dem. Sud. Ton. Lav.	31
" canker	Viola rostrata	Dem. Ton.	30
Virgin's bower	Clematis virginiana	Sti. Nor.	30
Wa-a-hoo bark	Euonymus atropurpureus	Cath. Exp.	40
Walnut bark	Carya alba	Ton. Sty.	25
Water cup	Sarracenia purpurea	Ner. Ton.	35
Water pepper	Polygonum hydropiper	Sti. Arc.	25
Whitewood bark	Liriodendron tulipifera	Sto. Aro.	25
Wickup	Epilobium angustifolium	A-sep. Eme. Ast.	
Wild turnip	Arum triphyllum	Sti. Exp. Acr. Nar.	35
Wild yam			30
Willow, pussy	Salix candida	A-sep. Ast.	31
" white	Salix alba	Ton. Ast.	
Wintergreen	Gaultheria procumbens	Sti. Diu. Sto. Emm.	25
Witch-hazel bark	Hamamælis virginica	Ton. Ast. Her.	25
" leaves	" "	"	25
Wormwood	Artemisia absynthium	Ton. Sti. A-bil.	22
Yarrow	Achillea millefolium	Ast. Sto. Det.	20
Zedoary			45

Pulverized Sweet Herbs, for culinary and other purposes, as Sage, Thyme, Summer Savory, Sweet Marjoram.

Hiera Picra	per lb.	$ 50
Homœopathic Globules	"	50
Honey, Strained	"	25
Hops, Loose	"	30
" Pressed, 1's ½'s and ¼'s	"	30
" " in ozs.	"	35
Hot Drops No. 6	"	75
ICELAND MOSS	"	12
Infusion of Opium	per doz.	2 50
" "	per lb.	2 50
Indigo, Bengal	"	2 25
" Gautemala	"	1 50
" Madras	"	1 25
" Manilla	"	1 15
Indigo-Carmine	"	1 50
Indigo-Compound	"	75
Ink, Arnold's Writing Fluid, qts.	per doz.	6 00
" " " " pts.	"	4 00
" " " " 1-2 pts.	per doz.	2 25
" " " " 2 oz.	"	60
" Black, 2 oz. Cone, 3 doz. in box	per box,	1 50
" Blue, " " " "	"	1 50
" Violet, " " " "	"	1 50
" Carmine	per doz.	1 25
" Indelible—(see Pat. Med. List, under Propr's name		
" India, Black, in Sticks	"	60
Insect Powder, Persian	per lb.	75
Iodine, Resub.	"	4 75
" " in 1 oz vials	per oz.	45
" Bromide, in 1 oz. vials	"	1 50
" Chloride, in 1 oz. vials	"	1 20
Iodoform, in 1-4 oz. vials	"	1 00
Irish Moss, Extra	per lb.	12
" " " by bbl.	"	10
Iron, Alum	"	1 10
" " in 1 oz. vials	per oz.	16
" Ammoniated	per lb.	40
" Ammonio Tartrate	"	1 20
" " " in 1 oz. vials	per oz.	18
" Arseniate, in 1 oz. vials	"	45
" Bromide, in oz. vials	"	50
" Carbonate Precip.	per lb.	25
" " Proto. (Vallet's Mass,) in 1-4 lb. jars.	"	50
" " Saccharated, Proto. Carb.	"	1 00
" Chloride, Tinct. of	"	35
" Citrate	"	1 40
" " in oz. vials	per oz.	18
" " and Ammonia	per lb.	1 50
" " " " in 1 oz. vials	per oz.	18

Iron, Citrate, and Magnesia, in 1 oz. vials	per oz.	$ 20
" " " Quinine	per lb.	8 50
" " " " in 1 oz. vials	per oz.	60
" " " Strychnine, in 1 oz. vials	"	45
" Ferrocyanide	per lb.	1 30
" " in 1 oz. vials	per oz.	18
" Filings	per lb.	12
" Hypophosphite, in 1 oz. vials	per oz.	50
" Iodide, in 1 oz. vials	"	60
" " Syrup of	per lb.	60
" Lactate, in 1 oz. vials	per oz.	25
" Oxide, Black	per lb.	45
" " Per Hydrated	"	45
" Per Chloride	"	1 20
" " in 1 oz. vials	per oz.	20
" Per Nitrate Solution	per lb.	45
" Per Sulphate	"	1 10
" " in 1 oz. vials	per oz.	16
" Phosphate	per. lb.	70
" Pyrophosphate	"	1 40
" " in 1 oz. vials	per oz.	18
" Quevenne's, by Hydrogen	per lb.	1 75
" " " " in 1 oz. vials	per oz.	18
" Chloride, Sol.	per lb.	40
" " Dry, in 1 oz. vials	per oz.	20
" Sulphate Crude—(see Copperas.)		
" " Pure	per lb.	9
" " Powd. Exsic.	"	20
" Sulphuret, or Sulphide	"	35
" Tannate, in 1 oz. vials	per oz.	50
" Tartrate and Ammonia, in 1 oz. vials	"	18
" " " Potass. Gran.	per lb.	1 00
" " " " Plates	"	1 25
" " " " " in 1 oz. vials	per oz.	18
" Valerianate, in 1 oz. vials	"	75
Isinglass, American, Fish	per lb.	1 75
" Cooper's Sheet	"	1 00
" " " by 12 lb. boxes	"	95
" " Shred	"	1 00
" " " by 12 lb. boxes	"	95
" Russian Sheet	"	5 50

JAMES'S POWDER	"	1 00
Japonica	"	10
" by bale	"	8¼
Juniper Berries	"	8
Jewellers' Rouge	"	2 00
Juices. (Pure Fruit.)—(see page 280.)		
Catawba Grape	per bot.	1 50
Blackberry	"	1 50

Juices. (Pure Fruit.)—		
Qiunce	per oz. bot.	$1 50
Raspberry	"	1 50
Strawberry	"	1 50
Cherry	"	1 50
Wild Cherry	"	1 50

KAMALA POWDER	per oz.	50
Kermes Mineral	per lb.	2 00
Kousso	"	1 25
" Powd.	"	1 50
" " in 1 oz. vials	per oz.	75

LAC, DYE	per lb.	45
Lac, Sulphur, English	"	18
" " P. & W.	"	30
Lactucarium, in 1 oz. vials	per oz.	50
Lapis Calaminaris	per lb.	12
Lead, Acetate (Sugar of)	"	25
" " C. P.	"	65
" Carbonate, Dry, Pure	"	12
" Chloride, in 1 oz. vials	per oz.	35
" Iodide, in 1 oz. vials	"	50
" Nitrate	per lb.	25
" Sol. Sub. Acet. (Goulard's Ext.)	"	30
" Tannate, in 1 oz. vials	per oz.	45
Leaves, Bay, Fresh	per lb.	12
" Buchu, Long	"	40
" " Short	"	14
" Laurel	"	12
" Matico	"	60
" Patchouly	"	2 00
" Rose, Red, Eng.	"	2 00
" Senna, Alex.	"	20
" " " Powd.	"	28
" " E. I.	"	20
" " Tinnevelly	"	25
Lime, Carbonate, Precip.	"	25
" Chloride, in jars	"	6
" " by cask	"	4½
" Hypophosphite	"	3 25
" " in 1 oz. vials	per oz.	30
" Iodide, in 1 oz. vials	"	65
" Phosphate, Precip.	per lb.	40
" Sulphite (for Cider)	"	25
" " " in cartoons	per doz.	1 50
Lint, Taylor's	per lb.	1 65
" " in ozs.	"	2 00
Liquorice, Calabria, P. & S.	"	42

Liquorice, Calabria, P. & S. Powd.	per lb.	$ 60
" " Imit. in 6 lb. boxes	"	28
" Refined Pipe Stem	"	38
" Sicily, in 6 lb. boxes	"	20
" Root, Select, in bundles	"	13
" " " Powd.	"	15
" " " Cut	"	22
Lithia, Bromide, in 1-4 oz. vials	per oz.	1 25
" Carbonate, " "	"	90
" Iodide, " "	"	1 50
Litmus	per lb.	1 10
" Paper	per quire,	1 20
Liver, Sulphur	per lb.	35
Lupuline	"	85
Lycopodium	"	75
Lye, Concent. in 4 doz. cases, Penn. Salt Co.	per case,	8 00
" " " "	"	8 00

MACE	per lb.	1 60
" Ground	"	1 75
Madder, Dutch	"	14
" " by bbl.	"	13
" Compound	"	25
Magnesia, Calcined, American	"	55
" " English, in 1 lb. bots. incl.	"	95
" " Jenning's, " "	"	1 25
" " Heavy, P. & W.	"	1 75
" " " "	per doz.	2 50
" Carbonate, Eng. in 1-8 lb. papers	per lb.	28
" " " 1-4 lb. "	"	27
" " Jenning's, in 1-8 lb. papers	"	38
" " " 1-4 lb. "	"	37
" " " S. S.	"	65
(For other Proprietary Magnesias, see Pat. Med. List.)		
" Hypophosphite, in 1 oz vials	per oz.	55
" Muriate, or Chloride	per lb.	1 00
" Nitrate, in oz. vials	per oz.	25
" Sulphite	per lb.	1 20
Manganese, Arseniate, in 1 oz. vials	per oz.	
" Black Oxide	per lb.	8
" Bromide, in 1 oz. vials	per oz.	1 25
" Carbonate, " "	"	45
" Hypophosphite, " "	"	65
" Iodide, " "	"	1 20
" Phosphate, " "	"	65
" Sulphate, " "	"	28
Manna, Flake, large	per lb.	1 25
" " small	"	65
" Sorts	"	40

Marble Dust, in bbls.	per bbl.	$2 75
Mercury	per lb.	1 90
" Acetate, in 1 oz. vials	per oz.	1 25
" Bichloride, (Corrosive Sublimate)	per lb.	2 00
" Bromide, in 1 oz. vials	per oz.	1 35
" Chloride, (Calomel)	per lb.	2 10
" " English, in 1 lb. bottles	"	2 85
" Cyanide, in 1 oz. vials	per oz.	60
" Iodide, Per (red), in 1 oz. vials	"	60
" " Proto (green), in 1 oz. vials	"	60
" Nitrate Sol., in 1 oz. vials	"	35
" Oxide, black, in 1 oz. vials	"	40
" Red, (Cinnabar)	per lb.	2 50
" Sub Sulph., (Turpeth Mineral)	"	2 25
" Sulphuret (Ethiop's Mineral)	"	1 75
" With Chalk, (Hydg. c. Creta)	"	1 00
Milk, Sugar of, Powd.	per lb	55
Morphia, Acetate, in ⅛ oz. vials	per oz.	5 65
" Muriate, " "	"	5 65
" Sulphate, " "	"	5 65
" Valerianate, " "	"	8 00
Mortars, (see page 150		
Moss, Iceland	per lb.	12
" Irish, select	"	12
" " by bbl.	"	10
Mucilage, 3 oz. (1 doz. in box)	per doz.	1 25
Mustard, Coleman's Durham, in 18 lb. kegs	per lb.	40
" " " in 4, 6, and 10 lb. cans	"	44
" " " in ¼ lb. cans	per doz.	1 75
" " " in ½ lb. cans	"	3 25
" " " in 1 lb. cans	"	6 00
" " Imitation, in 4, 6, and 10 lb. cans	per lb.	33
" Taylor's, in 4, 6, and 10 lb. cans	"	65
" " S. F., in ¼'s and ½'s	"	70
" " D. S. F., in ¼'s and ½'s	"	75
" Wilde's B. B., in 4, 6, and 10 lb. cans	"	35
Musk, Chinese, Canton, in 1 oz. caddies	per oz.	75
" True Grain	"	28 00
" " Pod	"	11 00

NUTMEGS, No. 1	per lb.	1 35
Nux Vomica	"	12
" Powd.	"	35

OAT MEAL, Coarse	per lb.	5
" " Fine	"	5
" " Robinson's Scotch	per doz.	1 75

OILS, Essential, we purchase in original packages of Importers, or the Manufacturers, and our effort is to furnish the best obtainable.

Oil, Almonds, Essent., Allen's	per lb.	$10 00
" " " "	per oz.	75
" " Sweet "	per lb.	45
" Amber, Crude	"	45
" " Rect.	"	60
" Animal, Dippels	per	50
" Anise	per lb.	3 50
" Arnica, in 1 oz. vials	per oz.	75
" Bay	"	1 00
" Benne (Sesame)	per gall.	2 00
" Bergamot, Sanderson's	per lb.	8 50
" Cade	"	60
" Cajeput	"	1 10
" Cantharides	per oz.	75
" Capsicum	"	75
" Caraway, German	per lb.	2 75
" Cassia	"	1 50
" Castor, American, White	"	
" " " by bbl.	"	
" " " common (machinists)	per gall.	
" " East India, in square *cans* (free)	per lb.	
" " in 2 oz. bottles	per doz.	75
" " " 3 "	"	1 00
" " " 4 "	"	1 50
" " " 8 "	"	2 50
" Cedar	per lb.	40
" " Red	"	65
" Chamomile, German, in 1 oz. vials	per oz.	3 00
" " Roman, " "	"	2 25
" Cherry Laurel	per lb.	85
" Cinnamon, Ceylon	per oz.	2 25
" Citronella, Native	per lb.	1 25
" " Winter's	"	1 50
" Cloves	"	3 75
" Cocoa Nut	"	18
" Cod Liver, Bulk, Brown, Norwegian	per gall.	2 00
" " " White	"	1 75
" " Proprietary, (see Patent Medicines.)		
" Cognac, Green, Opt.	per oz.	5 00
" " Straw	"	3 50
" " Essence	per lb.	10 00
" Cologne, Glenn's, bot. incl.	"	6 00
" " Lundborg's, in 1 lb. cans	"	7 50
" Copaiba	"	2 25
" Coriander	per oz.	1 75
" Croton, English	per lb.	3 00
" " " in 1 drm. vials	per doz.	1 50
" Cubebs	per lb	2 00

Oil, Cummin	per oz.	$	50
" Ergot	"		25
" Erigeron, (Fleabane)	per lb.	2	00
" Fennel	"	3	00
" Fern, Male	per oz.		75
" Fireweed	per lb.	2	00
" Fusel, P. & W.	"	1	50
" Geranium, Rose, Chiris's	per oz.	1	25
" Hemlock	per lb.		45
" Horsemint	"	3	50
" Jasmin, Chiris's	"	4	00
" Juniper Berries	per oz.		25
" " Wood	per lb.		75
" Laurel, Expressed	"		85
" Lavender Flowers	"	2	75
" " Garden	"	1	35
" " Spike	"	1	25
" Lemon, Sanderson's	"	5	00
" " Grass, Winter's	"	3	00
" Lobelia	per oz.	1	00
" Mace, Expressed	per lb.	2	50
" Melissa	"	7	50
" Musk, in 1 oz. vials	per oz.		75
" Mustard, Essential	"	2	00
" Myrrbane	per lb.		75
" Neroli, Bigarade, Chiris's	per oz.	4	50
" Nutmegs	"		50
" Olive, American, Loubon, qts. 1 doz. in case	per doz.	3	50
" " " " pts. 2 " "	"	2	25
" " " " ½ pts. 2 doz. "	"	1	50
" " Malaga	per gall.	1	60
" " Marseilles	"	3	50
" " " qts. 1 doz. in case	per doz.	6	00
" " " pts. 2 " "	"	3	25
" " Union Salad	per gall.		90
" " Vierge d'Aix, qts. 1 doz. in case	per doz.	10	50
" " " " pts. 2 " "	"	5	75
" " " " ½ pts. 2 " "	"	4	00
" Orange, Bitter	per lb.	6	50
" " Flowers, (Concent. Extract)	"	4	50
" " Sweet	"	4	00
" Origanum, Pure	"		75
" " Com.	"		50
" Palm	"		15
" Patchouly	per oz.	3	50
" Pennyroyal	per lb.	2	00
" Peppermint	"	5	50
" Pimento	per oz.		60
" Rhodium	"		50
" Rose, Commercial	"	4	50
" " Kissanlik, Pure	"	8	75

Oil, Rose, Geranium, Chiris's	per oz.	$ 1 25
" " Palma	per lb.	18 00
" Rosemary	"	1 00
" Rue	per oz.	25
" Rum, Jamaica	"	1 00
" " St. Croix	"	1 00
" Sandal Wood	"	1 00
" Sassafras	per lb.	90
" Savin	"	1 50
" Seneka	"	20
" Spearmint	"	4 50
" Spike	"	15
" Spruce	"	45
" Stillingia	per oz.	90
" Stone	per lb.	20
" Tanzy	"	3.25 to $5
" Tar	"	15
" Tobacco	per oz.	2 50
" Valerian	"	1 10
" Whisky, Bourbon	"	1 00
" " Irish	"	1 00
" " Monongahela	"	1 00
" " Rye	"	1 00
" " Scotch	"	1 00
" Wine, Light	per lb.	2 75
" Wintergreen	"	5 25
" Wormwood	"	5 00
" Wormseed	"	3 25
Orange Apple	"	15
" Peel, Bitter	"	25
" " Sweet	"	25
" " " Ground	"	28
" " " Powd.	"	40
Ointment, Basilicon	"	35
" Citrine	"	60
" Iodine	"	2 00
" Mercurial, ½ (Fort.)	"	1 25
" " ⅓ (Mitis)	"	1 00
" Savin	"	
" Stramonium	"	75
" Tar	"	50
Ox Galls, Inspissated, in 1 oz. jars	per oz.	22
PAPER, Dutcher's Fly	per ream	8 50
" Fly	"	7 50
" Wrapping, (see Van Schaack, Stevenson & Reid's list of Druggists' paper, page 106		
" Filtering and Filters, page 157		
Pearl-Ash	per lb.	15
Pepper, Black, Whole	"	28
" " Ground	"	35

Article	Unit	$	Cts.
Pepper, Cayenne, Pods	per lb.	$	35
" " Powdered	"		40
" White	"		50
Peppers, Long	"		50
Phloridzin, in 1 oz. vials	per oz.	4	00
Phosphorus, in 1 lb. cans	per lb.	1	75
" in 11 lb. cans	"	1	25
Pimento	"		18
" Ground	"		25
Pipe Clay	"		6
Piperine, in 1 oz. vials	per oz.	1	25
Plaster, Aconite, in ⅙ lb. rolls	per lb.	1	25
" Adhesive, "	"		40
" " Spread, Ellis's, in one and 5 yd. rolls	per yd.		20
" " " Eng. "	"		20
" Ammoniac, in ½ lb. rolls	per lb.	1	20
" " Cum Hydrg, in ½ lb. rolls	"	1	35
" Anodyne, in ⅙ lb. rolls	"	2	75
" Arnica, "	"	1	25
" Assafœtida, "	"	1	00
" Belladonna, "	"	1	25
" Cantharides, U. S. P.	"	1	10
" Calefacious, in ⅙ lb. rolls	"		60
" Court, (see Sundries, page —.)			
" Diachylon, Comp., in ⅙ lb. rolls	"		40
" " Simple, "	"		40
" Galbanum, "	"		75
" " Comp., "	"		65
" Hydrarg, Fort., "	"	1	25
" " Mite., "	"	1	00
" Irritating, Beech's, "	"		35
" Logani, "	"		50
" Olivæ, "	"		40
" Opii, "	"	2	75
" Oxycroc., "	"		45
" Plumbi, "	"		40
" Roboran's, "	"		40
" Saponis, "	"		45
" Simplex, "	"		50
Plasters, Proprietary, (see Patent Medicines.)			
Potash, Acetate	"		55
" American, 4 doz. in case	per case	8	75
" Arseniate, in 1 oz. vials	per oz.		50
" Babbitt's, 4 doz. in case	per case	8	75
" Bicarbonate, Crystals	per lb.		30
" Bichromate	"		25
" Carbonate, (Sal Tartar)	"		22
" Caustic, White	"		75
" " Cum Calc., (Vienna Caustic)	"		85
" Chlorate, English	"		35
" " " Powd.	"		45

Potash, Chlorate, French	per lb.	$ 50
" Citrate	"	1 50
" Crude	"	10
" " by cask	"	7½
" Hypophosphite, in 1 oz. vials	per oz.	30
" Permanganate, Med.	per lb.	1 00
" " Pure	"	2 00
" " " in 1 oz. vials	per oz.	25
" Phosphate	per lb.	2 75
" Prussiate, Red	"	85
" " Yellow	"	40
" Sulphate, Cryst.	"	15
" " Powd.	"	18
" Sulphite	"	60
" Tartrate	"	95
" " and Soda, (Rochelle Salt)	"	38
Potassium, in ¼ oz. vials	per oz.	4 00
" Bromide	per lb.	60
" " in 1 oz. vials	per oz.	15
" Cyanide, Fused	per lb.	75
" " Gran.	"	1 25
" Iodide, English	"	
" " "	"	3 50
" " " in 1 oz. vials	per oz.	35
" Sulphuret, (Liver of Sulphur)	per lb.	30
Propylamin, in 1 oz. vials	per oz.	1 75
" Chloride, in 1-8 oz. vials	"	6 50
Pumice Stone, Lump, Select	per lb.	6
" " Powd.	"	6

SELECT POWDERS, WARRANTED STRICTLY PURE.

Powder, Aromatic U. S. P.	per lb	$2 25
Powdered Aconite Root	"	38
" Angelica Root	"	35
" Anise Seed	"	25
" Assafœtida	"	45
" Bitter Root	"	45
" Black Cohosh	"	25
" Blood Root	"	18
" Blue Cohosh	"	30
" Blue Flag	"	35
" Boxwood	"	40
" Calamus	"	32
" Caraway Seed	"	30
" Cardamon "	"	3 00
" Colchicum Root	"	30
" " Seed	"	45
" Colocynth	"	40
" " Ext. Comp.	"	5 50
" Columbo	"	25

Powdered Comfrey	per lb.	$	35
" Coriander Seed	"		25
" Cranesbill	"		35
" Culver's Root	"		30
" Curcuma	"		13
" Dandelion Root	"		40
" Digitalis	"		35
" Elecampane	"		30
" Euphorbium	"		45
" Galangal	"		25
" Goldenseal	"		30
" Hellebore, Black	"		35
" " White	"		30
" Henbane	"		40
" Indian Hemp, White	"		40
" Indian Turnip	"		35
" Jalap Ext.	"	6	50
" Laurel, Leaves	"		35
" Lobelia, Herb	"		35
" Mandrake	"		20
" Olibanum	"		50
" Opium	"		
" Pareira Brava	"	1	00
" Peppermint	"		35
" Pink Root	"		45
" Pleurisy Root	"		35
" Poke Root	"		25
" Poplar Bark	"		20
" Rhatany	"		55
" Sage	"		30
" Salep	"	1	35
" Sandal Wood	"	1	00
" Sarsaparilla, Hond.	"		65
" Scullcap	"		65
" Skunk Cabbage	"		35
" Soap, White	"		60
" Stramonium Seed	"		35
" Summer Savory	"		40
" Sweet Marjoram	"		55
" Thyme	"		45
" Tragacanth	"	1	25
" Unicorn Root	"		40
" Wormseed, Levant	"		35
" Wormwood	"		35

All other Powders will be found under their respective drugs.

QUASSIA Chips	per lb	$ 7
Quicksilver. (See Mercury, page .)		
Quinine, Acetate, in 1 oz. vials	per oz	3 75
" Alkaloid, "	"	4 50
" Arseniate, "	"	6 00
" Cincho, (Nichols's,) in 1 oz. vials	"	1 60
" Citrate, "	"	3 75
" " and Iron, "	"	60
" Ferrocyanide, "	"	3 75
" Hypophosphite, "	"	5 00
" Iodide, "	"	4 00
" Muriate, "	"	3 50
" Sulphate, P. & W. "	"	2 25
" Sweet, (Stearns's), "	"	1 15
" Tannate, "	"	2 75
" Valerianate	"	6 50
Quercitron. (See Dye Woods, page 105.		

RED PRECIPITATE	per lb	2 30
Rice Flour, in 1 lb. pkges	"	14
Root, Aconite, German	"	30
" " " Powd.	"	38
" Alkanet	"	18
" Althea, Cut Square	"	28
" " Powd.	"	38
" Blood	"	12
" " Powd	"	18
" Calamus, White, Opt.	"	60
" " Peeled	"	25
" " Powd.	"	35
" Colchicum, English	"	25
" " Powd.	"	35
" Columbo, English	"	15
" " " Powd.	"	25
" Galangal	"	12
" " Powd.	"	25
" Gentian	"	11
" " Ground	"	15
" " Powd.	"	18
" Ginger, E. I.	"	17
" " " Ground	"	20
" " Jamaica		30
" " " Powd.	"	35
" Ginseng	"	1 85
" Goldenseal	"	
" " Powd.	"	32
" Ipecac	"	1 25
" " Powd.	"	1 40
" Jalap	"	35

Root, Jalap, Powd.	per lb.	$ 40
" Liquorice, Select, in Bundles	"	13
" " " Cut	"	22
" " " Powd.	"	15
" Orris	"	25
" " in Fingers	"	60
" " Powd.	"	28
" Pink	"	35
" " Powd.	"	45
" Rhatany	"	45
" " Powd.	"	60
" Rhei, E. I., Select	"	1 15
" " " Powd.	"	1 25
" " Turkey, Trimmed	"	7 50
" " " Powd.	"	8 00
" Sarsaparilla, Hond.	"	50
" " " Ground	"	58
" " " Powd.	"	65
" Seneka	"	1 10
" " Powd.	"	1 50
" Serpentaria	"	40
" " Powd.	"	50
" Squills	"	15
" " Powd.	"	50
" Valerian, American, (Ladies' Slipper)	"	25
" " " " Powd.	"	35
" " English	"	40
" " " Powd.	"	50
" " German	"	30
Rosin, Strained, by bbl. 280 lbs. gross	per bbl.	4 50
" " Pale, "	"	6 00
" " White	"	7 50
Rotten Stone, Lump	per lb.	7
" Powd.	"	8

SAFFRON, American	per lb	60
" Spanish	per oz.	75
Sago, Pearl	per lb.	10
Sal Acetosella	"	50
" Ammoniac	"	16
" " Powd.	"	25
" Epsom	"	4
" " by bbl.	"	3
" Glauber	"	3½
" " by bbl.	"	2½
" Nitre, Crude	"	14
" " " by bag, 250 lbs.	"	10½
" " Pure	"	15
" " " by keg, 100 lbs.	"	14
" " " Double (Large Crystals)	"	18

Item	Unit	$	¢
Sal Nitre, Pure, Double, by keg, 100 lbs.	per lb.	$	16
" " " Powd.	"		20
" " Refined, Hudson	"		12
" " " " by keg, 100 lbs.	"		10
" Prunelle	"		45
" Rochelle, P. & W.	"		38
" Soda	"		4
" " by cask	"		2½
" " by keg, 112 lbs.	"		2¼
" Tartar	"		22
Salacine, in 1 oz. vials	per oz.		55
Santonine, "	"		70
Sassafras Pith	per lb.	1	50
Seed, Anise, Italian	"		18
" " " Powd.	"		25
" " Star	"		65
" Burdock	"		25
" Canary	"		15
" Caraway	"		14
" " Powd.	"		30
" Cardamon, Malabar	"	2	00
" " Powd.	"	3	00
" Celery	"		55
" Colchicum	"		40
" " Powd.	"		45
" Coriander	"		14
" " Powd.	"		25
" Cummin	"		25
" Dill	"		35
" Flax	"		6
" " Ground	"		7
" Fœnugreek	"		12
" " Ground	"		12
" Fœnnel	"		18
" " Powd.	"		25
" Hemp	"		7
" Lobelia	"		40
" " Powd.	"		45
" Millet	"		6
" Mustard, Black	"		15
" " White	"		8
" Pumpkin	"		25
" Quince	"	1	50
" Stramonium	"		25
" " Powd.	"		35
" Watermelon	"		40
" Worm, American	"		15
" " " Powd.	"		25
" " Levant	"		25
" " " Powd.	"		35
Seidlitz Mixture, P. & W.	"		32

Seidlitz Powders	per doz.	$ 2	75
Silver, Bromide, in 1 oz. vials	per oz.	2	25
" Chloride, "	"	1	50
" Cyanide, "	"	2	25
" Iodide, "	"	2	50
" Nitrate, in 1 lb., ½ lb. and ¼ lb. bottles	per lb.	16	50
" " in 1 oz. vials	per oz.	1	10
" " Lunar Caustic, No. 1, in 1 oz. vials	"	1	10
" " " No. 2, "	"		85
" Oxide, in 1 oz. vials	"	1	75
Silvering Paste	per doz.	3	75
Silver Sand	per lb.		5
Snuff, Maccaboy, in bladders	"		75
" " in jars	"		75
" Scotch, in bladders	"		70
" " in jars	"		70
" " Garrett's, in 2 oz. papers	per doz.	1	50
" " " in 6 oz. bottles	"	4	50
Soap, Castile, Mottled, American	per lb.		
" " " " by box	"		
" " " Genuine	"		15
" " " " by box	"		11½
" " White, "Conti."	"		22
" " " " by box	"		18
" " " Powd.	"		60
Soda, Acetate	"		50
" Ash	"		6
" " by cask	"		3¾
" " Bi. Carb., (Allhusen's)	"		8
" " " by 112 lb. keg	"		6¾
" Bi. Sulphite	"	1	25
" Caustic, by drum, 600 lbs.	"		6½
" Chlorate, in 1 oz. vials	per oz.		25
" Chloride Liquor, C. S. bot.	per doz.	2	50
" " Gen. Labarraque's	"	9	00
" Citrate	per lb.	1	35
" Hypophosphite	"	3	25
" " in 1 oz. vials	per oz.		30
" Hyposulphite	per lb.		8
" Lactate Solution	"	6	00
" Nitrate, Purified	"		25
" Permanganate	"	1	35
" Phosphate	"		35
" Pyrophosphate	"	1	35
" " in 1 oz. vials	per oz.		18
" Silicate	per lb.		25
" Sulphite, Crystals	"		30
" " Gran.	"		60
Sodium, Metallic, in 1 oz. vials	per oz.		75
" Bromide, "	"		25
" Iodide, "	"		65

Spermaceti	per lb.	$ 35
Spirits Ammonia	"	45
" " Arom.	"	45
" Lavender	"	45
" " Comp.	"	45
" Nitre, 3f.,	"	35
" " 3f., by carboy	"	32
" " 4f.	"	40
" " 4f., by carboy	"	37
" " Nichols's	"	50
" " Squibb's	"	1 00
Starch, all kinds, at lowest market rates.		
Stove Polish, Dixon's	per doz.	60
" "	per gross	6 50
Strontia, Muriate	per lb.	35
" Nitrate	"	25
Strychnia, Acetate, in 1/8 oz. vials	per oz.	3 75
" Citrate, Iron, "	"	45
" Crystals, "	"	2 50
" " Powd., "	"	2 25
" Muriate, "	"	5 50
" Nitrate, "	"	4 00
" Sulphate, "	"	2 75
" Fleming's Solution, in 4 oz. vials	per bot.	50
Styrax, Liquid	per lb.	60
Sulphur, Flower	"	6
" " by bbl.	"	5
" Iodide, in 1 oz. vials	per oz.	65
" Lac, English	per lb.	18
" " P. & W.	"	30
" Roll, (Brimstone)	"	5
" Vivum	"	15
Sumac, Sicily, (Lead Seal)	"	10
Svapnia, in 1/8 oz. vials	per oz.	2 75
Syrup, Blackberry	per lb.	75
" Buckthorn	"	85
" Ginger	"	35
" Hydrate Chloral	"	1 00
" Iodide Iron	"	60
" Ipecac	"	65
" Morphia	"	50
" Orange Peel, Sweet	"	45
" Pepsin	"	60
" Poppies	"	65
" Rhei	"	65
" " Arom.	"	50
" Sarsaparilla	"	40
" " Comp.	"	50
" Seneka	"	65
" Squills	"	35
" " Comp., (Hive Syrup)	"	50

Syrup, Stillingia	per lb	$	50
" " Comp.	"		50
" Tolu	"		35
" Wild Cherry	"		35
All other Syrups furnished at lowest market rates.			
TAMARINDS	per lb.		12
Tannin	"	1	85
" in 1 oz. bots., vials included	per oz.		22
Tapers, 3 months	per doz		50
" 6 months	"		75
Tapioca	per lb.		10
" by bag	"		8
Tar, Barbadoes	"		18
" North Carolina	per bbl	5	50
" " in 1 gall. buckets	per doz	5	50
" " " ½ "	"	4	00
" " in quart cans	"	2	75
" " in pint cans	"	1	75
Tartar Emetic, Crystals	per lb.	1	00
" Powd.	"		85
Tea, Breast, German	"		50
Tin Foil, Heavy	"		38
" " Medium	"		45
" " Thin	"		55
" Grain	"		60
" Muriate, Crystals	"		40
" " Solution	"		18
Tincture Aconite Leaves	"		50
" " Root	"		75
" Aloes	"		35
" Aloes and Myrrh, (Elixir Pro)	"		60
" Arnica, U. S. P.	"		50
" Assafœtida	"		60
" Benzoin, Comp.	"		75
" Buchu	"		50
" Calabar Bean	"	2	00
" Camphor, U. S. P.	"		50
" Cannabis Indicæ	"	1	50
" Cantharides	"		50
" Capsicum	"		50
" " and Myrrh	"		60
" Cardamon	"		50
" " Comp.	"		75
" Cascarilla	"		50
" Castor	"	1	00
" Catechu	"		50
" Cimicifuga, (Black Cohosh)	"		50
" Cinchona, Red	"		60
" " Comp.	"		60

Tincture	Colchicum Seed	per lb.	$ 50
"	" Root	"	60
"	Columbo	"	50
"	Conium	"	50
"	Cubebs	"	50
"	Digitalis	"	50
"	Ergot	"	75
"	Galls	"	65
"	Gelseminum	"	75
"	" Merrill's	"	
"	Gentian, Comp.	"	50
"	Ginger, Jam.	"	60
"	Guaiac	"	65
"	" Ammoniated, (Dewey's)	"	1 00
"	" Comp.	"	1 00
"	Hellebore	"	60
"	Hops	"	50
"	Hyoscyamus	"	50
"	Iodine	"	1 00
"	" Comp.	"	1 50
"	Iron, Acet.	"	1 00
"	" " Ether	"	1 25
"	" Muriate	"	35
"	Jalap	"	1 00
"	Kino	"	50
"	Lobelia	"	50
"	Musk, Com.	"	3 50
"	" True	"	8 50
"	Myrrh	"	60
"	Nux Vomica	"	60
"	Opium, U. S. P., (Laudanum)	"	1 50
"	" Camph., (Paregoric)	"	75
"	" Deodorized	"	2 75
"	Quassia	"	35
"	" Comp.	"	90
"	Rhatany	"	60
"	Rhei	"	50
"	" and Aloes	"	75
"	Sanguinaria	"	50
"	Seneka	"	65
"	Serpentaria	"	50
"	Soap, (Camphorated, Opodeldoc)	"	75
"	Squills	"	50
"	Stramonium	"	50
"	Tolu	"	60
"	Valerian	"	50
"	" Ammon.	"	80
"	Veratrum Viridi	"	75
Tripoli,	Mount Eagle	per doz.	60
"	Soap	"	1 10
Turmeric		per lb.	13

Turpentine, Venice	per lb.	$ 35
" White Pine, Hard	"	8
" " Soft	"	10
Tutty, Prepared	"	1 00
Twines, Cotton, Sea Island, assorted colors	per doz.	1 00
" " White, 4 lb. ply	per lb.	35
" Hemp, Coarse	"	40
" Linen, Pink	"	1 00
" " Variegated	"	50
" " White	"	1 00
" Paper	"	25
URANIUM, Chloride, in 1 oz. vials	per oz.	1 15
" Nitrate, "	"	1 15
Oxide, "	"	1 15
VERATRIA, in 1-8 oz. vials	per oz.	3 00
Verdigris, Lump	per lb.	44
" Powd.	"	55
Vermillions. (See Paints, page 87.)		
WAFERS, Medicinal	per doz.	1 25
Washing Crystal	per gros	3 00
Water, Cherry Laurel	per lb.	45
" Cologne, (see Perfumery list under Sundries.		
" Distilled	per gal.	25
" Orange Flower, (Chiris's)	per lb	40
" " " Triple, in ½ pt. bots.	per doz.	3 00
" " " " in pt. bots	"	4 50
" " " " in qt. bots.	"	5 50
" Rose, Distilled, (Chiris's)	per lb.	45
" " Triple, " in ½ pt. bots.	per doz.	3 00
" " " " in pt. bots.	"	4 50
" " " " in qt. bots.	"	5 50
WATERS, Chalybeate and others.		
Congress, (C), Pints, 4 doz. in case	per case	8 50
" Quarts, 2 doz. in case	"	7 00
Empire, (E), Pints, 4 doz. in case	"	7 50
" Quarts, 2 doz. in case	"	6 00
Gettysburg, Quarts, 2 doz. in case	"	8 00
Geyser, Saratoga, Pints, 4 doz. in case	"	9 00
" " Quarts, 2 doz. in case	"	7 50
Hawthorn, Pints, 4 doz. in case	"	8 75
" Quarts, 2 doz. in case	"	7 00
Kissingen, (Hanbury Smith's), 3 doz. in case	"	7 50
St. Louis Magnetic Spring, Quarts, 2 doz. in case.	"	8 00
Vichy, (Hanbury Smith's,) 3 doz. [illegible]	"	7 50
Granular Effervescent Salts (H. S. & H.), (see p. 316), per doz.		
Wax Bayberry	per lb.	75

Wax, Bottle, Red, Green, Black and Yellow	per lb.	$ 25
" Fruit, Red	"	5
" Paraffine	"	35
" Sealing, Express	"	75
" " Irish Harp	"	1 00
" " Superfine, Black	"	75
" " " London Red	"	75
" White, Opt.	"	65
White Precipitate	"	2 50
WINES, MEDICATED—		
Wine, Antimony	per lb.	75
" Colchicum Root	"	75
" " Seed	"	75
" Ergot	"	75
" Ipecac	"	75
" Opium	"	2 75
" Pepsine	"	1 25
Wood, Naphtha	"	60

ZINC, Acetate	per lb.	75
" Bromide, in 1 oz. vials	per oz.	50
" Carbonate	per lb.	35
" Chloride	"	1 00
" " in 1 oz. vials	per oz.	18
" " Solution	per lb.	25
" Cyanide, in 1 oz. vials	per oz.	40
" Ferrocyanide, in 1 oz. vials	"	40
" Iodide, in 1 oz. vials	"	75
" Lactate, "	"	50
" Oxide, True	per lb.	30
" Phosphate	"	2 40
" Sulphate	"	9
" Tannate, in 1-4 oz. vials	per oz.	60
" Valerianate, in 1-4 oz. vials	"	75

☞ **See next page for addenda to this list.**

In shipping by water, we do not insure unless you so direct.

Filtering Racks, of Galvanized Wire.

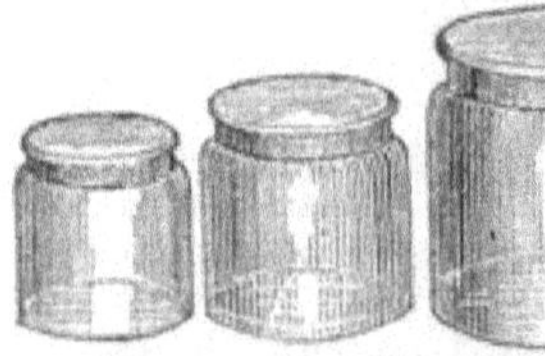

Ointment Pots, Glass, Metallic Covers.

Addenda to Chemical and Drug List.

Annattoine	per lb.	$ 1 65
Bark, Fringe Tree	"	65
Brandy Caps	per gross,	1 00
Broma	per lb.	45
Citrate Magnesia 1 s	"	1 25
Croton Chloral	per oz.	3 00
Flavine (See Dyes, page 105)		
Glycerine, Price's (bottle included)	per lb.	1 00
Herbs—Cotton Root	"	35
Hoffman's Anodyne	"	45
Ink Powders	per doz.	1 25
Iron Hydrocyanate	per oz.	85
Iron Muriate Solution (for making Tinct.)	per lb.	
Leaves, Buchu, powd	"	40
" Uva Ursi	"	13
Leeches	per doz.	1 00
Lime Sulpho. Carb	per oz.	35
Morphia, Bromide	"	7 50
Oil, Black Pepper	per lb.	1 35
" Neroli, Petit Grain	per oz.	1 75
" Olive, Garnier Fils', 5 pint bottles	each,	3 25
" " " " quart "	per doz.	9 00
" " " " pint "	"	5 25
" " Mottet's quart bottles	"	12 00
" " " pint "	"	6 50
" Thyme, White	per lb.	2 50
Plaster Paris, Dental	"	5
" " Common	"	3½
Potash, Caustic, Com	"	65
" Liquor	"	20
" Silicate	"	30
Quinine Bromide	per oz.	4 25
Root, Pareira Brava	per lb.	75
Saucers, Pink	per doz.	75
Seed, Poppy	per lb.	25
" Rape	"	10
" Sun Flower	"	40
Silver Nitrate, in cones	per oz.	1 50
Soda Phosphate and Ammonia	per lb.	1 50
" Sulpho. Carbolate	per oz.	35
St. John's Bread	per lb.	15
Tartrate Borax	"	1 75
Vaccine Virus (See page 81.)		
Wax, Bees, Prime Yellow	"	45
Wicks, No. 0	per gross,	40
" " 1	"	50
" " 2	"	90
" " 3	"	1 75
Zinc, Impure, (Tutty)	per lb.	25

See Addenda to this list, page 73.

Patent Medicines

AND

PROPRIETARY ARTICLES.

It will be our endeavor to maintain our reputation of carrying the Largest and most complete Stock in the West. *Handling this class of goods in largest quantity lots, and having the agency for the Northwest of many of the* leading *proprietary articles, we are prepared to make* special prices to jobbers. *In re-classifying our list, we have, as far as possible, done so under the* manufacturers' *name, as being more convenient for reference.*

VAN SCHAACK, STEVENSON & REID.

	Retail.	Wholesale. Per Doz.
ALBESPEYERE'S Blistering Tissue, large	$1 50	$12 00
" " " small	1 00	7 25
Alexander's Kid Glove Cleaner	25	1 90
Allcock's Porous Plasters	25	1 45
Allen's Lung Balsam (see page 331)	1 00	7 20
" Nerve and Bone Liniment	25	1 75
" (Mrs.) Hair Restorer	1 00	6 75
" " Zylobalsamum	50	3 50
Allport's Dentifrice	50	3 75
American Med. Co.'s Blood and Liver Pills	25	1 50
" " " Tonic of Health, large	1 00	6 00
" " " " " small	50	3 50
Anderson's Dermador, large	50	3 80
" " small	25	1 90
Apiol Capsules	1 00	8 00
Arnold's Bilious Pills	25	1 50
" Porous Plasters	25	1 50
" Soothing and Quieting Cordial	25	2 00
Arnold's (Seth) Balsam (See page 359)	25	2 00
" " Cough Killer, large	1 00	8 00
" " " " medium	50	4 00
" " " " small	25	2 00

PATENT MEDICINES—*Continued.*	Retail.	Wholesale Per Doz
Ashbaugh's Wonder of the World, large	$1 00	$ 8 00
" " " " small	50	3 75
" German Hair Renewer (	1 00	8 00
" Plant and Root Bitters, liquid	1 00	8 00
" " " " " package	1 00	8 00
" King of Malaria Pills	1 00	8 00
Atwood's Bitters, Jaundice	50	2 50
Horse Medicine, H. H. H., large	1 00	7 50
" " " small	50	3 75
" Quinine Tonic Bitters	1 00	9 00
Austin's Ague Drops	50	4 00
Ayer's Ague Cure	1 00	7 75
" Cherry Pectoral	1 00	7 75
" Hair Vigor	1 00	6 75
" Pills (see page 303)	25	1 50
" Sarsaparilla	1 00	7 75
BADEAU'S Blood Maker	1 00	7 50
" Bronchial Rock	35	2 50
" Liver Cure	1 00	7 50
Baker's Cod Liver Oil	1 00	6 50
" Citrate Magnesia Solution	35	2 75
" Pain Panacea, large	1 00	7 00
" " " medium	50	3 50
" " " small	25	1 85
Bally's Antidote	1 50	13 00
Barnes's Magnolia Water	1 00	7 50
Barrell's Indian Worm Confections	25	1 50
" " Liniment	35	2 25
Barrett's Hair Restorer	1 00	6 00
Barr's Ague Medicine	1 00	7 50
" Pectoral Elixir, large	1 00	7 50
" " " small	50	3 50
Barry's Black Hair Dye (see p.)	1 00	8 00
" Safe " "	1 00	8 00
" Pearl Cream	50	4 00
" Marfilina*	50	4 00
" Tricopherous	50	3 50
Batchelor's (W. A.) Alaska Seal Oil	1 00	8 00
" " Curative Ointment	50	4 00
" " Dentifrice	75	6 50
" " Hair Dye, large	3 00	27 00
" " " " small	1 00	8 00
Bateman's Drops (stamped)	15	75
Baxter's Worm Lozenges	25	1 75
Bazin's Camphor Ice	25	1 60
" Depilatory Powder	50	2 75
Becker's Eye Balsam	25	1 90

PATENT MEDICINES—*Continued.*	Retail.	Wholesale. Per Doz.
Bell's Specific Pills	$1 00	$ 7 50
Belloc's Charcoal	1 50	9 00
" " Lozenges	1 00	6 50
Benedict's G. W. Compound	1 00	7 00
" Bronchial Lozenges	25	1 50
Bennett's Root and Plant Pills	25	1 75
Benson's Asthma Cure	2 00	13 50
Benton's Pine Tree Tar Troches	25	1 50
Bernard's (Dr. W. V.) Vegetable Chill Cure	1 00	7 50
Bicknell's Dysentery Syrup	50	4 00
Bininger's London Dock Gin	1 00	8 25
Biokrene	1 00	8 00
Biroth's Powdered Pepsin	1 00	8 00
Blancard's Pills, large	1 50	10 50
" " small	1 00	6 00
Blackman's Balsam, large	1 00	7 00
" " small	50	3 25
" Red Salve	25	1 75
Blair's Gout and Rheumatic Pills	1 25	10 00
Bliss, Keene & Co.'s F. E. Cundurango, large	2 00	16 50
" " small	1 00	8 50
Blumm's F. & A. Pills	75	6 50
Bœrhave's Holland Bitters	1 00	8 00
Boker's Stomach "	1 50	14 00
Borden's Ext. Beef	50	3 75
Boschee's German Syrup (see page)	75	5 50
Boudault's Pepsin (Powder)	2 00	15 00
Boyd's Medicated Cream	50	3 50
Bragg's Arctic Liniment	25	1 75
" Mustang Liniment, large	1 00	7 40
" " " medium	50	3 70
" " " small	25	1 85
Brandreth's Pills	25	1 50
Brant's (J. W.) Condition Powders	25	1 50
" " Dexter Liniment (see page 302)	50	3 50
" " Pulmonic Balsam	1 00	9 00
" " Purifying Extract	1 00	9 00
" " Soothing Balm	35	2 25
" " Turkish Ointment	25	1 75
" " Yankee Salve	20	1 00
Brewer's (H. & J.) Home Made Beer	25	2 00
Bristol's Sarsaparilla	1 50	13 00
British Oil, (stamped)	15	75
Brou's Injection	1 50	9 00
Browen's Rat Killer	25	1 50
Browne's Chlorodyne	1 00	5 50
Brown's Blistering Tissue	2 00	13 50
" Camphorated Saponaceous Dentifrice	25	1 75
" Ext. Jam. Ginger	50	3 75

PATENT MEDICINES—*Continued.*	Retail.	Wholesale, Per Doz.
Brown's Red Horse Powders, large	$ 40	$ 3 00
" " " small	20	1 50
" (Dr. J. H.) Expectorant Syrup	1 00	7 50
" (N. K.) Ess. Jam. Ginger	50	3 50
" " Ext. Buchu, (see p. 429.)	1 00	7 50
" " Teething Cordial	25	1 75
" (O. P.) Assimilant	2 00	15 00
" " Acacian Balsam	1 00	8 00
" " Blood Purifier	1 00	8 00
" " Floral Bloom	1 00	8 00
" " Herbal Ointment	50	4 00
" " Liver Invigorator	1 00	8 00
" " Renovating Pills	50	4 00
" " Vermifuge	50	4 00
" " Woodland Balm	50	4 00
" Shakers' Valerian	50	3 25
" Troches, large (see page 311)	1 50	10 75
" " medium	75	5 50
" " small	35	2 60
" Worm Comfits	25	1 60
Bryan's Life Invigorator	1 00	7 75
" Pulmonic Wafers	50	3 25
Buckingham's Whisker Dye	50	3 75
Buchan's Carbolic Balm Ointment	25	1 75
" Hungarian Balsam	1 00	8 00
Budd's Liniment	25	1 85
" Columbian Ointment	25	1 85
Bull's Cedron Bitters	1 00	8 00
" Concentrated Extract Buchu	1 00	8 00
" Sarsaparilla (see page 358)	1 00	8 25
" Worm Confections	25	1 45
Burdsall's Arnica Liniment	50	3 25
Burnett's Cocoaine (see p. 309)	1 00	7 25
" Cod Liver Oil, (stamped)	1 00	8 00
" Florimel	1 00	7 50
" Kalliston	1 00	7 50
" Oriental Tooth Wash	1 00	7 50
Butt's Liniment, or Excelsior Medicine	50	3 50
" Pills	25	1 50
CALABAR Grains	50	3 75
Calder's Dentine	25	1 90
Carlsbad Salts	1 00	8 50
Carter's Ext. Smart Weed, large	1 00	7 00
" " small	50	3 75
Cary's Cough Cure (see page 288)	25	1 50
" Toothache Drops	25	1 50
Castle's Magnolia Catarrh Snuff	25	1 85
Catlin & Freese's Am. Mender, Cement	25	1 50

PATENT MEDICINES—*Continued.*	Retail.	Wholesale. Per Doz.
CASWELL, HAZARD & CO.'S PREPARATIONS (see p. 360)		
Camphor Ice with Glycerine	$ 25	$ 1 50
Chlorate of Potash Lozenges	25	1 50
Cod Liver Oil, (Hazard & Caswell's) stamped	1 00	6 50
" " " Iodo-Ferrated	1 50	10 00
" " " with Iodine, Brom. and Phosph.	1 50	10 00
Cold Cream, (Rose)	35	2 50
Dentine, (Tooth Powder)	50	3 00
Eau Aromatique, (Mouth Wash)	1 00	7 00
Elixir Phosphate Iron, Quinine and Strych.	3 00	22 00
" Calisaya Bark, simple	1 25	9 00
" Pepsin, Bismuth and Strychnia	3 00	22 00
" Taraxacum Comp.	1 50	10 50
Ferro-Phosphorated Elixir of Calisaya Bark	1 50	10 50
Ferro-Phosph. Elix. Calisaya Bark, with Bismuth	1 50	10 50
" " " " " Strychnia	1 50	10 50
Formodenta, (Tooth Paste)	75	5 00
Glyceria, (Hair Dressing, without oil), 8 oz.	1 00	8 00
Juniper Tar Soap, (for chapped hands, etc.)	50	2 50
Lotus Balm, (Hair Dressing, with oil), 4 oz.	50	4 00
Oil Philocome, (for the Hair), 2 oz.	75	5 00
Toilet Cologne, 4 oz.	1 00	6 00
" " 8 oz., with ill'mted heraldic label	1 50	10 50
Caylus's (Math.) Capsules, C.	1 00	7 50
" " " C. & C. (stamped)	1 00	8 00
" " " C. C. & M.	1 00	8 00
" " " C. C. Cit. Iron	1 00	8 00
" " " Injection	1 50	9 00
Centaur Liniment (J. B. Rose & Co.) Horse, large	1 00	7 50
" " " small	50	3 75
" (see page 289) " Family, large	1 00	7 50
" " " small	50	3 75
Chalfant's Coco Cream	50	3 25
Chamberlain's Pills, (see page 322)	25	1 35
" Relief, large	50	3 25
" " small	35	2 25
Chapin's Catarrh Inhalers	2 00	10 50
" " Inhalant	50	3 50
Charles's Gin, quarts	1 50	10 50
" " pints	75	6 50
Chaussier's Empress	1 00	7 00
Chevalier's Life for the Hair	1 00	7 00
Cheeseman's Arabian Balsam	50	4 00
" Female Pills	1 00	7 50
Christie's Ague Balsam	1 00	7 50
Christadoro's Hair Dye, small	1 00	8 00
" " Preservative	50	4 00
Clark's Anti-Bilious Compound	1 00	7 75
" Indelible Pencils	35	2 25

PATENT MEDICINES—*Continued.*	Retail.	Wholesale. Per Doz.
Clark's (Sir James) Female Pills, (genuine only)	$1 00	$ 7 50
" Peruvian Syrup, large	2 00	17 00
" " " small	1 00	8 50
Cloverine	25	1 50
Coddington's Capsicum Plaster large	2 00	12 00
" " " small	1 00	6 00
Coe's Cough Balsam, large	1 00	7 50
" " small	50	2 75
" Dyspepsia Cure	1 00	7 50
Coffeen's Chinese Liniment	25	1 85
Cooper's Magnetic Balm	35	2 00
Cook's Balm of Life	1 00	8 75
" Pills, (De La Cour's)	25	1 25
Comstock's Rational Food	75	6 00
Cosmoline, (Houghton's), Plain	50	4 00
" " Camphorated	50	4 00
" " Carbolated	50	4 00
" " Cerate (see page 301)	50	4 00
" " Pomade	50	4 00
" " Rose	50	4 00
" Veterinary, 1, 5 and 10 lb. cans		
Costar's Bed Bug Exterminator	25	1 50
" Corn Solvent	25	1 50
" Electric Powder	25	1 50
" Rat Exterminator	25	1 50
Covell's Cleaning Powder	25	1 75
Cram's (Dr. J. S.) Fluid Lightning	1 00	7 50
Crook's Benzoin Elixir	50	3 75
" Citron Balsam	25	1 75
" Comp. Syrup Poke Root	1 00	7 50
" Wine of Tar	1 00	7 20
Crossman's Specific	1 00	8 00
Croppi Oil (Olio Ricino)	35	3 00
Crumpton's Strawberry Balsam	50	3 00
Curtis & Brown's Household Panacea	35	2 75
Curtis's Mamaluke Liniment	50	3 50
Cutler's Inhalers (see page 291)	2 00	12 00
" Catarrh Inhalant	50	3 50
" Pulmonary Balsam, large	1 00	8 00
" " " small	50	4 00
DALLEY'S Horse Salve	50	4 00
" Pain Extractor	25	1 85
Darley's Heave Remedy	25	1 75
Davis's (Perry) Pain Killer, large, (see page 331	1 00	7 40
" " " medium	50	3 70
" " " small	25	1 85
Dean's Rheumatic Pills	50	4 00

PATENT MEDICINES—*Continued.*	Retail.	Wholesale. Per Doz.
Delamarre's Specific Pills	$1 00	$ 8 00
Delight's Spanish Lustral	1 00	7 00
Dell's (Farm's) Camphor Ice	25	1 50
" Santonine Lozenges	25	1 50
" Lola Montez Tooth Powder	25	2 00
Delluc's Eau Angelique	1 00	8 00
" Biscotine	75	5 50
Denton's Balsam, large	50	4 00
" " small	25	2 00
Deshler's Anti-Periodic or Fever and Ague Pills	1 00	7 75
Desmond's Samaritan Gift, for males	2 00	16 50
" " " " females	3 00	24 00
" " Root and Herb Juices	1 25	10 00
" " Wash (see p 328)	50	4 00
Devine's Pitch Lozenges	25	1 50
DeBing's Pile Remedy	1 00	8 00
DeGrath's Electric Oil	50	4 50
DeJongh's Cod Liver Oil	1 25	10 00
DeLoss's Pearl Drops, (White and Pink)	50	3 50
Diamond (Dr. Evory's) Catarrh Remedy	50	4 00
Dickman's Arnica Strengthening Plaster	25	1 50
" Court Plaster, Arnica, (assorted colors)	10	75
" " " Imperial	10	35
" " " India Rubber	10	75
" " " Mechanics', (on leather)	10	65
" " " Taffeta	10	75
" " " Tablets	25	1 75
" (Littlefield's) Isinglass Plaster	50	4 75
" " Pungents, (screw cap)	25	2 00
" " " (glass ")	25	2 00
Dixon's Aromatic Blackberry Carminative	25	2 00
" (Joseph) Stove Polish	10	60
Dodd's German Cough Balsam, large	1 00	7 50
" " " " medium	50	3 75
" " " " small	25	1 90
" Diarrhœa Cure (see page 326)	35	2 00
" Nervine	1 00	7 50
" Liver Pills	25	1 50
Down's Vegetable Balsamic Elixir, large	1 00	8 00
" " " " small	35	2 85
Drake's Plantation Bitters	1 00	8 25
Duane's Cough Cordial, large	1 00	7 50
" " " small	35	2 75
" Dentition Syrup	25	1 75
" Liver Pills	25	1 50
" Santonine Worm Lozenges	25	1 50
Du Barry's Revelenta Arabica Food	1 50	12 00
Dundas Dick & Co.'s Capsules Copaiba	75	5 50
" " " and Oil Cubebs	1 00	7 50

PATENT MEDICINES—*Continued.*	Retail.	Wholesale. Per Doz.
Dundas Dick & Co.'s Capsules C. C. and Matico	$1 00	$ 7 50
" " Oil Sandal Wood	2 00	15 00
" Matico Injection,	1 00	8 00
Duponco's Golden Female Pills	1 00	7 00
Durno's Catarrh Snuff	25	2 00
Dutcher's Dead Shot	25	1 85
" Lightning Fly Killer......quire	5	45
Dwyer's Bitter Wine of Iron, (see page ... doz.	1 00	8 00
" Cholera Remedy ...... "	50	4 00
" Pectoral Syrup, large ...... "	1 00	7 00
" " " small ...... "	50	3 50
Dyson's Fragrant Dentilave	25	1 75
" Oriental Balm	50	3 50
" Liquid Hair Dye	50	4 00
EDE'S Diamond Cement	35	2 00
" Pungents	1 00	5 50
Edey's Carbolic Baby Wash	1 00	8 00
" " Troches	25	1 85
Eilert's Daylight Pills	25	1 25
" Extract Tar and Wild Cherry	1 00	6 75
Ellis (Chas.) Son & Co.'s Adhesive Plaster, 1 yd. rolls,	50	2 50
" " " 5 "		
" Cit. Magnesia, granular, doz. ..	50	4 00
" " " " 1 lb. bots. lb.	2 00	1 25
" Prepared Citrate (powder)	40	3 00
" Willow Charcoal, doz.	25	1 90
Elgin's Phantom Toilet Powder, large	50	3 50
" " " " small	25	2 00
Elmore's Pills	25	1 50
Englehard 's Vegetable Liver Pills	25	1 50
English Granulated Salts, (in elegant glass screw cap blue bottles.)		
English Granulated Citrate Magnesia	50	[illegible]
" " Kissingen	50	4 50
" " Seltzer	50	4 50
" " Vichy	50	4 50
Ennis's Fragrant Tooth Soap	25	1 50
Espic's Cigarettes	1 00	7 50
Eureka Hair Restorer	1 00	6 50
FAHNESTOCK'S (B. A.) Vermifuge	25	1 50
" (B. L.) "	25	1 50
" " Lung Syrup	25	1 75
" " Worm Pastiles	25	1 50
Fahrney's Blood Cleanser (Panacea)	1 25	10 00
" Uterine Tonic	1 25	10 00

PATENT MEDICINES—*Continued.*	Retail.	Wholesale. Per Doz.
Farnham's Asthma Remedy	$2 00	12 00
“ Tooth Lozenges, large	50	3 50
“ “ small	25	1 75
Farrell's Arabian Liniment	25	1 85
Fayard & Blain's Rheumatic Paper	50	2 75
Fellow's Hypophosphites	2 00	16 50
Fennimore's Cough Syrup, large	50	3 50
“ “ “ small	35	2 00
“ Liniment, large	70	3 50
“ “ small	35	2 00
Fetridge's Balm Thousand Flowers	75	6 00
Filkins's Pills	25	1 50
Fink's Magic Oil	25	1 75
Fitch's Biliary Corrector, large,	1 00	8 25
“ “ “ small	50	4 25
“ Catarrh Liniment	50	4 25
“ “ Specific	50	4 25
“ Cathartic Pills	25	1 90
“ Cough Curer	1 00	8 25
“ Heart Corrector, large	1 50	12 75
“ “ “ small	75	6 50
“ Nervine	50	4 00
“ King and Queen's Toilet	1 00	8 00
“ Pulmonary Balsam	1 00	8 25
“ “ Expectorant	1 00	8 25
“ “ Liniment	50	4 25
“ Universal Tonic	50	4 25
Fitler's Rheumatic Remedy	1 25	9 50
Flagg's Cough Killer, large	1 00	8 00
“ “ “ small	50	4 00
“ Relief, large	1 00	7 75
“ “ small	50	3 75
Foord's Pectoral Syrup, large	1 25	9 00
“ “ “ small	75	5 00
“ Tonic Cordial	35	2 75
Ford's (Dr.) R. & N. Cure	1 00	8 00
Forsha's Alterative Balm, large	1 50	12 00
“ “ “ medium	75	6 00
“ “ “ small	50	3 00
Fosgate's Anodyne Cordial	50	2 75
Fougera's Comp. Dragees Santonine, (see page 342).	50	4 00
“ “ Iceland Moss Paste	50	4 00
“ “ Iodinized Cod Liver Oil	1 50	11 50
“ Ready-made Mustard Plaster	50	4 00
Foutz's Condition Powders	25	1 65
Fowle's Pile and Humor Cure	1 00	7 50
Francis' Water of Happiness	25	1 50
Frank's Grains de Sante	50	4 00
Fronfield's Cattle Powders	25	1 65

PATENT MEDICINES—*Continued.*	Retail.	Wholesale. Per Doz.
Frost's Egypt Salve	$ 25	$ 1 25
" Liniment	35	1 75
" London Pills	25	1 25
Fuller's (Mrs.) Rheumatic Pills	75	5 50
Furgeson's Wonderful Oil	50	3 75
GARDNER'S (Mrs.) Indian Balsam of Liverwort	50	4 00
Gardiner's Rheumatic Compound	1 00	8 50
Gaudichaud's Comp. Ext. Sandalwood, (see page 351)	1 00	8 00
Gayetty's Medicated Paper	50	4 50
Gibson's Injection		
Godfrey's Cordial, (stamped)	15	75
Goodale's Catarrh Remedy	1 00	7 50
Gouraud's Medicated Soap, (see page 347)	50	4 50
" Oriental Cream	1 50	12 00
Graefenberg Uterine Catholicon, (see page 362)	1 50	8 75
" Children's Panacea	50	3 50
" Consumptive's Balm	3 00	18 00
" Dysentery Syrup	50	3 50
" Eye Lotion	25	1 50
" Fever and Ague Remedy	50	3 50
" Green Mountain Ointment	25	1 50
" Health Bitters	25	1 50
" Manual of Health	50	3 50
" Pile Remedy	1 00	6 50
" Sarsaparilla Compound	1 00	6 50
" Vegetable Pills	25	1 50
Graham's Barber Soap, bars, 12 bars in box, per lb.	35	2 75
" Hair Dye, (Black and Brown), per doz.	50	4 25
Granvil's Pile Salve	1 00	7 50
Gray's Ointment	25	1 85
Green's August Flower, (see page 315)	75	5 50
" Oxygenated Bitters	1 00	7 75
Gregg's Constitution Water	1 00	8 00
" " Life Syrup	1 50	10 50
Gregory's Instant Cure of Pain	25	1 75
Grimault's Injection	1 50	10 50
Griswold's Salve	25	1 85
Guffroy's Cod Liver Oil, Dragees, large	2 00	16 50
" " " " " medium	1 25	11 00
" " " " " small	75	6 50
Guild's Green Mountain Asthma Cure	2 00	12 00
Guysott's Sarsaparilla and Yellow Dock	1 00	7 50
HAGAN'S Magnolia Balm	75	5 25
Hale's Cough Cordial	1 00	7 50
Hall's Balsam	1 00	7 50
" Vegetable Sicilian Hair Renewer (see page 335)	1 00	6 75

PATENT MEDICINES—*Continued.*	Retail.	Wholesale, Per Doz.
Ham's Oil of Gladness	$ 25	$ 2 25
Hamburg Tea	25	1 00
" " Breast	25	1 75
Hambleton's Hair Stain	50	4 00
Hamlin's Wizard Oil, large	1 00	7 00
" " " small	50	3 50
" Cough Balsam	50	3 50
Hanbury Smith, Hazard & Co., (see page 376)		
Hardy's Eye Balm	25	1 50
Harlem Oil, Genuine (stamped)	15	60
Harris's Cedar Camphor, large	1 50	11 50
" " " small	1 00	7 75
" " " unpressed	50	4 00
Harrison's Peristaltic Lozenges, large	60	4 75
" " " small	30	2 50
Harter's Fever and Ague Specific	75	4 50
" " " " Pills	75	4 50
" German Worm Confection	25	1 50
" Iron Tonic	1 00	7 50
" Liniment	50	3 50
" Liver Pills	25	1 50
" Lung Balm	75	4 75
" Soothing Drops	25	1 75
Harvell's Condition Powders	25	1 50
Hasting's Comp. Syrup Naphtha	1 00	8 00
Hawley's Pancreatic Emulsion	1 50	10 50
" " " Cod Liver Oil	1 50	10 50
" Am. Pepsin Powder	1 00	8 00
Hay's Pile Ointment	1 00	0 50
Hegeman's Benzine, (see page 324)	25	1 60
" Camphor Ice	25	1 75
" Chlor. Potash Lozenges	25	1 75
" Cod Liver Oil	1 00	7 00
" Elixir Calisaya	1 50	12 50
" " " Ferrated	1 25	10 50
Heimstreet's Hair Coloring	75	4 25
Heiskell's Tetter Ointment	50	3 75
Helmbold's Comp. Fluid Buchu	1 25	9 50
" Catawba Grape Pills	50	3 50
" Fluid Extract Compound Sarsaparilla	1 25	9 50
" Improved Rose Wash	75	5 25
Henley's California I X L Bitters, Drug, (see p. 383	1 00	8 00
" " " " Family	1 00	8 00
Henry's Carbolic Salve	25	1 75
" Magnesia, Gen.	1 50	10 50
" Root and Plant Pills	25	1 75
" Vermont Liniment	50	3 75
" World Tonic and Blood Purifier	1 00	8 00
" Worm Confections	25	1 50

PATENT MEDICINES—*Continued.*	Retail.	Wholesale, Per Doz.
Herrick's Pills	$ 25	$ 1 50
" Plasters	25	1 50
Hibbard's Pills	25	1 50
Hill's Balsam, Honey	25	1 00
" Hair Dye	50	4 50
Hilton's Cement	25	1 75
Hobensack's Worm Vermifuge	25	1 85
Hodgson's Diamond Cement	25	1 75
Hoff's Malt Extract, (see page 329)	50	4 50
Hofmann's Hop Pills, (see page)	50	4 00
Holmes's Mexican Fluid Ext. Buchu	75	6 00
Holloway's Arnica Plaster, large	50	2 25
" " " medium	35	1 75
" " " small	25	1 25
" Ointment, (stamped), large	1 00	7 75
" " " medium	62	4 50
" " " small	35	1 75
" Pills, (stamped), large	1 00	7 75
" " " medium	62	4 50
" " " small	25	1 75
" Worm Confections	25	1 50
Hoofland's German Bitters	1 00	8 00
" " Tonic	1 50	11 75
" Greek Oil	50	4 00
" Podophyllin Pills	25	1 75
Hooper's Female Pills, Genuine	1 00	7 50
" " " Imitation	25	75
" Aromatic Cachous, metal	25	1 50
" " " paper	25	1 25
" Ladies' "	25	1 50
" Pastiles	25	1 75
Horsford's Acid Phosphate	1 00	6 00
Hostetter's Stomach Bitters, (see page 345)	1 00	8 25
Houghton's Pepsin Liquid	1 00	7 00
" " Powder	1 00	7 00
Hovey's Hair Balm	50	4 00
Howe's Ague Tonic	1 00	7 50
" Concentrated Tonic	1 00	7 50
" Tonic Bitters	1 00	7 50
" (S. D.) Arabian Tonic	1 00	8 50
" " " Milk Cure	1 00	8 50
Hubbel's Prepared Wheat	50	4 00
Hueyk's Chlorate Potassa Lozenges	25	1 50
Humbolt's Bitters	1 00	7 50
Humphrey's Specifics (Nos. 1 to 15 inclusive)	25	1 75
" " " 16 to 34 " exc'pt 28, 32 & 33	50	3 50
" " " 28, 32 and 33	1 00	7 00
" " " 28, (see page) ... pkg.	5 00	48 00
" " Anchor	2 00	14 00
" " Crescent	5 00	36 00

PATENT MEDICINES—*Continued.*	Retail.	Wholesale. Per Doz.
Humphrey's Specific, Star	$5 00	$36 00
Hunnewell's Eclectic Pills	25	1 85
" Opal Cement	30	2 25
" Tolu Anodyne	50	3 75
" Universal Cough Remedy, large	50	3 75
" " " " small	25	1 85
Hunt's Liniment	50	3 75
Huntley's Benzine	15	1 25
Hurd's Liver Stimulant	1 00	8 00
Husband's Calcined Magnesia	50	3 25
" Isinglass Adhesive Plaster	1 00	7 20
Hutchins's Headache Pills	25	1 50
Hyatt's Infallible Life Balsam	1 00	9 00
Hyde's (Rob't H.) Chiropodin	25	
IMPORTED French Botanical Bitters	1 00	8 00
Imperial Granum, large	1 25	10 00
" " small	75	6 00
Ingraham's Macedonia Oil	50	4 00
" Worm Candy Tablets	25	1 50
" Blood and Liver Pills	25	1 50
Isaacsen's Sure Pop, (see page 318)	25	1 50
JACKSON'S Home Stomach Bitters	1 00	7 00
Jayne's Ague Mixture, (see page 341)	1 00	7 20
" Alterative	1 00	7 20
" Carminative Balsam, large	50	3 75
" " " small	35	2 75
" Expectorant	1 00	7 20
" Hair Tonic	1 00	7 20
" Liniment, or Counter-Irritant	50	3 75
" Liquid Hair Dye	1 00	8 50
" Sanative Pills	25	1 60
" Tape-Worm Specific	2 00	19 00
" Tonic Vermifuge	35	2 75
Jaques's Worm Cakes	25	1 50
Jew David's Hebrew Plaster	25	1 50
Jewsbury & Brown's Oriental Tooth Paste, large	1 25	8 50
" " " " small	75	5 50
Johnson's Anodyne Liniment	35	3 00
" Condition Powder	25	1 50
" Rheumatic Compound	1 00	7 50
" Vigor of Life, large	1 00	6 50
" " " small	50	3 25
Jouven's Glove Cleaner	25	2 00
Judkin's Specific Ointment	25	1 50
Judson's Mountain Herb Tea	25	1 50
" Pills	25	1 50

PATENT MEDICINES—*Continued.*	Retail.	Wholesale. Per Doz.
KALCKOFF (Dr. A. F.) & Co.'s Mexican Fever Cure Bitters	$1 00	$ 8 00
Kalckoff (Dr. A. F.) & Co.'s Nerve Soother	1 00	8 00
Karoly's Catarrh Remedy, Nos. 1 and 2	1 00	7 50
Kasirine, (for the Hair)	1 00	8 50
Kay's Coaguline	25	2 00
Kearney's (F. E.) Buchu	1 00	8 00
" Rose Wash	50	4 00
Keir's Genuine Petroleum	50	3 50
Kellogg's Worm Tea	25	1 75
Kendall's Amboline	1 00	6 50
Kennedy's (Donald) Hair Tea, (see page 294)	1 00	6 75
" " Healing Ointment	50	3 85
" " Medical Discovery	1 50	12 75
" " Rheumatic Dissolvent	1 50	12 75
" " " Liniment	50	3 85
" " Salt Rheum Ointment	50	3 85
" " Scattering "	1 00	7 75
" " Scrofula "	1 00	7 75
" (S. H.) Hemlock Plasters	25	1 50
" " " Ointment	50	4 00
" " " Liniment	1 00	7 50
" " " Pinus Candensis	2 00	18 00
Kerr's System Renovator	1 25	9 00
Kibbe's (S. C.) Cardamons	10	75
" " Flag Root	10	75
Kiefer's Taraxine or Liver Invigorator	1 00	8 00
Kidder's (Mrs.) Cordial	1 50	8 00
" Asthmatic Pastiles, (advanced)	50	2 75
" Indelible Ink	25	2 00
" Tonic Powders		
Kimm's Holland Worm Cakes	25	1 50
King's Prepared Prescription	1 00	7 50
Kinsman & Wardner's "Take Out"	25	1 50
Kromer's Hair Dye, (Upham's)	50	4 00
" Anti-Bilious Pills		
LAIRD'S (G. W.) Bloom of Youth	75	5 50
Lallemand's Gout Specific	2 00	12 50
Lamberson's Nerve and Bone Liniment	25	2 00
Langley's Root and Herb Bitters, quarts	1 00	8 00
" " " " pints	75	6 25
" " " " ½ pints	35	3 50
Lanman & Kemp's Florida Water	1 00	6 75
Laville's Gout Liquor	5 00	42 00
Leamon's Aniline Dyes, the best in use, (see page 319)	25	1 75
Lee's (Windham) Pills	25	1 50
Liebig's Texan Extract of Beef, ¼s	75	4 50
" " " " ½s	1 25	8 50

PATENT MEDICINES—*Continued.*	Retail.	Wholesale. Per Doz.
Liebig's Texan Extract of Beef, ½s	$2 00	$15 00
" " " " 1s	3 50	27 00
Lightner's Sticky Fly Paper		
Lilly & Phelan's Aromatic Liquid Pepsin	1 75	15 00
Lindsey's Blood Searcher	1 00	7 50
Littlefield's Constitutional Catarrh Remedy	1 00	8 00
Louden's Sanative Pills	25	1 85
Low's Electric Liniment	50	3 50
Ludlum's Gon. Specific	1 00	7 00
Lunt's Eye Salve	50	3 00
" Worm Candy	25	1 50
Lyon's French Periodical Drops	1 50	8 50
" Kathairon	50	3 75
" Magnetic Insect Powder	25	1 85
" Tooth Tablets	50	3 75
MANN'S Ague Balsam	1 00	7 50
" Hair Luxuriant	1 00	6 50
Mamaluke (Original) Liniment	25	1 75
Marchisi's Uterine Catholicon	2 00	16 00
Marshall's Catarrh Snuff	25	2 00
" Catholicon (see Graefenberg)	1 50	8 75
Marsden's Pectoral Balm, large	1 00	8 00
" " " small	50	4 00
Marvin's Cod Liver Oil	1 00	6 50
Mason's Challenge Blacking, large	15	70
" " " medium	10	45
" " " small	5	35
" Hair Dye, large	50	3 50
" " small	25	1 75
Mathew's Venetian Liquid Hair Dye	75	5 75
McAllister's All-healing Ointment	25	1 65
McGuire's Extract Benne Plant	75	5 75
McKay's Chinee Oil	50	4 00
McLain's Bon Bons—Vermifuge (see page)	25	1 50
" Candid Castor Oil	25	1 75
" Cod Liver Oil	1 00	6 00
McLane's (Dr. C.) Celebrated Vermifuge	25	1 50
" " Liver Pills, (see page 309)	25	1 50
McLean's (Dr. J. H.) Candy Vermifuge	25	1 50
" " Strengthening Cordial & Blood Purifier	1 00	8 50
" " Universal Pills	25	1 50
" " Volcanic Oil Liniment, large	1 00	7 00
" " " " medium	50	3 50
" " " " small	25	1 75
" " Wonderful Healing Plaster	25	1 50
McMunn's Elixir Opium	50	3 75

PATENT MEDICINES—*Continued.*	Retail.	Wholesale. Per Doz.
Mead's Catarrh Cure	$ 50	$ 3 50
Melvin's (Mitchell's Nov.) Adhesive Plaster, ½ yd. rolls, pr yd.	75	37
" " " " 1 " "	75	40
" " " Ready cut, ⅜ in. per doz.	15	1 12
" " " " ½ " "	20	1 37
" " " " ¾ " "	25	1 62
" " " " 1 " "	30	2 25
" " " " 2 " "	50	3 50
" Porous Plaster, 5 yd. rolls, each	3 00	2 00
" " per doz.	25	1 25
" Surgeons' Isinglass Plaster, on silk, (black, flesh, and white), per yard	1 00	65
" Surgeons' Companion, (walnut box), each	2 00	1 25
Merchant's Celebrated Gargling Oil, large	1 00	7 40
" " " " medium	50	3 75
" " " " small (see p 377)	25	1 85
" " " " (Family) small	25	1 85
" Vegetable Worm Tablets, (see page).	25	1 50
Merrill's Eclectic Veg. Irritating Plaster, in ¼ lb. rolls	25	35
" " " " " in tin boxes, doz.	25	1 75
" Neutralizing Cordial	1 00	8 00
" Syrup Sarsaparilla Comp.	1 00	8 00
" " Stillingia Comp.	1 00	8 00
Metcalf's Great Rheumatic Remedy	1 00	7 50
Miller's Antheo (for the complexion)	50	3 75
" Hair Dye	50	4 50
" Magnetic Balm	25	1 90
" Peerless Blacking, large	10	70
" " " small	5	35
" Sewing Machine Oil	25	1 50
Mishler's Herb Bitters	1 00	8 00
Mitchell's Syrup Ipecac	50	3 25
" Universal Liniment, large	1 00	8 00
" " " small	50	4 00
Moffatt's Life Pills	25	1 50
" Phœnix Bitters	1 00	7 50
Moller's Norwegian Cod Liver Oil	1 00	6 75
Moore & Taylor's Pastilles de Paris	50	2 75
Moore's Essence of Life	25	1 90
Morehead's Magnetic Plaster	25	1 85
Morgan & Son's Sapolio	25	1 25
Morrison's Pills, No. 1	75	5 50
" " No. 2	75	5 50
Morse's Indian Root Pills	25	1 50
Mortimer's Medicated Paper	50	2 25
" Jones's Packets	10	60
Morton's Chemical Eraser	25	1 65
Mother Noble's Syrup, large	1 00	8 40
" " small	50	4 20

PATENT MEDICINES—*Continued.*	Retail.	Wholesale. Per Doz.
Mott's Chalybeate Pills	$ 50	$ 2 75
" Liver Pills	25	1 50
Muchmore's Glue Pots	40	2 50
Murison's (Jas.) Pectoral Cough Candy	10	60
Murray's Fluid Magnesia	75	4 75
NESTLE'S Mother's Milk Substitute	1 00	8 00
Neuril	1 00	8 00
Neuriline	1 00	8 00
Nichols's Elixir Val. Ammonia	50	3 25
" " Bark and Iron	1 00	8 00
" Granular Citrate Magnesia	35	2 75
" Liebig's Food	25	2 25
" Medicinal Cod Liver Oil	1 00	6 75
" Syr. Sarsaparilla and Iod. Lime	1 25	8 50
Nowill's Honey of Liverwort	35	2 75
Norwood's Tincture Veratrum Viridi	1 50	13 00
OSBORN'S Golden Ointment, large	50	3 50
" " " small	25	1 50
Osgood's Indian Chologogue	1 50	12 75
Owen's Extract Buchu	1 00	7 50
PAGE'S Arnica Oil	25	1 75
" Climax Salve	25	1 85
" Mandrake Pills	25	1 50
Painter's Diamond Cement	25	1 75
Palmer's Cosmetic Lotion	1 00	8 00
" Invisible Powder	25	2 00
Parmelee's Biliary Corrector	1 00	8 00
" Vegetable Pills	25	1 50
Parrish's Liquid Rennet	35	2 00
Parson's Purgative Pills	25	1 50
Patchin's Magnetic Oil	35	2 75
Payson's Indelible Ink	35	2 00
" " " Combination	50	4 50
Pearls's White Glycerine, large	50	12 00
" " " small	75	6 00
Pearson's Cream Blanche	75	4 50
Peckham's Healing Balsam	50	2 75
Peleg White's Salve	25	85
Perrin's Fumigator, for Catarrh, large	1 00	7 75
" " " small	25	1 65
Perry's Comedone. (see p. 316)	2 00	14 50
" Dead Shot Vermifuge	25	1 90
" Moth and Freckle Lotion	2 00	14 50
Petit's Canker Balsam	25	1 75
" American Eye Salve	25	1 50

PATENT MEDICINES—*Continued.*	Retail.	Wholesale. Per Doz.
Phalon's Hair Invigorator, large (see p. 349)	$1 50	$11 50
" " " small	75	6 00
" Magic Hair Dye	1 00	8 00
" Paphian Lotion	1 00	8 00
" " Soap	25	2 00
" Snow White Enamel	50	4 00
" Vitalia	1 00	8 00
Phelps's Cathartic Lozenges	35	2 50
Phillip's Milk of Magnesia (see p. 378)	50	4 00
Phinney's (Dr.) Family Pills	25	1 50
Pierce's Favorite Prescription	1 50	11 00
" Golden Medical Discovery	1 00	7 50
" Purgative Pellets	25	1 50
" Nasal Injection	50	4 00
Piso's Cure for Consumption, large (see page 334)	1 00	7 75
" " " small	50	4 00
Pitcher's Castoria (Castor Oil Substitute)	35	2 85
Planten's Gelatine Capsules (See page 344)	1 00	8 00
" " " Castor Oil, No. 1	25	1 40
" " " " No. 3	50	3 50
" " " Cod Liver Oil, No. 3	50	3 75
" " " Copaiba, No. 1	25	1 10
" " " " No. 2	35	1 90
" " " " No. 3	50	2 75
" " " and Cubebs, No. 1	25	1 75
" " " " No. 2	40	3 25
" " " " No. 3	65	4 50
" " " Empty, Nos. 1 and 2	75	5 00
" " " Oil of Turpentine	50	3 25
Poland's Humor Doctor	1 00	7 50
" White Pine Compound	1 00	7 50
Pond's Extract, large (see page 379)	1 75	15 00
" " medium	1 00	8 00
" " small	50	4 00
Poor Man's Bitters	25	2 00
Poor Richard's Eye Water	25	2 00
Popham's Asthma Specific	1 00	7 50
Porter's (Madam Zadoc) Cough Balsam, large	75	6 50
" " " " " medium	50	4 00
" " " " " small	25	2 00
" " " Stomach Bitters	25	2 00
Pozzoni's Chemical Hair Balm (see page)	1 00	8 00
" Med. Complexion Powder, pink	50	3 00
" " " " white	50	3 00
Pratt's Abolition Oil, large	1 00	7 50
" " " small	50	3 75
Pridham's Asthma Specific	1 00	8 00
" Quinine Wine	1 00	7 50

PATENT MEDICINES—*Continued.*	Retail.	Wholesale. Per Doz.
Proctor's Nature's Hair Restorer	$1 00	$ 6 00
Punderson's Condition Powders	25	1 50
QUERU'S Cod Liver Oil Jelly	1 00	8 00
RADWAY'S Railroad Pills	25	1 50
" Ready Relief	50	3 75
" Sarsaparillian Resolvent	1 00	8 50
Radcliffe's (Dr.) Great Remedy or Seven Seals or Golden Wonder, large (see page 308)	1 00	7 50
Radcliffe's (Dr.) Great Remedy or Seven Seals or Golden Wonder, small	50	3 75
Rand's Sea Moss Farine	35	2 20
Ransom's Hive Syrup and Tolu, large	50	4 00
" " " small	35	2 50
" King of Blood	1 00	7 50
Red Horse Condition Powder, large	50	3 00
" " " small	20	1 50
Red Jacket Stomach Bitters	1 00	6 50
Redding's Russia Salve	25	1 85
Reeves's Ambrosia	1 00	7 00
Reid's Sewing Machine Oil	25	1 50
Renne's Pain Killer Magic Oil	25	1 75
Reynold's Gout Specific	3 00	24 00
Richardson's Sherry Wine Bitters	1 00	8 00
Ridge's Food for Infants, No. 1	35	2 75
" " " " " 2	65	5 20
" " " " " 3	1 25	10 00
" " " " " 4	1 75	14 00
Rigollett's Mustard Leaves, flat	50	4 50
" " " in rolls	50	4 50
Ring's Rose Injection	50	4 00
" Vegetable Ambrosia	1 00	6 25
Roback's Stomach Bitters	1 00	8 00
" Scandin, Blood Purifier	1 00	7 50
" Blood Pills, plain	25	1 40
" " sugar-coated	25	1 50
Robinson's Groats	25	1 50
" Oatmeal	25	1 75
" Patent Barley	25	1 50
Rogers's Eff. Citrate Magnesia	50	3 00
" Liverwort Tar and Canch.	1 00	7 50
" Silver Polish	25	2 00
Roman Eye Balsam	25	1 75
Root's (John) Bitters	1 00	8 00
" German Ointment	25	1 85
Rosadalis (Clement & Co.'s)	1 50	10 50

PATENT MEDICINES—*Continued.*	Retail.	Wholesale. Per Doz.
Rowand's (Dr.) Comp. Syrup Blackberry Root	$ 35	$ 1 90
Rushton's (F. V.) Cherry Pectoral Troches	35	2 25
" " Cod Liver Oil	1 00	7 00
SAGE'S Catarrh Remedy	50	3 50
Sand's Clove Anodyne	25	1 75
Sanford's Liver Invigorator	1 00	9 00
Sapoliene	50	4 00
Sarjent's Blackberry Cordial	50	4 00
" Elixir Calisayæ Ferratum	1 50	12 00
" Pain Extractor	50	3 50
Saunders's Tooth Powder	50	3 50
Savory & Moore's Pancreatic Emulsion	1 25	10 50
Sawyer's Fluid Extract Bark, (see page)	1 00	8 00
" (Miss) Salve	50	4 00
Scarpa's Acoustic Oil	1 00	7 00
Schenck's Mandrake Pills, (see page 317)	25	1 60
" Pulmonic Syrup	1 50	12 00
" " Candy	25	1 75
" Sea Weed Tonic	1 50	12 00
Schneeberger's Snuff	10	35
Scovil's Blood and Liver Syrup	1 00	7 50
" Worm Killer	25	1 50
Seely's Catarrh Remedy	1 00	7 00
" Pile Ointment	50	3 50
Seller's Imperial Cough Syrup (see page 325)	50	2 75
" Liver Pills, S. C.	25	1 50
" Vermifuge	25	1 50
Senier's Asthma Remedy	50	4 00
Shallenberger's Fever and Ague Antidote (Pills)	1 50	9 50
Shamrock Oil	50	3 50
Shaw's Voxopines, (for the throat)	25	1 75
Shepard's (C. N.) Compound Wahoo Bitters	50	4 00
Sheridan's Cavalry Condition Powders	25	1 75
" Condition Powders	25	1 75
Sholl's Infallible Chill and Fever Cure	50	4 00
Shoshonee's Remedy	1 00	8 00
Simmon's Liver Regulator, liquid	1 00	7 50
" " " powder	1 00	7 50
Simpson's Hyssop Oil	1 00	7 50
Singer's Gravel Paper, (for birds)	25	2 00
Sloan's Condition Powders, large	50	2 70
" " " small	25	1 35
" Hoof Ointment	1 00	7 50
" Instant Relief	50	3 75
" Ointment, Horse, large	50	3 50
" " Family, small	25	1 75
Smolander's Extract Buchu	1 00	7 50

PATENT MEDICINES—*Continued.*	Retail.	Wholesale. Per Doz.
Smith's Golden Liniment	$ 50	$ 3 50
" Green Mountain Renovator	1 00	7 50
" Old Style Bitters	1 00	8 00
" Tonic Syrup	1 00	8 00
" Window Cleaner	25	1 50
" (Dr. Wm. Manlius) Saccharated Pepsin	1 00	9 00
Soule's Sovereign Balm Pills	25	1 50
Spalding's Cephalic Pills	25	1 50
" Prepared Glue	25	1 75
" Throat Confections	25	1 50
Speer's Port Grape Wine	1 50	9 00
" Standard Wine Bitters	1 00	8 50
Stafford's Olive Tar Inhalers	1 00	6 00
" Iron and Sulphur Powders	1 00	7 50
" Olive Tar	50	3 85
Star Medicated Paper, (best in use)	50	3 00
Stearns's Coco Oleine	75	5 50
Steer's Opodeldoc, (solid) stamped	25	1 37
Stephens's All-right Corn Salve	25	1 75
Sterling's Ambrosia	1 00	8 00
Stevens's Eye Salve	25	1 50
" Vegetine	1 25	10 50
Stone's Cough Syrup	50	3 50
Storm's Cough Candy	10	60
Streeter's Magnetic Liniment	50	3 75
Strickland's Aromatic British Brandy	1 00	6 00
" Cough Balsam	50	3 00
" English Cordial Gin	1 00	6 00
" Kentucky Bourbon Whisky	1 00	6 00
" Liquors, in flasks	50	3 00
" Old Jamaica Rum	1 00	6 00
" Pile Ointment	1 00	6 50
" Stomach Bitters	1 00	6 00
" Vegetable Pills	25	1 50
" Wine of Life (see page 363)	1 00	6 00
St. John's Cough and Consumption Syrup, large	1 00	8 00
" " " " " small	60	4 50
" Condition Powders	25	1 50
" Liniment	30	2 50
Suire's Effervescing Crab Orchard Salts	1 00	8 00
Swaim's Panacea	2 00	18 00
Sweet's Liniment	50	3 25
" Strengthening Cordial	1 00	7 50
Sykes's Atmospheric Insuflators	50	3 50
" Catarrh Cure, dry, in packages	1 00	7 50
" " " liquid	1 00	8 00
Sylvester's Deodorized Benzine	25	1 20

PATENT MEDICINES—*Continued.*	Retail.	Wholesale. Per Doz.
TANNER'S German Ointment	$ 25	$ 1 75
Tarrant's Extract Cubebs and Copaiva	1 00	8 00
" Indelible Ink	50	3 00
" Seltzer Aperient, (see page 329)	1 00	8 25
Taylor's Liver and Stomach Corrector	1 00	8 00
Tebbett's Hair Regenerator	1 00	6 00
Thompson's Aromatic Tooth Soap	25	1 75
" Eye Water, (Troy)	25	1 35
Thorn's Extract Cop. and Sarsaparilla, genuine	1 50	11 25
" " " " imitation	1 00	5 50
Thurston's Tooth Powder, large	50	3 50
" " " small	25	1 75
Tilden's Bromo Chloralum, a powerful deodorizer and disinfectant (see p. 364)	50	3 75
" Fluid Extract Buchu, U. S. P.	1 00	7 50
" Ferrated Wine of Wild Cherry	1 00	8 50
" Sarsaparilla, with Iod. Pot. & Pyro. Iron	1 00	6 50
Tobias's Horse Liniment	1 00	9 50
" Venetian " large	1 00	8 50
" " " small	50	4 25
Topping's Syrup	1 00	8 50
Tourtelot's Extract Beef, large	1 50	13 00
" " " solid, small	75	6 00
Townsend's (Old Jacob) Sarsaparilla	1 25	10 50
" (S. P.) "	1 25	10 50
Townsley's Toothache Anodyne	25	1 60
Trask's (Dr. A.) Magnetic Ointment, large	40	2 85
" " " " small	25	1 85
Tubb's Universal Pain Eradicator	35	3 00
Turlington's Balsam, (stamped)	15	75
Turner's Neuralgia Pills	1 00	7 50
UPHAM'S Asthma Cure	50	4 00
" Depilatory Powder	1 00	7 50
" Fresh Meat Cure	1 00	8 00
" Freckle, Tan and Pimple Banisher	50	4 00
" Hair Gloss and Curling Fluid	50	4 00
" Japanese Hair Stain, black and brown	50	4 00
" Vegetable Electuary for Piles	1 00	7 50
Usquebaugh (Old Settlers') Bitters	1 00	8 00
VALENTINE'S Preparation Meat Juice	1 25	10 00
Van Buskirk's Fragrant Sozodont	75	6 25
Van Deusen's Worm Confections	25	1 50
Van Schaack, Stevenson & R.'s Arnica Plasters, large	35	1 75
" " " " med.	25	1 50
" " " " small	20	1 25

PATENT MEDICINES—*Continued.*

	Retail.	Wholesale. Per Doz.
Van S., S. & R.'s Alaskine Camphor Ice with Glyc'ne	$ 25	$ 1 50
" " Arnica and Bellad. Plasters, No. 2	25	1 50
" " " " " " 3	20	1 25
" "		
" " Belladonna Plasters, No. 1	50	2 25
" " " " " 2	35	1 75
" " " " " 3	25	1 50
" " Poor Man's " on canvas ...	20	1 00
" " " " on kid	25	1 50
" " " " on paper	15	75
" " Porous " (the best made)	25	1 25
" " Warming " No. 2	25	1 50
" " " " " 3	20	1 25
" " Benzine	25	1 50
" " Seidlitz Powders	40	2 75
Van Wagenen's Horse Salve	1 00	7 50
Vaughn's Lithonthriptic Mixture	1 00	8 00
Velpau's Diarrhœa Remedy	50	4 00
" French Female Pills	1 00	6 50
WADSWORTH'S Dry Up	1 00	8 00
Wakefield's Ague and Fever Pills	1 00	7 00
Wakefield's Ague Specific, ...	1 00	6 75
" Blackberry Balsam	35	2 25
" Cathartic Pills	25	1 50
" Cough Syrup, large	70	3 50
" " small	35	2 15
" Egyptian Liniment, large	50	3 25
" " " small	25	1 75
" " Salve	25	1 25
" Eye Salve	25	1 50
" Liver Pills	25	1 50
" Magic Pain Cure	50	3 50
" Nerve and Bone Liniment	25	1 60
" Rheumatic Plaster	25	1 50
" Wine Bitters	75	5 00
" Worm Destroyer	25	1 50
Walker's California Vinegar Bitters	1 00	8 15
Ward's Excelsior Hair Dye, large	50	3 50
" " " " small	25	1 75
" Rosaline	1 00	6 00
Warner's Cough Balsam	1 00	6 50
" Dyspepsia Tonic	1 00	6 50
" Emmenagogue	1 00	6 50
" English Gin (see page 363)	1 00	6 00
" Pile Remedy	1 00	6 50
" Wine of Life	1 00	6 00
Watson's Neuralgia King	1 00	7 75

PATENT MEDICINES—*Continued.*	Retail.	Wholesale. Per Doz.
Wayne's Diuretic Elixir	$1 00	$ 8 00
Weaver's (Dr. S. A.) Cerate	35	2 75
" Canker and Salt Rheum Syrup	1 25	10 50
" Extract Fire Weed	1 00	7 50
Welchman's (E.) Gamgee Powder	25	1 25
" Mamaluke Liniment	25	1 75
Wells's Carbolic Tablets	25	1 75
" Ext. Jurubeba	1 00	8 50
Wheaton's Itch Ointment	50	3 50
Whitcomb's Asthma Remedy	1 50	11 50
" Soothing Syrup	25	1 75
White's (Dr. N. G.) Elixir	35	3 00
" Specialty for Dyspepsia	1 00	8 50
Whittlesey's Ague Cure	50	4 00
" Cough Granules	25	2 00
" Dyspepsia Cure	1 00	7 00
Wilbor's Cod Liver Oil and Phosphate Lime	1 00	9 00
Wilder's Pills	25	1 75
" Sarsaparilla	1 00	8 25
" Wild Cherry	1 00	8 25
" Worm Syrup	25	1 75
Wilhoft's Anti-Periodic Remedy	1 50	8 75
Williams's Asthma Remedy	75	5 00
" (Dr.) Catarrh Specific	2 00	12 00
" (J. L. & Sons) Rheumatic Compound		
Wilson's Headache Pills	25	2 25
Wilson's (E. A.) Preparation of Hypophosphites and Blodgetti	3 00	24 00
" Neuropathic Drops	50	4 25
Winchell's Teething Syrup	25	1 65
Winchester's genuine preparation Hypophosphites, lge	2 00	16 00
" " " " sml	1 00	8 00
" " " " with Manganese	1 00	8 50
Winslow's (Mrs.) Soothing Syrup	35	2 60
Wishart's Dyspepsia Pills	1 00	7 50
" Pine Tree Tar Cordial, qts.	1 25	10 50
" " " pts.	75	5 25
" Worm Drops	25	1 90
Wistar's Balsam Wild Cherry	1 00	7 50
" Cough Lozenges, (stamped)	25	1 00
Wolcott's Pain Annihilator, large	1 00	7 50
" " " medium	50	3 75
" " " small	25	1 85
" " Paint, large	1 00	7 50
" " " medium	50	3 75
" " " small	25	1 85
Wolfe's Celebrated Schiedam Schnapps, qts.	1 50	12 00
" " " " pts.	75	6 50

PATENT MEDICINES—*Continued.*	Retail.	Wholesale. Per Doz.
Wallace's Green Mountain Ointment	$ 50	$ 3 00
" Pile Panacea, (see page 362)	1 00	8 00
" Tonic Stomach Bitters	1 00	7 50
" Wizard Salve	25	1 50
Warner's White Wine and Tar Syrup, (see page 371)	1 00	7 50
West's (S. C.) Cathartic Pills, (see page 356)	25	1 25
" " Dead Shot Itch Cure	50	3 00
" " Instant Cure of Pain, large	50	3 50
" " " " small	25	1 75
" " Jenny Lind Liniment, large	50	3 75
" " " " small	25	2 00
" " Pulmonary Balsam, large	1 00	7 00
" " " " medium	50	3 75
" " " " small	25	2 00
" " Vegetable Liver Pills	25	1 25
White's Salve		
Wigg's (W. B.) Buckeye Salve		
" " Persian Tooth Polish		
Willson's Carbolated Cod Liver Oil	1 00	7 50
Wood's Improved Hair Restorative	1 00	5 50
Woodruff's Chain Lightning Fly Paper, (see p. 350) quire	5	45
Wright's Indian Vegetable Pills, plain	25	1 40
" " " " sugar-coated	25	1 50
" " " Syrup	35	2 00
Wyeth Bros.' Elegant Preparations, (see page 366)		
Wynkoop's Iceland Pectoral	50	3 75
YOUATT'S Condition Powders, large	50	2 50
" " " small	25	1 25

☞ We request our customers to send in their orders for Dr. Gouraud's Oriental Cream, (and other proprietary articles likely to *freeze*,) before *severe cold weather*, as no care we can use in packing is sure to secure against freezing.

ADDENDA.

	Retail.	Wholesale Per Doz
Andersen's Pure Norwegian Cod Liver Oil	$1 00	$6 50
Barton's Dewberry Cordial	1 00	8 00
Bartholic's Mothers' Relief	2 00	16 50
Bell's (G. L.) Balsam of Alpine Moss		
Benedict's Prairie Herb Tonic	1 00	7 00
Bliss' Asthma Remedy, large	2 00	15 00
" " " small	1 00	8 00

ADDENDA—*Continued.*	Retail.	Wholesale. Per Doz.
Brewer's Granular Effervescent Tartrate Soda.......		
" Guarana, Powder.......................lb.		
" " Elixir, (see page 347).......4 oz.		
" Wine of Wild Cherry Bark and Iron......		
Bowen's Catarrh Cure..............................	$ 1 00	$ 8 00
Brigg's Alleviator, for Catarrh, etc................	1 00	7 50
" Corn and Chilblain Remedy..	50	4 00
" Pile Remedy..............................	1 00	7 50
" Throat and Lung Healer.................	1 00	7 50
Brunn's Pure Norwegian Cod Liver Oil......... ...	1 00	6 50
Collins' (Dr. S. B.) Female's Friend, (see page 368)..		
" " Liquor Antidote...............		
Crab Orchard Springs Salts, (see page 369)........	1 00	8 50
Crumb's Ague Killer, (see page 354)...............	1 00	8 00
" Comp. Pills..............................	25	1 50
" Inhalers.................................	2 00	11 50
" Peruvian Alterative	1 00	8 00
" Sweet Castor Oil.........................	25	2 00
" Worm Tablets............................	25	1 90
Daniels' Tantamiraculous and Tonic Bitters, (see page 372)	1 00	8 00
" " " small, "	50	4 00
Delmonico Little's Syrup, large, (see page 355).....	1 00	8 00
Engelhard's Remedy..............................	50	3 50
Forest Tar, for Consumption, etc., (see page 375....	1 00	7 50
" " in Solution, for inhaling, for Catarrh, etc.	1 00	7 50
" " Troches, for Bronchitis, etc.	25	1 50
" " Salve, for purifying, etc.	25	1 50
" " Inhalers, for inhaling the Solution......	50	3 00
" " Soap, cleansing and purifying..........	25	1 50
Gardner's Idaho Gum Crystals, (see page 353)......	25	1 75
Gifford & Tomlinson's H. H. H. Horse Medicine, large	1 00	7 50
" " " (see p. 323) small	50	3 75
Giles' Iodide of Ammonia Lin'nt, Family, large, (see p. 373)	1 00	8 00
" " " " " small	50	4 00
" " " " Anim'l, large	1 00	8 00
" " " " " small	50	4 00
Granger's (late Halleck's,) Lacto Phosp'd Pepsin....		
Hale's Honey of Horehound and Tar, large........	1 00	8 00
" " " " " small........	50	4 00
Hall's Galvano-Electric Plaster, (see page 370).....	50	4 00
Harbridge's (F.) Cough Balsam....................	75	
Haven's Positive Cure for Dyspepsia...............	1 00	
" " " Catarrh.................	1 00	
Hood's Excelsior Liniment, large..................	1 00	8 00
" " " medium	75	6 50
" " " small	35	2 75
" " Troches	25	2 00
" Vegetable Blood Purifier.................	1 00	8 00
Houdan's Chicken Powder........................	25	

ADDENDA—*Continued.*

	Retail.	Wholesale. Per Doz.
Houdan's Chicken Powder	$ 50	
Japanese Hair Dye		
Kennedy's Prairie Weed	1 00	$ 7 75
Koenig's Hamburg Breast Tea	25	1 75
" " Drops, (see page 365)	50	3 50
" " Hamburger Salve	25	1 75
Leibig's Extract of Malt		
Loring's Vegetable Specific—Dyspepsia, etc.	1 00	8 00
Merriam's (Mrs. Wm.) Cough Syrup		
Miller's Baby Syrup	35	2 50
Pierce's Extract of Smartweed	50	4 00
Pike's Toothache Drops	25	1 75
Piso's Remedy for Catarrh, (see page 334)	1 00	7 75
Pratt's New Life	1 00	8 00
Price's Liver Invigorator	1 00	8 00
Renne's Pain Killing Magic Oil, large, (see page 374)	1 00	8 00
" " " " " medium	50	4 00
" " " " " small	25	2 00
Rush's Bitters	1 00	8 75
" Buchu and Iron	1 00	8 25
" Lung Balm	1 00	8 25
" Mandrake Pills	25	1 65
" Remedy	2 00	18 00
" Restorer	1 00	8 25
" Sarsaparilla and Iron	1 00	8 50
Scheffer's Pepsin	1 00	8 00
Sloan's (Dr.) Soothing Compound	25	1 75
Stafford's Florida Water	50	4 50
Talcott's "Magic Cure" for Chills, etc.	1 00	8 00
Towne's Universal Catarrah Cure	1 00	7 50
Van Schaack, Stevenson & Reid's Pure Cod Liver Oil	1 00	6 50

SPECIAL NOTICE TO THE TRADE.

Our friends and customers will not be the losers by sending us their orders direct by mail; traveling salesmen from other Drug houses, offering to sell from OUR LIST, or duplicating from our prices current, do not always give you the BENEFIT OF ANY DECLINE IN THE MARKET, as we aim to do, regardless of the cost.

The injustice of using another's labor in this way is apparent to all, and our attention has been called to it by many of our business friends. We do not hesitate to send out a *priced* list, which we do to keep our friends advised in regard to the market.

We know those who have so long dealt with us must feel we are always ready, and the *first* to meet the market on *low figures*.

VAN SCHAACK, STEVENSON & REID.

Pure Wines and Liquors

FOR MEDICINAL USE.

Van Schaack, Stevenson & Reid's Catalogue

Fully appreciating the importance of this class of goods to the Drug Trade, we guarantee the following line in every way reliable. The imported goods have *age*, and we confidently recommend them as being the best imported, and selected *especially* for the wants of the Drug Trade.

ALES,	Imported,	Allsopp's Gen. English	pints,	per doz.,	$2 60
"	"	Byass's " "	"	"	2 60
"	"	Jeffry's, Edinburg	"	"	2 60
"	"	Muir's, "	"	"	2 60
"	"	McEwan's	"	"	2 60
PORTER,	"	Barclay's	"	"	2 60
"	"	Byass's	"	"	2 60
"	"	Guinnes's Dublin Stout	"	"	2 60
"	"	Barclay's	quarts,	"	4 50
"	"	Byass's	"	"	4 50
BRANDY,	Cognac, Hennesy, Pale, 1867			per gall.,	8 50
"	" " Dark, 1867			"	8 50
"	Otard, Dupuy & Co., 1869			"	8 00
"	Pinet, Castillon & Co., 1869			"	7 50
"	Pellevoisin, Rochelle			"	6 50
"	Hope, Seignette French			"	6 00
"	California Grape, Los Angelos, 1868			"	4 50
"	Catawba "			"	3 25
"	" Kelly Island			"	4 00
"	Blackberry, 1869			"	3 50
"	" 1870			"	3 00
"	Domestic, old			"	3 50
"	" new			"	2 50
GIN,	Meder Swan			"	5 00
"	Schiedam			"	3 75
"	Imperial			"	2 00
"	Von Wordragen's Aromatic and Medicinal Imperial Gin; this most superior article has become in most demand of any of the imported Gins; large bottles			per doz.,	7 50
RUM,	Jamaica, London Dock, High Proof			per gall.,	6 00
"	St. Croix			"	5 50

RUM, New England, Genuine, old	per gall.	$3 00
" Domestic	"	2 00
WHISKY, Bourbon, Very Old	"	4 75
" " Kentucky, 3 years old	"	3 50
" " " 2 years old	"	2 50
" " " 1 year old	"	2 25
" " Domestic, at Market Price.		
" Rye, Stuart's Old	"	5 00
" " 3 years old	"	3 50
" " 2 years old	"	3 00
" " 1 year old	"	2 25
" " Domestic, at Market Rates.		
WINE, Calif'nia, Angelica, Perkins, Stern & Co., 1869	"	2 75
" " " " 1868	"	3 00
" " " " 1867	per doz.	9 00
" " Port, Perkins, Stern & Co., 1869	per gall.	2 75
" " " " 1868	"	3 00
" " " " 1868	per doz.	9 00
" " Sherry, Crown, "	per gall.	2 75
" Madeira, Almedia	"	4 50
" " Sicily	"	2 75
" Malaga, Sweet	"	2 25
" Port, Pure Juice, Schiller, Old	"	5 00
" " Burgundy	"	2 50
" " Cadiz	"	3 75
" Sherry, Harmony Nephews	"	4 50
" " Sicily	"	2 50

PERKINS, STERN & CO.'S LIST.

	Vintage.	Per gall.	Per case.
Hock	1864,	Bulk.	$11 00
" Sonoma	1868,	$1 75	7 50
Claret, "	1868,	1 75	7 00
Port, Los Angelos, very fine	1864,	5 00	14 00
" "	1868,	3 00	9 00
" "	1869,	2 50	
Angelica, Los Angelos	1860,		16 00
" "	1868,	3 00	9 00
Muscatel, "	1868,	3 00	10 00
Brandy, "	1768,	8 00	18 00
" "	1866,	6 00	13 50
" "	1869.	4 50	
Wine Bitters		3 00	
Pacific Bitters			8 50
Natural Grape Champagne, a dry, pleasant Wine....(quarts)			18 00
" " " (pints, 2 doz.)			20 00
Sparkling Catawba, very fine (quarts.)			18 00
Crown Sherry per gall.		$2 75	9 00

RHINE AND MOSELLE WINES.

		Currency. Gallon.	Case.
1869.	Deidesheimer	$1 75	$7 00
1870.	Duerkheimer	2 25	7 75
1869.	Forster Traminer	2 75	9 00
1868.	Zeltinger Moselle	2 50	8 25
1868.	Schwarzhofberger Moselle	3 00	9 50
1868.	Brauneberger Moselle	4 00	12 00
1865.	Ruedesheimer	3 50	11 00
1869.	Hochheimer	4 00	12 00
1869.	Geisenheimer	2 25	8 00
1868.	Ruppertsberger	3 00	9 50
1870.	Niersteiner	2 50	8 25
1870.	Ungsteiner	2 25	7 75
1868.	Johannesberger	6 00	18 00
1870.	Assmannshauser, Red	3 50	11 00
1868.	Ingelheimer, Red	4 50	14 00

RHINE WINES IMPORTED IN CASES.

Niersteiner, from Josef Falck, in Mainz	qt.	$14 00
" " "	pt.	15 00
Ruedesheimer, " "	qt.	15 00
" " "	pt.	16 00
Liebfraumilch, " "	qt.	17 00
" " "	pt.	18 00
Hochheimer Dom, " "	qt.	18 00
" " "	pt.	19 00
Edelweis, " "	qt.	24 00
Brauneberger Moselle, " "	qt.	20 00
" " " "	pt.	21 00
Assmannshauser, Red, " "	qt.	19 00
" " " "	pt.	20 00
Steinwein in Bocksbeutel, from J. Oppman	qt.	24 00
" " "	pt.	26 00

CLARETS.

FRENCH WINES IMPORTED IN CASES.

	Qt.	Pt.
St. Julien	$ 4 50	
Chateau Bouillac	6 00	
St. Estephe	7 00	
Pauillac, of Flouch Freres	12 00	$13 00
St. Emilion, "	14 00	15 00
Pontet Canet, of Tenet & de Georges	16 00	17 00
Chateau Larose, of Cuzol, Fils & Co.	20 00	22 00
Chateau Lafite, "	20 00	
Chambertin Burgundy, "	36 00	37 50
Beaune, "	21 00	22 50
Haute Sauternes	13 00	14 00

CATAWBA WINE.

	Gallon.	Case.
Kelley's Island, 1871	$2 00	$ 8 00
" 1872	1 50	6 50
" 1873	1 25	6 50
Sparkling Imperial	quart,	13 00
"	pint,	15 00

SCOTCH AND IRISH WHISKY.

Scotch, Ramsay's	$ 4 50 to $6 00
" Caol Ila	4 50 to 6 00
" in cases	16 00
Irish Whisky, Mehan's	4 00 to 5 50
" " in cases	16 00

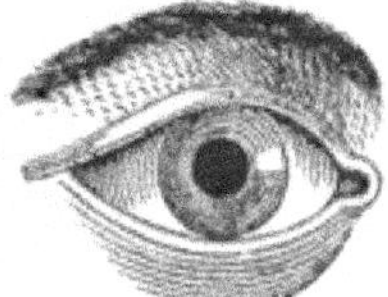

Artificial Human Eyes.

Of excellent construction, and of the very best manufacture, fully equal to the celebrated French, the enamel being of great purity, and not liable to break.

☞ Parties ordering from a distance, will please indicate whether right or left eye, the natural color of the eye, and, as nearly as possible, the size of the cavity, and the size of the stump.

Price List of Best Telegraph and Parlor Matches

Sold by Van Schaack, Stevenson & Reid.

SQUARE TELEGRAPH MATCHES.

No. 2.	Contains 3 gross, small paper boxes, 100 matches in each, packed in ⅛ gross caddies, $2.50 per gross; per case	$7 75
No. 6.	Contains 6 caddies, 24 boxes of 300 matches in each, $1.25 per caddy; per case	7 75
	Also in wood caddies (3 in a rack), 24 boxes of 300 matches in each, per caddy	1 25
No. 8.	Contains 144 boxes, 200 matches in each, per case	5 00
	Also in wood caddies (5 in a rack), 24 boxes of 200 matches in each, per caddy	85
No. 9.	Contains 144 boxes, same as No. 6—without the caddies—per case	7 50

ROUND MATCHES.

No. 4.	1-8 gross—contains 3 gross, same as above, $2.55 per gross; per case	7 50
No. 4½.	1-16 gross—contains 3 gross, 9 paper boxes in each box, $2.55 per gross; per case	7 50
No. 7.	Contains 3 gross, dime boxes, 300 matches in each, $2.50 per gross; per case	7 50
No. 16.	Contains 4½ gross—assorted boxes. 80 300s; 40 500s; 20 1000s; $2.50 per gross; per case	11 00

ROUND PARLOR MATCHES.

No. 10.	In 1, 2 and 3 gross cases, dime boxes, 300 matches in each, per gross	3 00
No. 11.	In 1, 2 and 3 gross cases, flat drawer boxes, 100 matches in each, per gross	3 15
No. 12.	In 2 gross cases, flat drawer boxes, 70 matches in each, $2 75 per gross; per case	5 50

As we go to press, a Bill is before Congress to remove stamps on matches. If it passes, we shall make the proper decline in prices.

Van Schaack, Stevenson & Reid.

VACCINE VIRUS.

Owing to the increased and widely extending demand for

ANIMAL VACCINE VIRUS

Stables have been established for its propagation in the vicinity of Chicago, and we are now prepared to distribute it while in the best possible condition.

For immediate use, Points are the most reliable of all forms of stored virus. They are charged on both sides of the pointed end. Crusts are less reliable, but can be kept longer.

Vaccine Virus must be kept *cool* and *dry*. Envelop it in tin-foil or tissue paper, and bury it in pulverized burnt alum, or Plaster of Paris, and keep in a tightly corked bottle.

This virus is the product of heifer inoculations, from the celebrated *spontaneous case* of *Cow Pox* discovered at *Beaugency*, in *France*, in 1866, *and has never passed through the human subject.*

Scarify thus in three places. The scratches should produce a very *slight* appearance of blood. Aim slightly to break up the cuticle. If blood flows, wait and wipe it off before applying the virus. The charged points should be slightly moistened by cold water from the point of the lancet, and immediately rubbed over and into the scratches. Time and patience are essential to success. One Point is sufficient for two insertions.

Crusts are to be dissolved into the consistency of pus, with cold water, and then spread abundantly over and worked into the scarified surface. Time must be given for the virus to be absorbed and the surface to dry, before the clothing is re-applied. No *plaster* or *dressing* must be used. In cases of young children, it is *always* best to have the arm *entirely freed from clothing* before commencing the operation. *Or moisten the crust well and* "POCKET" *it in at least three places.*

Points must be used within ten days after the package is received. A *total* failure in the results will entitle the party to a duplicate package without charge.

Cow Pox Virus is efficient, protective and safe, and is less liable to complications than humanized virus.

15 Large Ivory Points	**$2 00**
6 " " "	**1 00**
1 Crust of Single Vesicle	**2 00**
1 Crust Compound Vesicle	**3 00**
1 Large Typical Crust	**4 00**

The Trade supplied by

Van Schaack, Stevenson & Reid,

92 and 94 Lake Street, cor. Dearborn, Chicago.

For the convenience of our customers, we submit cuts showing sizes of Vial Corks, from Nos. 1 to 10 inclusive. In ordering, please specify whether X (Fine) or XX (Superfine) are wanted. Put up in 5 gross bags.

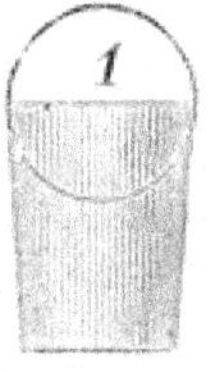

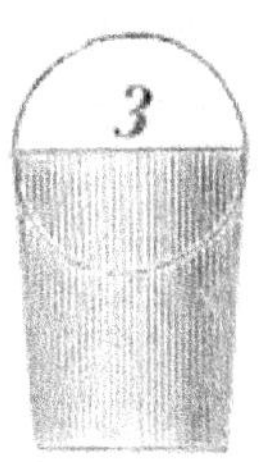

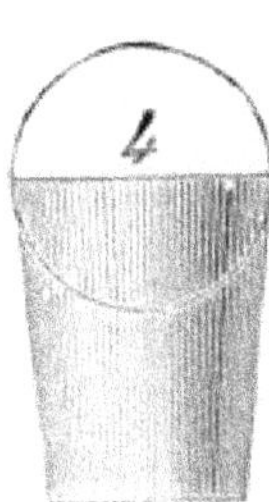

VAN SCHAACK,

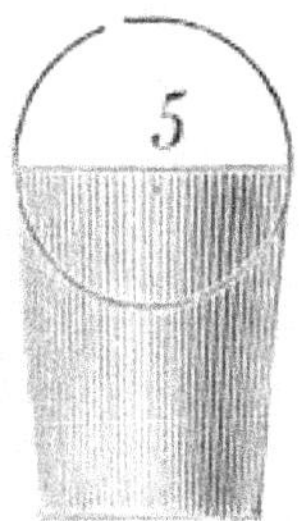

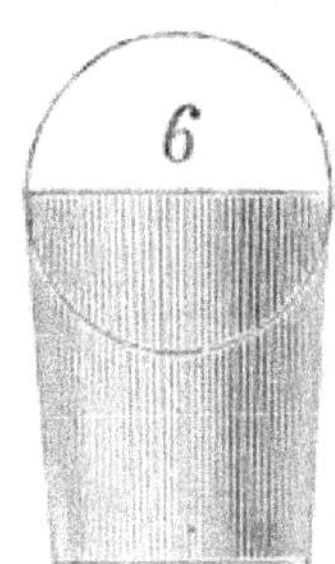

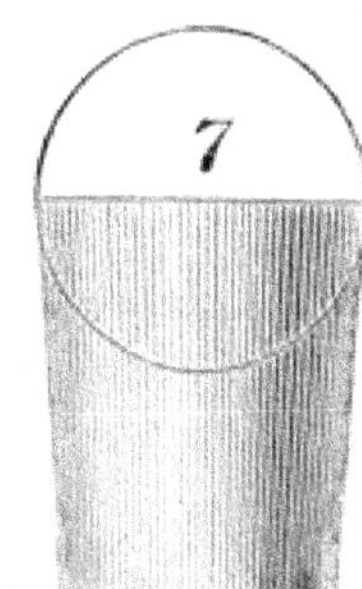

STEVENSON

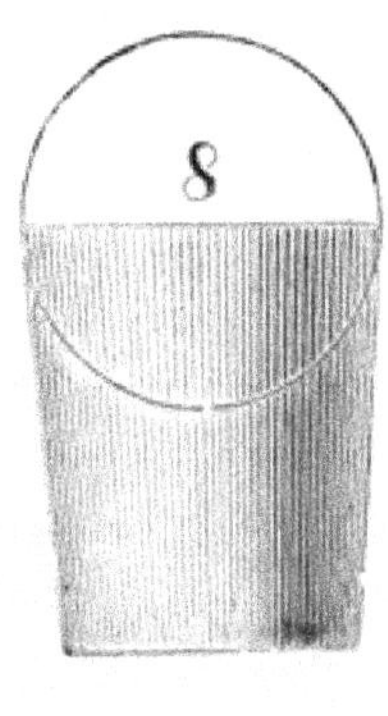

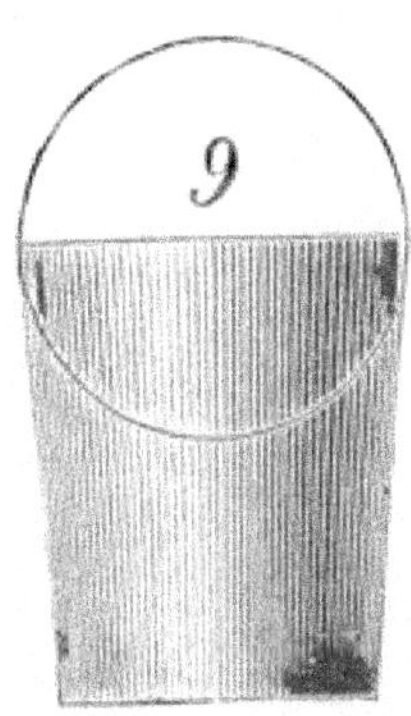

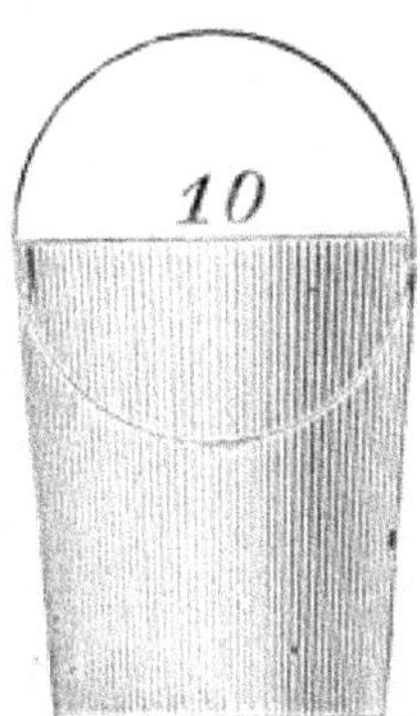

& REID,

WHOLESALE DRUGGISTS, CHICAGO.

MACHINE  CUT CORKS.

VAN SCHAACK, STEVENSON & REID'S

PRICE LIST,

For Druggists, Grocers, Liquor Dealers & Mineral Water Bottlers.

Superfine [XX] Straight or Taper Corks.

No.	$
No. 0	22
1	25
2	28
3	33
4	38
5	46
6	55
7	77
8	1 00
9	1 10
10	1 30
11	1 55
12	1 75
13	1 90
14	2 10
15	2 35
16	2 50
17	2 75
18	3 00
19	3 25
20	3 50

Fine [X] Straight or Taper Corks.

No.	$
No. 0	8
1	9
2	10
3	14
4	16
5	19
6	25
7	35
8	45
9	55
10	65
11	76
12	90
13	1 00
14	1 15
15	1 35
16	1 50
17	1 70
18	1 90
19	2 10
20	2 30

Superfine [XX] Wine Corks.

No.		$
No. 7—1½	in. long	1 20
8—1⅝	"	1 45
9—1½	"	1 70
10—1½	"	2 00

Short Taper [XX] Corks.

No.	$
No. 1	16
2	20
3	24
4	28
5	33
6	40
7	50
8	65
9	80
10	95
11	1 15

Common Flask Corks Tapered

No.	$
No. 8	33
9	38
10	44
11	50
12	56
Assorted, 6 to 12	35

Lager Beer Corks.

No.	$
No. 12	65
13	70
14	80
15	90
16	1 00
17	1 10
18	1 20
19	1 25
20	1 35

Soda Corks.

No.		$
No. 8—1¼	in. long	50
8—1⅜	"	55
8—1½	"	60
9—1¼	"	60
9—1⅜	"	65
9—1½	"	70
10—1¼	"	62
10—1⅜	"	70
10—1½	"	75

Assorted [X] Vial Corks.

Nos.	$
Nos. 1 to 6	14
3 to 7	25

Assorted [XX] Vial Corks.

Nos.	$
Nos. 1 to 6	40
3 to 7	50

Fine [X] Wine Corks.

No.		$
No. 7—1¼	in. long	85
8—1⅜	"	90
9—1½	"	1 00
10—1⅝	"	1 10

Long, Straight Vial Corks.

No.	$
No. 1	22
2	25
3	30
4	35
5	45
6	50

Short Taper [X] Corks.

No.	$
No. 1	7
2	9
3	12
4	14
5	16
6	18
7	25
8	30
9	36
10	45
11	50

Specie or Flat Corks.

Size		$
1	inch	26
1⅛	"	33
1¼	"	40
1⅜	"	46
1½	"	55
1⅝	"	63
1¾	"	75
1⅞	"	90
2	"	1 00
2⅛	"	1 15
2¼	"	1 25
2⅜	"	1 40
2½	"	1 55
2⅝	"	1 75
2¾	"	1 90
2⅞	"	2 15
3	"	2 30
3¼	"	2 75
3½	"	3 25
3¾	"	3 75
4	"	4 25

☞ Our Corks are cut the *full size*, and the quality is unsurpassed. For sizes of Taper Vial Corks, see preceding page.

Paints, Oils and Heavy Goods.

We have in store a large and varied stock, and shall *promptly meet any decline* by the manufacturers.

BENZOLE, Light, by bbl	per gall.	$	18
" Heavy, "	"		20
Black, Drop, Dry, American, No. 1	per lb.		15
" " " English,	"		20
" " in Oil, 1 lb. cans	"		20
" " Coach, in Oil, 1 lb. cans	"		20
" " India, in Oil, "	"		20
" Paint, "	"		12
" Ivory, Dry	"		07
" " in Oil, 1 lb. cans	"		20
" Lamp, Common, Assorted	"		10
" " " in ¼s	"		12
" " " in ½s	"		10
" " " in 1 s	"		08
" " Eddy's Refined, in ¼s	"		32
" " " " in ½s	"		30
" " " " in 1 s	"		28
" " " " Assorted	"		30
" " Fingers, (1 oz. papers)	per 100		1 25
Black, Lead, German	per lb.		07
" " East India	"		12
Blue, Prussian, Dry, pure	"		75
" " Pow'd	"		80
" " in Oil, (assorted cans)	"		65
" " All Qualities, dry and in Oil, from 40 cts. to 65 cts.	"		65
" Soluble, No. 1	"		85
" " Chinese, Dry	"		90
" Paris, Lump	"		85
" Ultramarine, (H. B. Brand)	"		25
" " in Oil, (assorted cans)	"		40
Bronzes. All shades and colors, Nos. 1000 to 6000, from $3 to $10 per lb.			
Brown, Spanish, by bbl	"		02
" " less than bbl	"		03
" Van Dyke, Dry	"		12
" " in Oil, (1 lb. cans)	"		22
DIAMONDS, Ebony, Keyed	each,		3 75
" " Plain	"		3 00
" Ivory, Keyed	"		6 00
" Pocket	"		6 00
Dryer, Patent, 1 lb. cans	per lb.		12

FLOCK, all Colors	per lb.	$ 90
Frosting, White	"	65
GLAZIER'S Points, (in packs)	per doz.	1 25
Gold Leaf, Deep XX	per pack,	8 75
" Pale	"	8 00
Graining Combs, Leather	per set,	75
" Steel, American	"	1 50
" " English, in tin cases	"	1 75
Green, Chrome, pure, dry, in 6 lb. boxes	per lb.	18
" " " in Oil, in 1 lb. cans	"	20
" " Dry, all grades, in 6 lb. boxes, from 13 cts. to 16 cts.	"	16
" " in Oil, all grades, assorted cans, from 14 cts. to 17 cts.	"	17
" Hampden, Dry, in 6 lb. boxes	"	22
" " in Oil	"	24
" Magnesia, Dry, (H & S.)	"	22
" " in Oil	"	24
" Marseilles, Wood & Son's, in Oil, in 1 lb. cans	"	25
" Paris, Dry, Pure	"	40
" " " Brand, (A.)	"	38
" " " " (B.)	"	35
" " " " (C)	"	32
" " in Oil from 28 cts. to 38 cts.	"	38
LEAD, White, Dry	"	12
Lead, White, in Oil, *Strictly Pure*, Van Schaack, Stevenson & Reid's, (forfeited if not found *Strictly Pure*,) in 25, 50, 100 and 500 lb. kegs	per hund.	10 00
Lead, White, *Pure*, in Oil, Van Schaack, Stevenson & Reid's, in 25, 50, 100 and 500 lb. kegs	"	9 00

N. B.—Error of Printer.—Lead list should read:

Lead, White, in Oil, *Strictly Pure*, Van Schaack, Stevenson & Reid's, (forfeited, if not found *Strictly Pure*,) in 25, 50, 100 and 500 lb. kegs,	per hund.	$9.75
Lead, White, in Oil, *Pure*, Van Schaack, Stevenson & Reid's, in 25, 50, 100, and 500 lb. kegs	" "	8.75
D. B. Shipman's *Strictly Pure White Lead*, (warranted)	" "	9 75

Paints, Oils and Heavy Goods.

We have in store a large and varied stock, and shall *promptly meet any decline* by the manufacturers.

BENZOLE, Light, by bbl	per gall.	$ 18
“ Heavy, “	“	20
Black, Drop, Dry, American, No. 1	per lb.	15
“ “ “ English,	“	20
“ “ in Oil, 1 lb. cans	“	20
“ “ Coach, in Oil, 1 lb. cans	“	20
“ “ India, in Oil, “	“	20
“ Paint, “	“	12
“ Ivory, Dry	“	07
“ “ in Oil, 1 lb. cans	“	20
“ Lamp, Common, Assorted	“	10
“ “ “ in ¼s	“	12
“ “ “ in ½s	“	10
“ “ “ in 1 s	“	08
“ “ Eddy's Refined, in ¼s	“	32
“ “ “ “ in ½s	“	30
“ “ “ “ in 1 s	“	28
“ “ “ “ Assorted	“	30
“ “ Fingers, (1 oz. papers)	per 100	1 25
Black, Lead, German	per lb.	07
“ “ East India	“	12
Blue, Prussian, Dry, pure	“	75
“ “ Pow'd	“	80
“ “ in Oil, (assorted cans)	“	65
“ “ All Qualities, dry and in Oil, from 40 cts. to 65 cts.	“	65
“ Soluble, No. 1	“	85
“ “ Chinese, Dry	“	

FLOCK, all Colors	per lb.	$ 90
Frosting, White	"	65
GLAZIER'S Points, (in packs)	per doz.	1 25
Gold Leaf, Deep XX	per pack,	8 75
" Pale	"	8 00
Graining Combs, Leather	per set,	75
" Steel, American	"	1 50
" " English, in tin cases	"	1 75
Green, Chrome, pure, dry, in 6 lb. boxes	per lb.	18
" " " in Oil, in 1 lb. cans	"	20
" " Dry, all grades, in 6 lb. boxes, from 13 cts. to 16 cts.	"	16
" " in Oil, all grades, assorted cans, from 14 cts. to 17 cts.	"	17
" Hampden, Dry, in 6 lb. boxes	"	22
" " in Oil	"	24
" Magnesia, Dry, (H & S.)	"	22
" " in Oil	"	24
" Marseilles, Wood & Son's, in Oil, in 1 lb. cans	"	25
" Paris, Dry, Pure	"	40
" " " Brand, (A.)	"	38
" " " " (B.)	"	35
" " " " (C)	"	32
" " in Oil from 28 cts. to 38 cts.	"	38
LEAD, White, Dry	"	12
Lead, White, in Oil, *Strictly Pure*, Van Schaack, Stevenson & Reid's, (forfeited if not found *Strictly Pure*,) in 25, 50, 100 and 500 lb. kegs	per hund.	10 00
Lead, White, *Pure*, in Oil, Van Schaack, Stevenson & Reid's, in 25, 50, 100 and 500 lb. kegs	"	9 00
Lead, White, in Oil, Fahnestock, Haslett & Schwartz, 25, 50 and 100 lb. kegs, (see page)	"	10 00
" " Challenge, 25 lb. cans, (see page).	"	9 00
" " Continental, 25 lb. tin pails	"	10 00
" " " 12½ lb. tin pails	"	10 50
" " Diamond, 25 lb. tin pails	"	8 00
" " Stirling, 25 lb. tin pails	"	7 00
" " " 12 ½ lb. tin pails	"	7 50
" " Shipman's, (D. B.) Str. Pure, in 25, 50 and 100 lb. kegs	"	10 00

This brand is greatly growing in favor.

Lead, White, Assorted, 1 to 5 lb. cans, (100 lb. cases)		per lb.	$ 10
Litharge		"	12
MINERAL Paint, Ohio,	by bbl.	"	2½
" "	less than "	"	3½
" Winter's,	by "	"	3
" "	less than "	"	4
OCHRE, Yellow, American, less than cask		"	2¼
" " French, in Oil, pure, 1 lb. cans		"	12
" " Havre, Dry, less than cask		"	3¼
" " Rochelle, Dry, "		"	4
Orange, Mineral		"	15
Oil, Carbon, by bbl.		per gall	Market
" Castor, American, white, by bbl.		"	Price.
" " " " less than bbl.		"	"
" " " common (Machinists')		"	"
" " E. I., (5 gall. *cans* free)		per lb.	"
" Lard, Extra,	by bbl	per gall.	"
" " No. 1,	"	"	"
" " No. 2,	"	"	"
" Linseed, Raw,	"	"	"
" " Boiled,	"	"	"
" Lubricating, (Machinery,) Golden,	"	"	65
" " " W. Virginia,	"	"	40
" " " "Natural,"	"	"	60
" Neats Foot,	"	"	90
" Sperm, Winter Bleached,	"	"	2 25
" Tanner's Bank,	"	"	55
" " Straits,	"	"	60
" Whale, Winter Bleached,	"	"	80
PAINT Mills,		each,	6 50
Paper, Emery		per ream, from $6 00 to	6 50
" Sand, Flint, Beader & A's		" " 4 50 "	5 50
" " " " ass'd		" " 4 50 "	5 50
" " " Abbot's		per ream,	3 50
" " " " assorted		"	3 50
Putty, in bulk, 100 lb. tubs		per lb.	3¼
" in bladders, by bbl.		"	3½
" " in 25, 50 and 100 lb. bxs.		"	4
" Knives, Riveted, Round and Square		per doz.	2 50
RED, India, English, Dry		per lb.	15
" " in Oil, pure, 1 lb. cans		"	20
" Lead, American, strictly pure		"	10½
" " English		"	12
" " Painters'		"	10
" " in Oil, 1 lb. cans, from 15 to 20 cts.		"	
" Venetian, American, by bbl.		"	2½
" " English, Cookson's, by bbl.		"	3¾
" " in Oil, pure, 1 lb. cans		"	13
Rose Pink, bright		"	16

SIENNA, Burnt, Dry	per lb.	$ 6
" " in Oil, pure, 1 lb. cans	"	22
" " " ......from 14 to 20 cts.	"	
" Raw, Dry	"	6
" " in Oil, pure, 1 lb cans	"	22
" " " ......from 14 to 20 cts.	"	
Smalts, Black, Coarse and Fine	"	10
" Blue, "	"	18
" Brown, "	"	10
" Green, "	"	12
" Red, "	"	15
" Vermillion, "	"	18
		00
TURPENTINE, by bbl	per gall.	Market
" less than bbl	"	Price.
UMBER, Burnt, Dry	per lb.	6
" " in Oil, pure, in 1 lb. cans	"	20
" " " from 14 cts. to 18 cts.	"	
" Raw, Dry	"	6
" " in Oil, pure, 1 lb. cans	"	20
" " " from 14 cts. to 18 cts.	"	
VARNISH, Asphaltum	per gall.	1 25
" Coach, No. 1	"	1 75
" " No. 2	"	1 50
" " Body, Extra	"	3 50
" Copal, No. 1	"	1 50
" Damar	"	1 75
" English, in 1 gall. cans, all kinds.	"	5 50
" Flowing, No. 1	"	2 50
" Rubbing Body	"	3 75
" Shellac, (Alcohol)	"	4 50
" Turpentine, Japan	"	1 00
" Wearing Body	"	4 50
Vermillion, American, Dry	per lb.	25
" " Hampden	"	23
" " in Oil	"	30
" Chinese, in 1 oz. pkgs	"	3 25
" English, (A. & B.), Deep	"	2 50
" " " Pale	"	2 50
" Trieste	"	2 25
WHITING, Spanish, by bbl	"	$1\frac{3}{4}$
" " less than bbl	"	$2\frac{1}{2}$
" Paris, by bbl	"	$3\frac{1}{2}$
" " less than bbl	"	4
White Cremnitz, Lump	"	25
" " Powd	"	30

YELLOW, Chrome, Dry, Extra	per lb. $	25
" " " M	"	20
" " " K	"	15
" " Oil, Extra	"	26
" " " M	"	20
" " " K	"	15

ZINC, Dry, N. J.	"	12
" " Red Seal	"	15
" In Oil, N. J. (25 lb. tin pails)	"	9
" " Vielle Montague, (25 lb. iron pails)	"	14
" " 1 to 5 lb. cans, (100 lb. cases)	"	18
" In Varnish, 1 to 5 lb. cans, (100 lb. cases)	"	25

DISTEMPER COLORS.

Burnt Umber,	1 lb. bottles	$	16
Raw "	" "		16
Burnt Sienna,	" "		18
Raw "	" "		18
Vandyke Brown,	" "		20
Drop Black,	" "		20

GRAINING COLORS.

Our Light Oak, Dark Oak and Walnut Graining Colors are prepared ready for use, and are put up in from 1 to 25 lb. cans.

Light Oak, asst. 1 to 5 lb. cans, 50 lb. cases	$	22
Dark " " " "		22
Walnut, " " "		22

Cases assorted, viz.: 16-1, 10-2, 3-3, 1-5 = 50 lbs.

JAPANNED SIGN PLATES.

Black Japan,	10x14 inch, Light, per doz.	$	1	75
" "	10x14 " Heavy, "		2	50
" "	12x17 " " "		5	00
" "	14x20 " " "		6	50
" "	18x24 " " "		10	00
" "	20x28 " " "		20	00
" "	24x30 " " "		25	00
" "	24x36 " " "		30	00
" "	28x56 " " "		45	00
" Jap. both sides,	10x14 " " "		3	35
" "	12x17 " " "		6	75
" "	14x20 " " "		8	75

LAKES AND DRY FINE COLORS.

In connection with Mander Bros.' Varnishes, we are importing fine colors required for Coach, Car and Fresco Painting, viz.:

Chatemuc Carmine Lake, per lb.	$		75
Carmine Lake, A Extra, "		5	00
" " A "		4	00
" " B "		3	00
Scarlet " A "		10	00
" " B "		7	00
" " C "		6	00
Florentine " A Light, "		2	50
" " A Deep, "		2	50
" " B Light, "		2	00
Munich " A "		15	00
" " B "		10	00
" " C "		6	00
" " D "		2	50

☞ For prices of Anilines and Carmines, see pages .. and ...
Colors in Oil, assorted cans, 1 cent less per lb.

We will fill orders for everything needed in the PAINT LINE at the very Lowest Market Price.

PAINTERS' BRUSH STOCK.

Subject to per cent. discount.

Except those noted as net.

In comparing prices, please notice that our LIST figures are LOWER than those of most manufacturers.

Brick Stripers.

	Net.
Per doz., - -	$1 25

Okatka Paint Brushes.

Every Brush has "D. White & Sons" Branded on the Handle.

All White Bristles—Wire Bound.

No.		
No. 1,	per doz.,	$13 50
0,	"	17 00
2-0,	"	20 50
3-0,	"	24 50
4-0,	"	29 50
5-0,	"	34 50
6-0,	"	38 50
7-0,	"	42 50
8-0,	"	46 50

Common Paint Brushes are wire bound, good mixed quality, gray centre, white outside, for general use.

No.		
No. 6,	per doz.,	$ 2 50
5,	"	3 00
4,	"	3 75
3,	"	4 50
2,	"	5 50
1,	"	6 50
0,	"	8 00
2-0,	"	9 50
3-0,	"	11 00
4-0,	"	12 50
5-0,	"	14 00
6-0,	"	16 00

Blind Brushes.

All white Russia Bristles.

No.		
No. 1,	per doz.,	$10 50
0,	"	12 00
2-0,	"	15 00
3-0,	"	18 00

Gloss Varnish Brushes.

All Finest White French Bristles.

Every Brush has "D. White & Sons" Branded on the Handle.

Oval. Wire Bound.

No.		
No. 6,	per doz.,	$ 3 00
5,	"	4 25
4,	"	5 00
3,	"	6 00
2,	"	8 00
1,	"	10 00
0,	"	13 00
2-0,	"	16 00
3-0,	"	19 00
4-0,	"	22 00
5-0,	"	25 00
6-0,	"	29 00
7-0,	"	33 00
8-0,	"	38 00

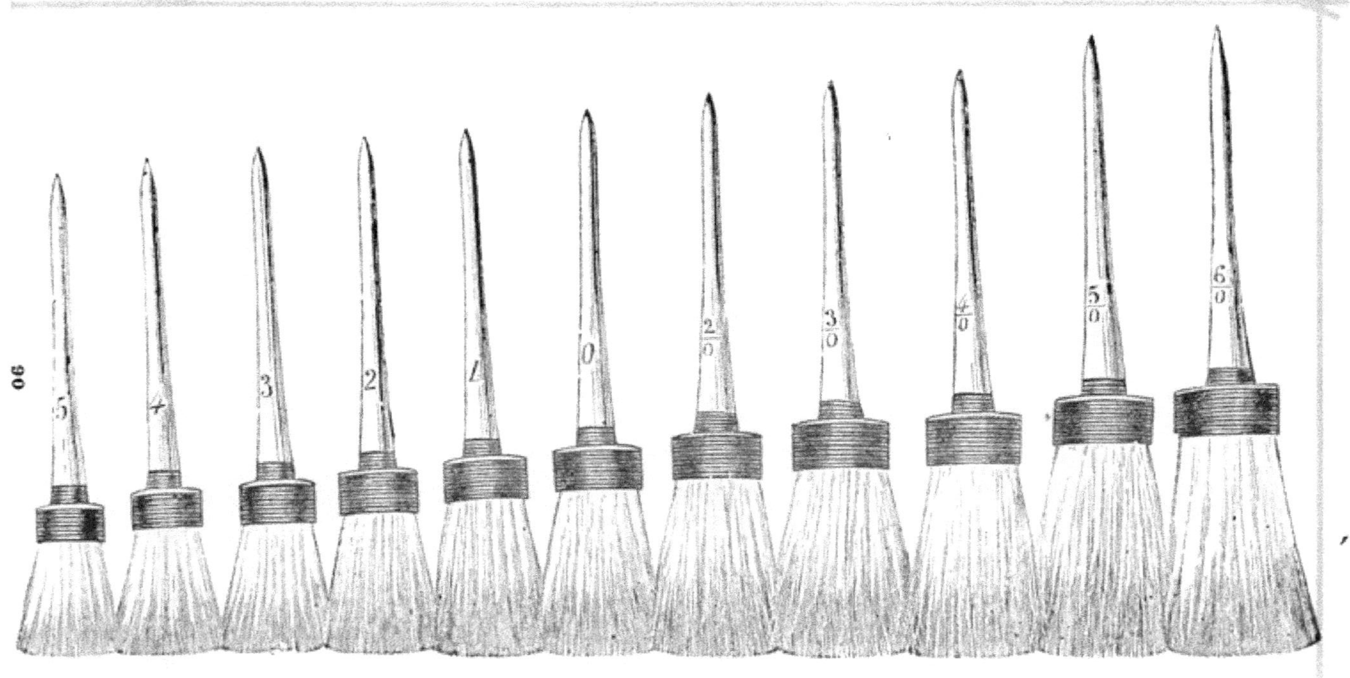

PAINT BRUSHES.

HORSE BRUSHES—Wood Back.

No. 10, per dozen,	$4 00	No. 12, per dozen, $5 50
" 11, "	4 50	

HORSE BRUSHES—Leather Back.

No. 100, per dozen,	$ 7 00	No. 115, per dozen,	$13 50
" 110, "	10 00	" 120, "	16 50

Genuine Emerson's Razor Strap.

[ESTABLISHED 1810.]

(See Fancy Goods List.)

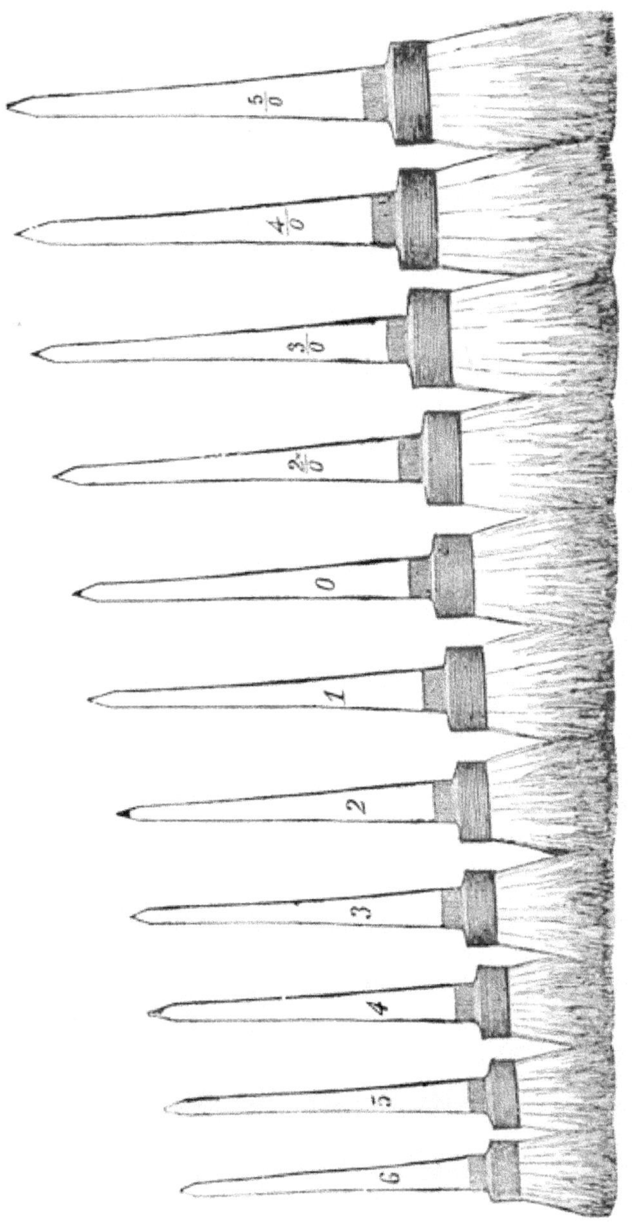

VARNISH BRUSHES.

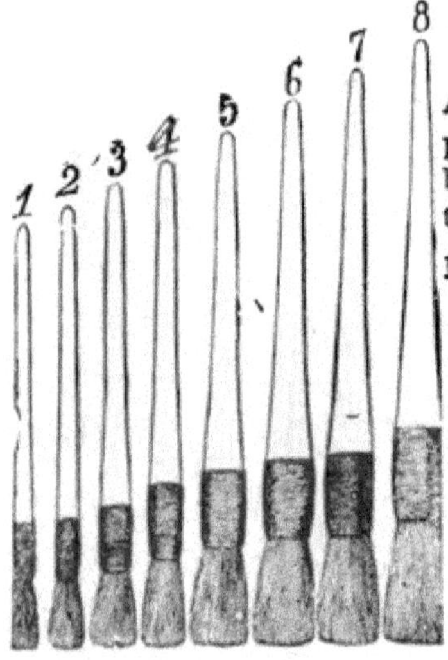

American Sash Brushes

Subject to per cent. discount.

Are twine bound, well fastened, pure, all white, superfine French bristles, specially adapted for practical painters' use.

No.		
No. 1,	per doz.,	$1 00
2,	"	1 30
3,	"	1 60
4,	"	1 90
5,	"	2 20
6,	"	2 50
7,	"	2 90
8,	"	3 20

French Sash Brushes.

Subject to per cent. discount.

Are wire bound, extra fastened, pure, all white, superfine French bristles, specially adapted to practical painters' use.

No.		
No. 1,	per doz.,	$1 20
2,	"	1 60
3,	"	2 00
4,	"	2 40
5,	"	2 80
6,	"	3 20
7,	"	3 70
8,	"	4 20

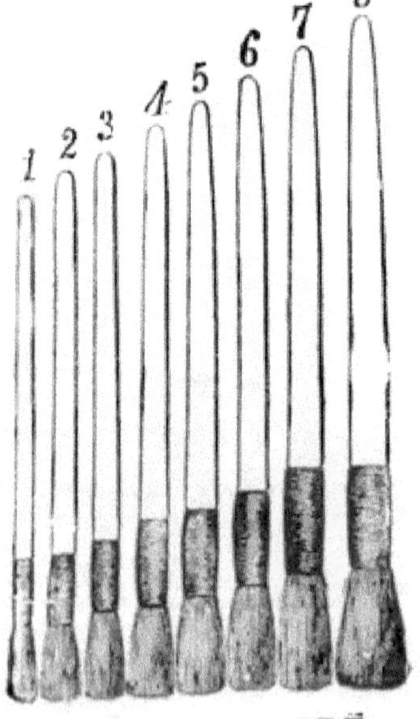

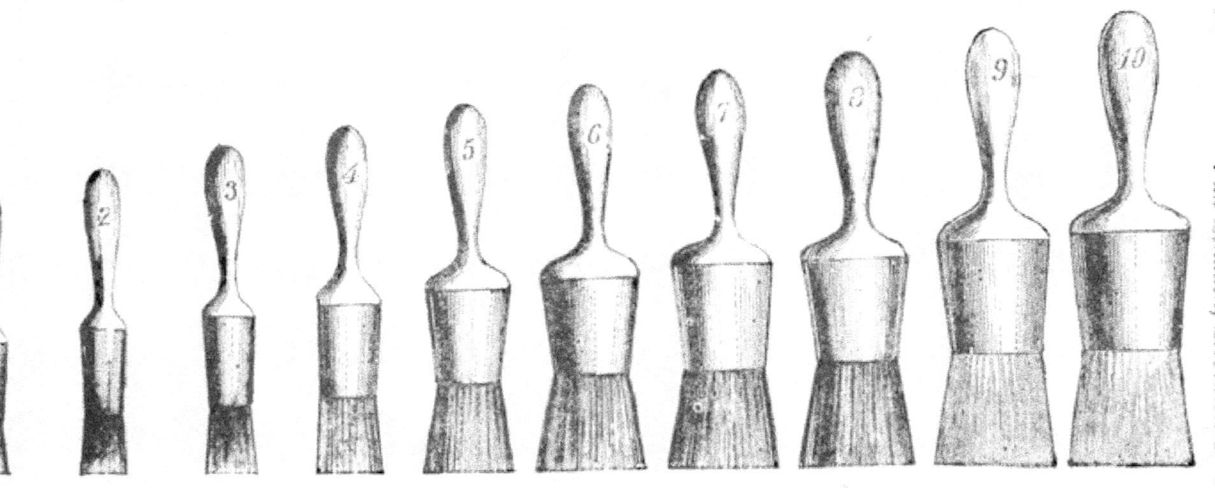

SHORT HANDLE STENCIL, TIN BOUND.

		Net.			*Net.*
No. 1,	per dozen,	$1 10	No. 6,	per dozen,	$2 25
2,	"	1 25	7,	"	2 75
3,	"	1 50	8,	"	3 50
4,	"	1 75	9,	"	4 00
5,	"	2 00	10,	"	5 00

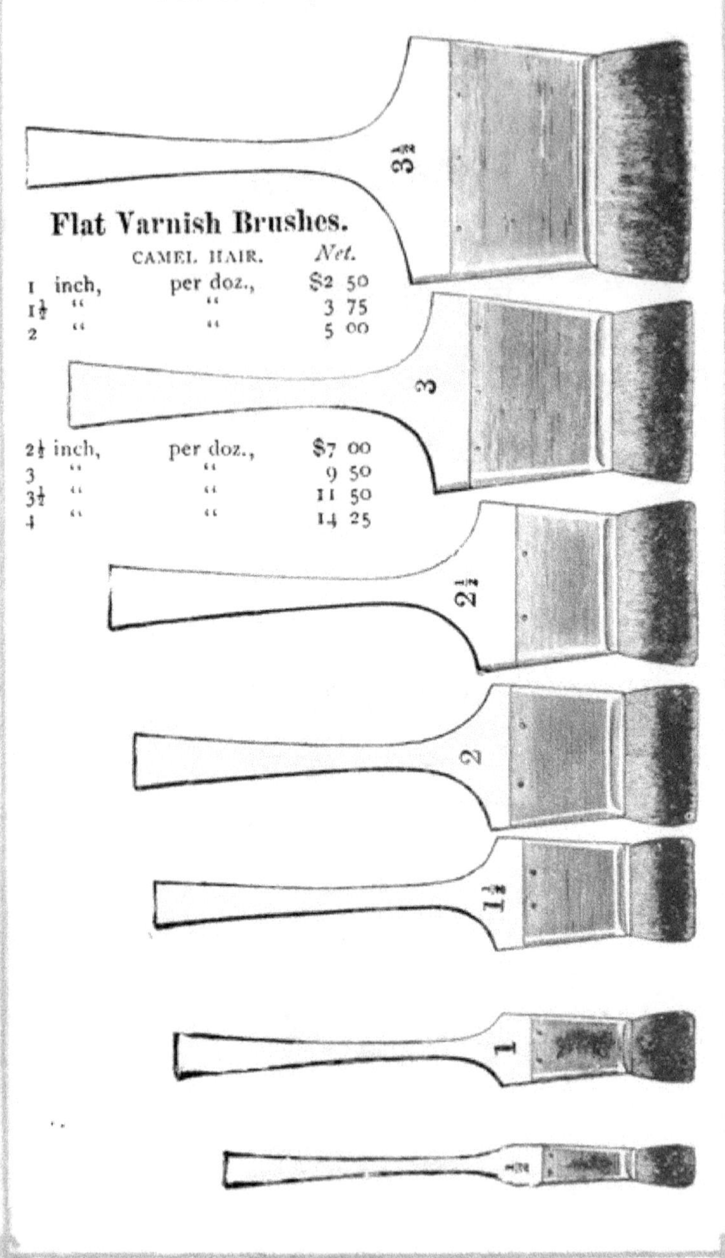

Flat Varnish Brushes.

CAMEL HAIR.

Size		*Net.*
1 inch,	per doz.,	$2 50
1½ "	"	3 75
2 "	"	5 00
2½ inch,	per doz.,	$7 00
3 "	"	9 50
3½ "	"	11 50
4 "	"	14 25

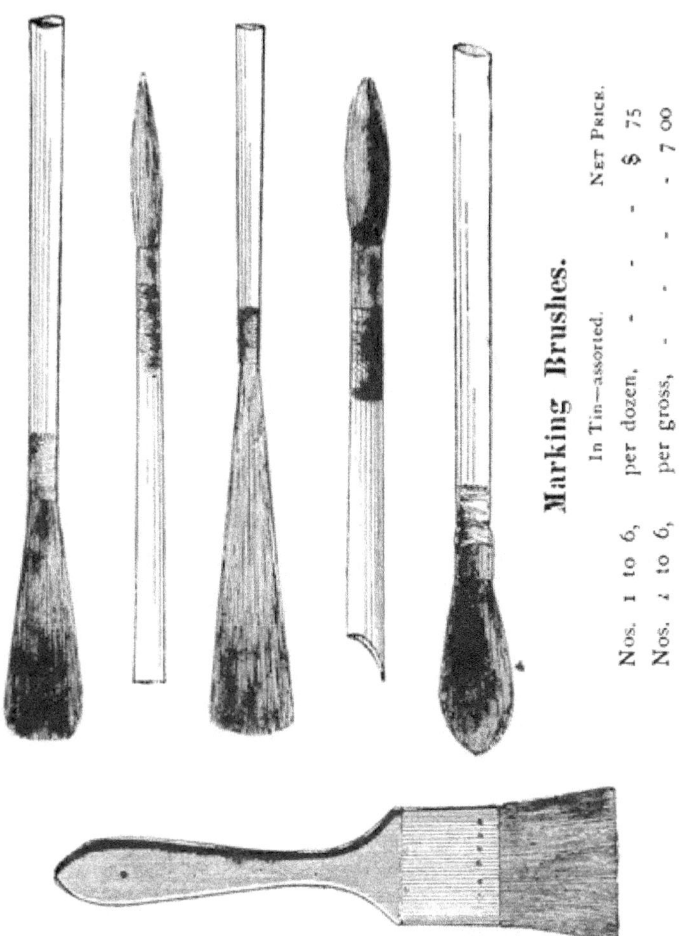

Marking Brushes.

In Tin—assorted.		Net Price.
Nos. 1 to 6,	per dozen, - - - -	$ 75
Nos. 1 to 6,	per gross, - - - -	7 00

Flat Varnish Brushes.

BRISTLE.		*Net.*
1 inch,	per dozen, - - - - - - -	$1 50
1½ "	" - - - - - - - -	2 25
2 "	" - - - - - - -	3 00
2½ "	" - - - - - - - -	4 00
3 "	" - - - - - - -	5 50
3½ "	" - - - - - - - -	7 00
4 "	" - - - - - - -	9 00

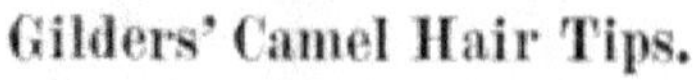

Gilders' Camel Hair Tips.

Net.

1 inch,	4 inches wide, each, 25 cts.	
1½ "		
2 "		
2½ "		

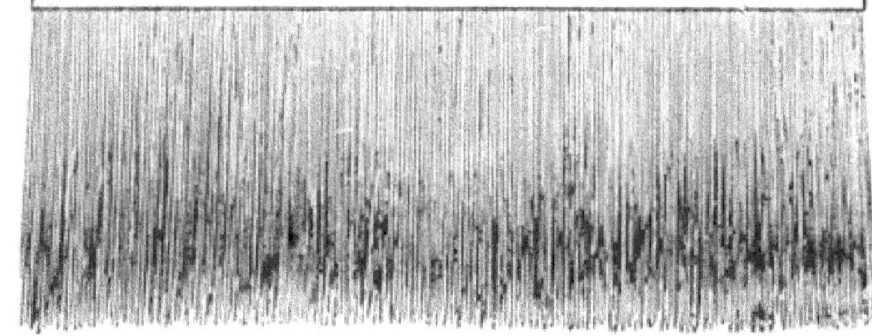

Painters' Dusters.

Subject to per cent. discount.

No. 14,	per doz.,	Black Mixed,	$ 9 00
15,	"	"	11 00
16,	"	"	13 00
40,	"	All Bristles,	14 00
50,	"	"	21 00
60,	"	"	27 00

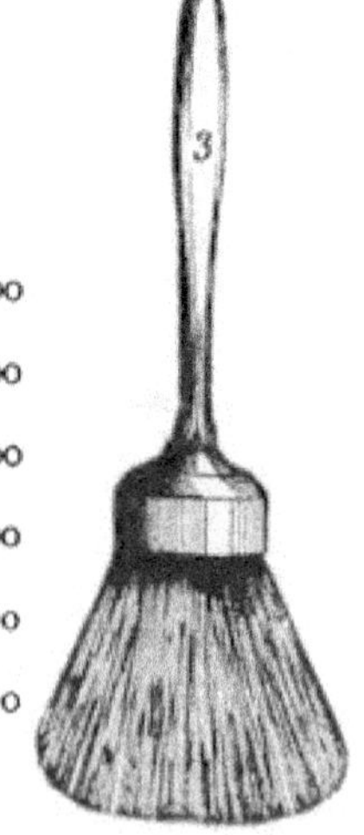

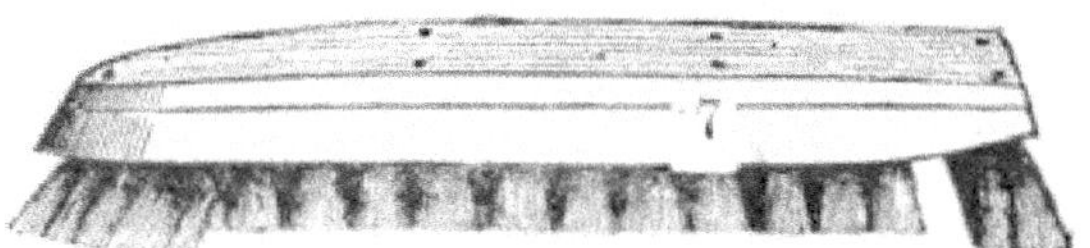

Scrub Brushes.

	Net.
No. 1, per dozen,	$1 50
7, "	2 25
13, " all bristles,	5 00
7, with handles, per dozen,	2 75
12, " " all bristles,	5 25
3,	2 00

Window Brushes.

	Net.
Each,	75 cts.

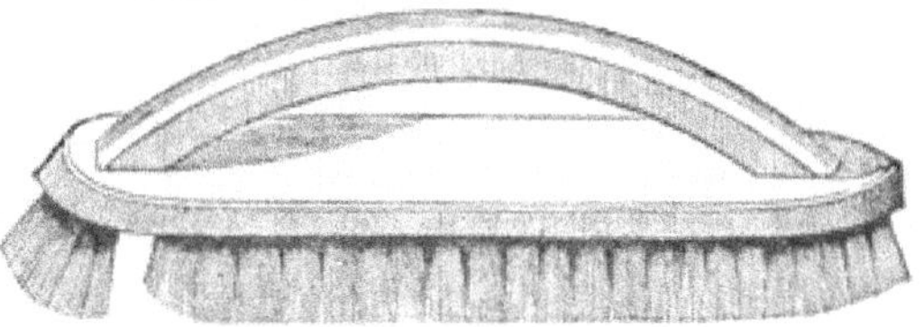

Stove Brushes.

	Net.
No. 1, per dozen,	$2 50
2, "	3 00
3, "	4 00

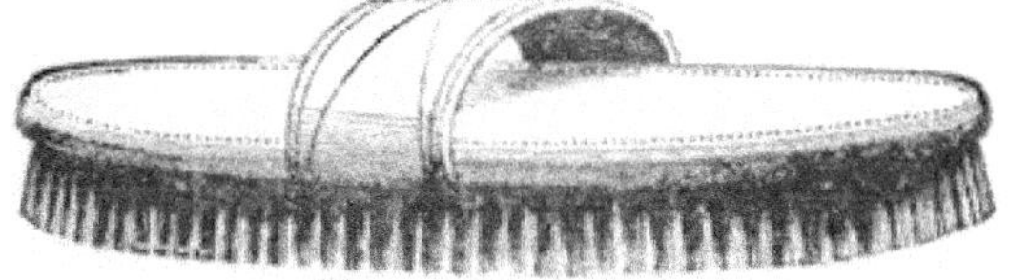

Horse Brushes.

For prices see page 91.

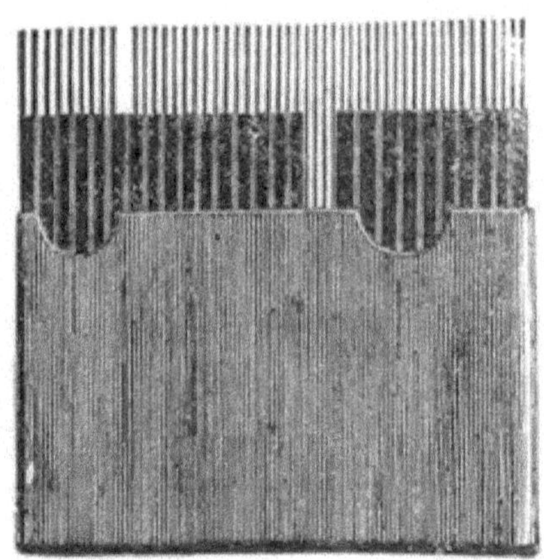

Graining Combs.

	Net.
Leather, per set of 6,	$ 75
American Steel, per set of one doz., -	1 25
English, per set, in cases, - -	1 75

Knotted Badger Blenders.

SET IN BONE.

Size	*Net.*
2 inch,	50 cents per inch.
2½ "	
3 "	
3½ "	
4 "	
4½ "	
5 "	
5½ "	
6 "	

Whitewash Heads.

Common. Whitewash Heads are leather bound, ordinary mixed quality, for general family use.

No. 5,	width 6 inches,	per doz.,	$5 00
6,	" 7 "	"	6 50
7,	" 7½ "	"	8 00
No. 8,	width 8 inches,	per doz.,	$ 9 50
9,	" 9 "	"	11 50
10,	" 9 "	"	14 00
12,	" 9 "	"	16 50
14,	" 9 "	"	20 00

Okatka Whitewash Heads are leather bound, all white Russia bristles, specially adapted for practical whitewashers' use.

No. 9,	Width, 9 inches,	per doz.,	$53 00
No. 11,	Width, 9½ inches,	per doz.,	$ 64 00
13,	" 9½ "	"	76 00
15,	" 9½ "	"	98 00

Bristle Dusters.

For Counters, etc. *Net.*

Common,	35 cents each.
Extra,	50 "
Fancy,	75 "

Subject to per cent. discount.

Super Paste Brushes.

OKATKA.

These Paste Brushes are leather bound, extra thickness, pure, all white bristles, specially adapted for floor painting and paper hangers' use.

3½	inches wide,	per doz.,	$12 00
4	"	"	16 00
4½	"	"	20 00
5	"	"	26 00

Kalsomine Brushes.

These Kalsomine Brushes are zinc bound, riveted, pure, all white Russia Bristles, superfine, extra heavy quality, specially adapted for finest work, for practical masons and finishers.

3	inches wide,	per doz.,	$18 00
4	"	"	23 00
5	"	"	33 00
6	"	"	48 00
7	"	"	61 00
8	"	"	90 00

Common Varnish, or Oval Paint.

All white Bristles, Wire Bound.

No. 6,	per doz.,	$ 3 00
5,	"	4 25
4,	"	5 00
3,	"	5 50
2,	"	7 00
1,	"	8 50
0,	"	10 00
2-0,	"	12 00
3-0,	"	15 00
4-0,	"	17 00
5-0,	"	19 00
6-0,	"	21 00
7-0,	"	24 00

The number of each Kalsomine Brush denotes its width in inches.

WINDOW GLASS.

PRICE LIST OF
VAN SCHAACK, STEVENSON & REID.

Subject to —— discount.

Size of Glass.	No. of Lights per box 50 ft.	Size of Glass.	No. of Lights per box 50 ft.	Size of Glass.	No. of Lights per box 50 ft.	Size of Glass.	No. of Lights per box 50 ft.
6 by 8	150	16 by 32	14	22 by 38	9	28 by 44	6
7 by 9	114	16 by 34	13	22 by 40	8	28 by 46	6
8 by 10	90	16 by 36	12	22 by 42	8	28 by 48	5
8 by 12	75	16 by 38	12	22 by 44	7	28 by 50	5
8 by 14	64	16 by 40	11	22 by 46	7	30 by 34	7
9 by 12	66	16 by 42	11	22 by 48	7	30 by 36	7
9 by 13	61	16 by 44	10	22 by 50	7	30 by 38	6
9 by 14	57	16 by 46	10	22 by 52	6	30 by 40	6
9 by 15	53	16 by 48	9	22 by 54	6	30 by 42	6
9 by 16	50	16 by 50	9	22 by 56	6	30 by 44	5
9 by 18	44	18 by 20	20	22 by 58	6	30 by 46	5
10 by 12	60	18 by 22	18	22 by 60	5	30 by 48	5
10 by 13	55	18 by 24	17	24 by 26	12	30 by 52	5
10 by 14	51	18 by 26	15	24 by 28	11	32 by 34	7
10 by 15	48	18 by 28	14	24 by 30	10	32 by 36	6
10 by 16	45	18 by 30	13	24 by 32	9	32 by 38	6
10 by 17	42	18 by 32	12	24 by 34	9	32 by 40	6
10 by 18	40	18 by 34	12	24 by 36	8	32 by 42	5
10 by 20	36	18 by 36	11	24 by 38	8	32 by 44	5
10 by 22	33	18 by 38	11	24 by 40	7	32 by 46	5
10 by 24	30	18 by 40	10	24 by 42	7	32 by 48	5
10 by 26	28	18 by 42	10	24 by 44	7	32 by 50	5
10 by 30	24	18 by 44	9	24 by 46	7	32 by 52	4
12 by 14	43	18 by 46	9	24 by 48	6	32 by 54	4
12 by 15	40	18 by 48	8	24 by 50	6	34 by 36	6
12 by 16	37	18 by 50	8	24 by 52	6	34 by 38	6
12 by 17	35	18 by 52	8	24 by 54	6	34 by 40	5
12 by 18	33	18 by 54	7	24 by 56	5	34 by 42	5
12 by 20	30	20 by 22	16	24 by 58	5	34 by 44	5
12 by 22	27	20 by 24	15	24 by 60	5	34 by 46	5
12 by 24	25	20 by 26	14	26 by 28	10	34 by 48	4
12 by 26	23	20 by 28	13	26 by 30	9	34 by 50	4
12 by 28	21	20 by 30	12	26 by 32	9	34 by 52	4
14 by 16	32	20 by 32	11	26 by 34	8	34 by 54	4
14 by 18	29	20 by 34	11	26 by 36	8	34 by 56	4
14 by 20	26	20 by 36	10	26 by 38	7	34 by 58	4
14 by 22	23	20 by 38	9	26 by 40	7	34 by 60	3
14 by 24	21	20 by 40	9	26 by 42	7	36 by 40	5
14 by 26	20	20 by 42	9	26 by 44	6	36 by 42	5
14 by 28	18	20 by 44	8	26 by 46	6	36 by 44	5
14 by 30	17	20 by 46	8	26 by 48	6	36 by 46	4
14 by 36	14	20 by 48	7	26 by 60	5	36 by 48	4
16 by 18	25	22 by 24	14	28 by 30	9	36 by 50	4
16 by 20	22	22 by 26	13	28 by 32	8	36 by 52	4
16 by 22	20	22 by 28	12	28 by 34	8	36 by 54	4
16 by 24	19	22 by 30	11	28 by 36	7	38 by 56	3
16 by 26	17	22 by 32	10	28 by 38	7	38 by 58	3
16 by 28	16	22 by 34	10	28 by 40	6		
16 by 30	15	22 by 36	9	28 by 42	6		

SIZES.	1st Quality.
6x 8 to 7x 9..	$6 75
7x10 to 8x10..	7 75
8x11 to 10x12..	8 25
8x14 to 10x15..	8 75
11x14 to 12x17..	10 00
12x18 to 16x24..	11 00
18 x 22 ..	12 00
15x26 to 20x30.	13 00
15x36 to 24x30..	14 00
26x28 to 24x36..	15 00
26x36 to 26x44..	16 00
26x46 to 30x50..	18 00
30x52 to 30x54..	
30x56 to 34x56..	
34x58 to 34x60..	
36x60 to 40x60..	

All sizes over 52 inches in length, and not making 81 united inches, will be charged in 84 united inches bracket.

An additional 10 per cent. will be charged for all Glass more than 40 inches wide.

We handle glass only from reliable manufacturers, who pay special attention to selecting and packing.

Special attention given to orders for odd sizes, double thick, etc.

FEATHER DUSTERS.

Van Schaack, Stevenson & Reid,

Manufacturers' Agents.

FULL CENTRES.

Ostrich Feather Dusters.

MANUFACTURERS' PRICE LIST.

Subject to —— per cent. discount on extra heavy full centre.

N. B.—Dealers will please notice that the *Bell and Picture Dusters* are numbered according to the length (in inches) of the Feathers.

BELL DUSTERS—Full Centres.

No.	Plain, per doz.	Length of Feathers.
No. 4	$ 2 00	4 inches
5	3 00	5 "
6	6 00	6 "
7	9 00	7 "
8	13 00	8 "
9	19 00	9 "
10	24 00	10 "
11	30 00	11 "
12	36 00	12 "
13	41 00	13 "
14	45 00	14 "
15	48 00	15 "
16	48 50	16 "
17	49 00	17 "
18	49 50	18 "
19	50 00	19 "
20	50 50	20 "
21	51 00	21 "
22	51 50	22 "
23	52 00	23 "
24	54 00	24 "

FEATHER DUSTERS—CONTINUED.

CORNICE DUSTERS.

	Plain, per doz.
No. 1	$ 50 00
2	52 00
3	54 00
4	58 00
5	60 00

CARRIAGE DUSTERS.

For Railroad Cars or Carriages, very heavy.

	Plain, per doz.
No. 1	$ 46 00
2	58 00
3	66 00
4	72 00
5	78 00

PICTURE DUSTERS.

Soft, Fine Feathers—for Pianos, Paintings and other Polished Surfaces.

	Plain, per doz.	Length of Feathers.
No. 5	$12 00	5 inches
6	15 00	6 "
7	20 00	7 "
8	24 00	8 "
9	30 00	9 "
10	36 00	10 "
11	42 00	11 "
12	48 00	12 "
13	54 00	13 "
14	60 00	14 "
15	66 00	15 "
16	72 00	16 "

COUNTER DUSTERS.

	Plain, per doz.
No. 1	$ 8 00
2	12 00
3	15 00
Made of soft, fine Feathers	18 00

DYE WOODS, DYE STUFFS, ETC.

Barwood	by bbl.	per lb.,	$3\frac{1}{2}$
Brazil Wood	"	"	$7\frac{1}{2}$
Camwood, commercial	"	"	7
" pure	"	"	10
Ebony Wood	"	"	$4\frac{3}{4}$
Flavine	"	"	75
Fustic, Cuba	"	"	$3\frac{3}{4}$
Hypernic	"	"	$4\frac{1}{2}$
Hachewood	"	"	$4\frac{1}{2}$
Logwood, "Campeachy	"	"	$2\frac{1}{2}$
Nicwood	"	"	$3\frac{1}{2}$
Orchill Paste	"	"	50
Peachwood	"	"	5
Quercitron	"	"	4
Red Saunders	"	"	$5\frac{1}{2}$
Sumac, Lead Seal	"	"	7
Turmeric	"	"	12

The quotations attached are merely approximations of our prices, at this date. When large quantities of goods are ordered, and there is an opportunity of doing better than quoted prices, customers shall have the advantage of it.

Manufacturers will find us fully supplied with all the adjuncts for dyeing purposes: Madder, Madder Compound, Tin Crystals, (which will be found in our Drug List,) and special prices sent on application. With a large experience in the *wants* of this branch of trade, particular attention is paid to handling those only of the *best quality*.

Van Schaack, Stevenson & Reid,
WHOLESALE DRUGGISTS, CHICAGO.

DRUGGISTS' PRESCRIPTION & WRAPPING

PAPER.

MANILLA PAPER BAGS.

¼ lb., for packages of 500		$ 50	10 lb., for packages of 500		$ 2 90
½ "	"	60	12 "	"	3 50
1 "	"	70	14 "	"	4 50
1½ "	"	80	16 "	"	4 75
2 "	"	1 00	20 "	"	5 25
3 "	"	1 15	25 "	"	5 75
4 "	"	1 40	30 "	"	7 00
5 "	"	1 75	35 "	"	7 75
6 "	"	2 10	50 "	" extra heavy	8 75
7 "	"	2 25	35 "	" "	9 75
8 "	"	2 50			

MANILLA PAPER.

Weight to Ream.	Size.	Per lb.
8 lb.	15 x 20	13 cents.
10 "	15 x 20	13 "
12 "	18 x 24	13 "
15 "	18 x 24	13 "
16 "	20 x 30	12½ "
20 "	20 x 30	12½ "
20 "	22 x 32	12½ "
25 "	22 x 32	12½ "
30 "	22 x 32	12½ "
40 "	22 x 32	12½ "
50 "	22 x 32	12½ "
20 "	24 x 36	12½ "
25 "	24 x 36	12½ "
30 "	24 x 36	12½ "
35 "	24 x 36	12½ "
40 "	24 x 36	12½ "
60 "	24 x 36	12½ "
70 "	24 x 36	12½ "

STRAW PAPER.

Size, 13 x 18	4 cents per lb
" 16 x 22	
" 18 x 26	
" 22 x 32	
" 26 x 36	
" 36 x 40	

Blue Seidlitz	19 x 24	per ream,	$ 4 00
White "	24 x 36	"	4 50
Fine Prescription Paper, Assorted Colors,	20 x 25	"	6 00
Prescription " " "	24 x 36	"	6 00

Other sizes and styles furnished upon application.

POCKET CUTLERY.

OUR FIRST BRAND

Is noted for HARD WEAR, GOOD QUALITY, STYLE AND FINISH; EVERY BLADE HAND FORGED AND WARRANTED GOOD. Can almost always supply any pattern ordered, including Budding, Corn, Cork Screw, Farrier's, Nail, Pruning, Pencil, Rule, Scissor, Smoking, Spaying and Sportsman's Knives. From $3.00 to $100.00 per dozen.

OUR SECOND BRAND,

"Manufacturers' Union Cutlery Co.," affords a greater variety of Price and Style than can usually be obtained in low priced goods, and will compare favorably with most of the makes in the market.

	List price per doz.
Boys' one blade Knives	$1 25 to 2 00
3 to 4 inch one blade Jack Knives	2 00 to 4 50
3 to 4 " " " " brass lined	4 00 to 6 00
3 to 4 " two blade " common	2 50 to 3 50
3 to 4 " " " " good	4 50 to 6 50
3 to 4 " " " " extra finish	6 00 to 7 00
3 to 4 " " " G. S. Bolster brass lined	7 00 to 8 00
3 to 4 " " " " fifty styles fancy	7 00 to 10 00
Three Blades, 3½ in. Ivory or Buffalo G. S. Bolster	6 00
" 3½ " " Stag or Buffalo G.S. Bolster, with Shield	7 00 to 9 00
" 3½ " Shell and Pearl " "	8 00 to 12 00
" 3½ to 4 inch, over 100 styles, extra finish	8 00 to 15 00
Four Blades, 3½ inch Ivory, Stag and Buff. Congress, with Shield	6 00 to 8 00
" 3½ " Pearl and Shell " " " "	8 00 to 12 00
" 3½ " Ivory, Stag and Buffalo, straight haft "	7 00 to 9 00
" 3½ " Pearl and Shell, " " "	9 00 to 12 00

PEN KNIVES.

2 or 3 inch Ladies' small Ivory, 1 blade	1 50 to 2 50
2 or 3 " " Pearl or Shell	2 50 to 4 00
2½ " " fine Ivory, 2 blades	4 00 to 6 00
2½ " " Shell or Pearl, "	5 00 to 8 00
Various Patterns in three or four "	6 00 to 12 00

In ordering by mail, state about what priced goods are wanted, and the quantities of each.

We also carry a line of low priced German Pocket Knives.

The Trade supplied by

Van Schaack, Stevenson & Reid,

92 and 94 Lake Street, Chicago.

STATIONERY, ENVELOPES, ETC.

—FOR SALE BY—

VAN SCHAACK, STEVENSON & REID.

RULED PAPERS.

FIRST CLASS, AND SURE TO GIVE SATISFACTION.

NAME.	Weight per Ream.		Price per Pound
Pearl Spring Mills, Billet	3	lb.	33 cts
Hudson River Mills, Billet	3	"	30 cts
Pearl Spring Mills, Octavo	4	"	33 cts
Hudson River Mills, Octavo	4	"	30 cts
Pearl Spring Mills, Note	4, 5, 6	"	33 cts
Hudson River Mills, Note	4, 5, 6	"	30 cts
Pearl Spring Mills, Packet Note	7, 8	"	33 cts
Hudson River Mills, Packet Note	7, 8	"	30 cts
Pearl Spring Mills, Letter	8, 9, 10, 12	"	33 cts
Hudson River Mills, Letter	8, 9, 10, 12	"	30 cts
Pearl Spring Mills, Foolscap	9, 10, 12, 14	"	33 cts
Hudson River Mills, Foolscap	9, 10, 12, 14	"	30 cts
Pearl Spring Mills, Legal Cap,	10, 12, 14, 16	"	33 cts
Hudson River Mills, Legal Cap,	10, 12, 14, 16	"	30 cts
Pearl Spring Mills, Bill Cap,	10, 12, 14	"	37 cts

ENVELOPES.

No.	Quality.	Size.	Price per 1000.	Size.	Price per 1000
280	Thick Manilla	5½	$1 05	6	$1 25
	Extra Thick Manilla	5½	1 15	6	1 35
430	Double Thick Manilla	5½	1 35	6	1 75
5394	Thick D. K. Buff	5½	1 75	6	1 85
7430	Extra Thick Corn	5½	1 75	6	2 00
7430	Double Thick Corn	5½	2 00	6	2 50
7450	Extra Thick Gold	5½	1 75	6	2 00
7450	Double Thick Gold	5½	2 00	6	2 50
7840	Extra Thick Canary	5½	1 75	6	2 00
7840	Double Thick Canary	5½	2 00	6	2 50
7850	Extra Thick Amber	5½	1 75	6	2 00
7850	Double Thick Amber	5½	2 00	6	2 50
6920	Extra Thick Orange	5½	2 25	6	2 50
6920	Doubie Thick Orange	5½	2 25	6	2 65
6940	Extra Thick Canary	5½	2 25	6	2 50
6940	Double Thick Canary	5½	2 50	6	2 75
4530	Thick White Wove	5½	2 25	6	2 50
5540	Double Thick White Wove	5½	2 50	6	2 75
	Extra Thick Pearl Wove	5½	2 25	6	2 50
	Extra Blue Laid	5½	2 50	6	2 75
	Double Pearl	2¾	2 25	4½	2 75
	Double Pearl	5¾	3 00		

CIGARS AND TOBACCO.

Sold by Van Schaack, Stevenson & Reid.

CIGARS—HAVANA FILLERS, HAND MADE.

	In Box.	Per M.
Flor del Fumar, extra size	50	$75 00
" Flounder, original	50	75 00
La Flor de Henry Clay, Concha Regalia	50	78 00
Old Reliable, Full Regalia	50	83 00
The Identical, Reina Victoria	100	65 00
Diamond Crown, Regalia	50	70 00
Rich and Rare, Concha	100	60 00
La Rosa Santiago, "	100	60 00
Old Reliable, "	100	60 00
Diamond Crown, "	100	50 00
Chicago Favorite, "	100	50 00
Figaro	100	50 00
Flor. de Cuba, Concha	100	43 00

CIGARS—SEED MIXED, HAND MADE.

	In Box.	Per M.
L'Africana, Media Regalia	100	$45 00
Silver Lake, Full Regalia	50	45 00
Concha Media	100	40 00
Magnolia, Concha	100	30 00
1876	100	30 00
First Lesson	100	30 00
Delight	100	30 00

FINE CUT CHEWING TOBACCO.

EUREKA.

60 lb. Packages	80c.	20 lb. Packages	81c.
40 "	80c.	10 lb. Pails	82c.

GRAY EAGLE.

60 lb. Packages	80c.	20 lb. Packages	81c.
40 "	80c.	10 lb. Pails	82c.

SWEET CLOVER.

60 lb. Packages	76c.	20 lb. Packages	77c.
40 "	76c.	10 lb. Pails	78c.

JAMESTOWN.

60 lb. Packages	74c.	20 lb. Packages	75c.
40 "	74c.	10 lb. Pails	76c.

GOLDEN WEST.

60 lb. Packages	66c.	20 lb. Packages	67c.
40 "	66c.	10 lb. Pails	68c.

ANCHOR.

60 lb. Packages	66c.	20 lb. Packages	67c.
40 "	66c.	10 lb. Pails	68c.

HUNTSMAN.

60 lb. Packages	55c.	20 lb. Packages	56c.
40 "	55c.	10 lb. Pails	57c.

RYE BLOOM.

60 lb. Packages	51c.	20 lb. Packages	52c.
40 "	51c.	10 lb. Pails	53c.

We make no charge for Packages containing Fine Cut Chewing.

Druggists' and Stationers' Elastic Bands and Rings.

ELASTIC BANDS.

Light.	Per Gross.
½ inch wide, assorted, lengths 2¼ to ¼ inch	$1 20
⅝ " " " "	1 70
1 " " " "	2 35
1½ " " " "	3 60
Heavy.	
X ¼ inch wide, 2 inches long	50
X ½ " 2 "	1 20
X ¾ " 2 "	1 55
XX ¼ " 2½ "	60
XX ½ " 2½ "	1 35
XX ¾ " 2½ "	1 85
XXX ¼ " 3 "	75
XXX ½ " 3 "	1 55
XXX ¾ " 3 "	2 15
XXXX ¼ " 3½ "	90
XXXX ½ " 3½ "	1 80
XXXX ¾ " 3½ "	2 40
⅛ inch wide, 2 inches long, No. 30	30
" 2½ " " 31	35
" 3 " " 32	40
" 3½ " " 33	45
" 1½ " " 51	25

ELASTIC RINGS.

Light, No. 1, 2½ inches inside diameter	$1 20
" 2, 1⅝ " "	90
" 3, ⅝ " "	60
" 50, 1 " "	20
Heavy, No. 1, 2½ " "	2 00
" 2, 1⅝ " "	1 25
" 3, ⅞ " "	75
" 42, ⅞ " "	60
" 43, 1⅛ " "	75
" 44, 1⅜ " "	1 00
" 45, 1⅝ " "	1 35
Asst'd Sizes, Nos. 42, 43, 44, 45, one-half gross in a box	80
o Assorted	95
oo "	1 10
ooo "	1 25
¼ "	70
½ "	1 50
Small Rings, No. 8, ¾ inches long, per gt. gross	1 00
" 10, 1⅛ " "	1 10
" 11, 1⅜ " "	1 15
" 13, 1¾ " "	1 35
" 14, 2 " "	1 40
" 16, 2½ " "	1 60
" 18, 3 " "	1 90
" 19, 3½ " "	2 10
" 20, 1½ " "	1 35

RUBBER GOODS.

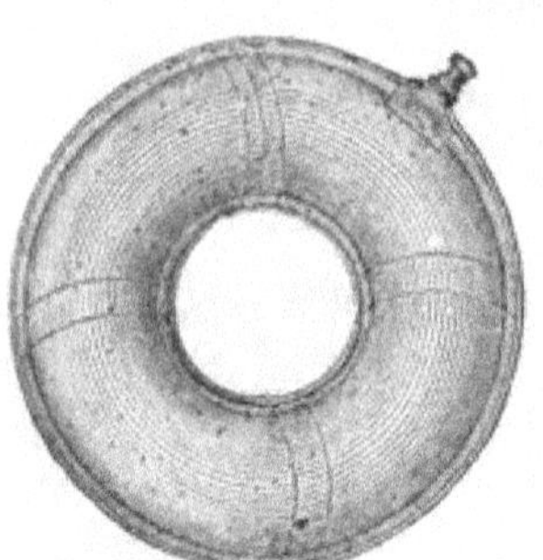
BED PANS.

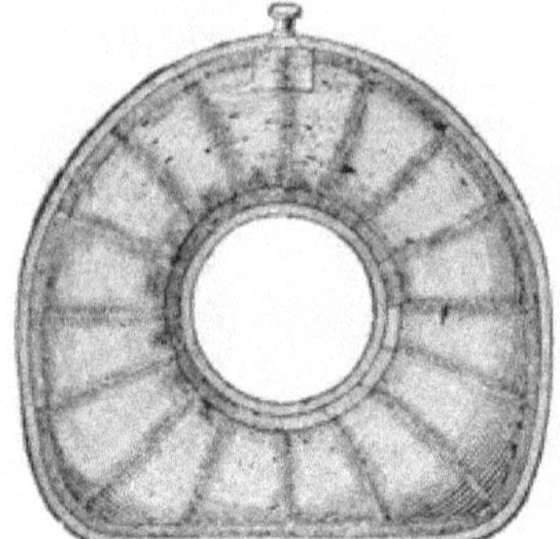
CHAIR CUSHIONS.

BATH TUBS.

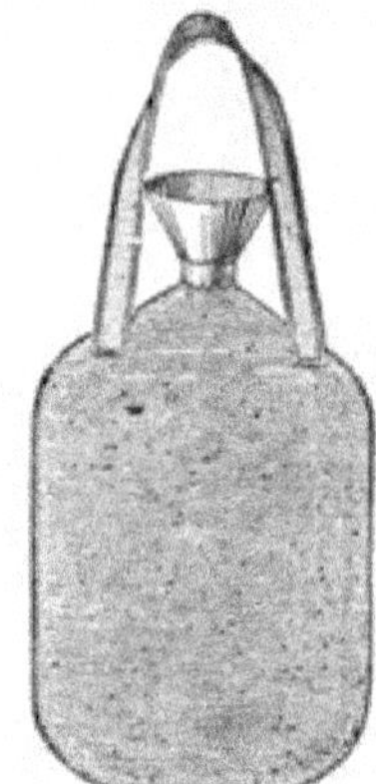
WATER BAG.

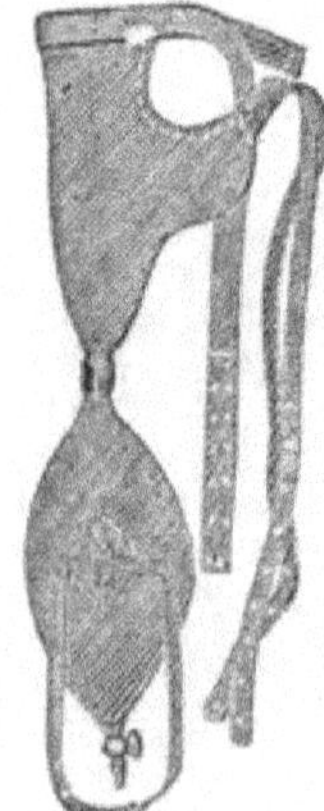
Urinal, Male (A).

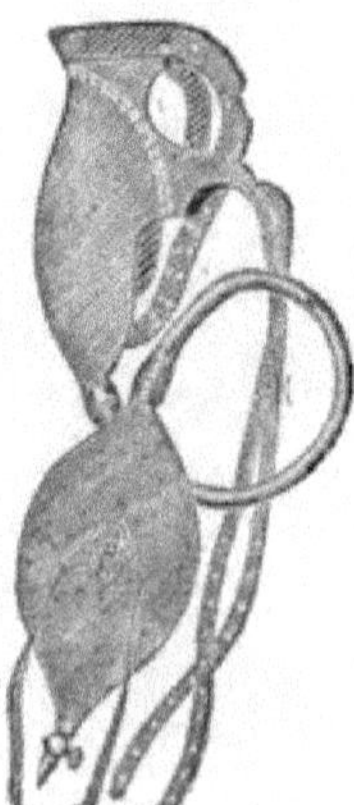
Urinal, Male (B).

RESERVOIR SYRINGE.

DRUGGISTS' CONFECTIONERY.

Sold by Van Schaack, Stevenson & Reid.

Item	Packing		Price
A. B. Gum Drops, Lemon	per pound, 5 lb. boxes,		17
" " " Liquorice	"	"	17
" " " Rose	"	"	17
" " Stick, Rose, Lemon, and Liquorice	"	"	18
Apple Jelly, Bon Bons	"	"	30
" Cakes	"	"	30
" Rings	"	"	30
Assorted Kisses	"	"	18
Barber Pole	"	"	18
Brandy Drops	"	"	25
Brandy and Wine Bottles, Assorted	"	"	25
Cachous, for the breath	"	"	30
Caramels, all flavors	"	"	30
Caraways, Pearled	"	"	26
Chocolate Cream Drops	"	"	25
" " Stick	"	"	25
Chocolate Ice	"	"	35
Cloves, Pearled	"	"	30
Cocoa Bar	"	"	28
" Ice	"	"	35
Cream Bar	"	"	18
Cream Candy, all flavors	"	"	25
Cream Dates	"	"	30
Cream Bon Bons	"	"	25
" Assorted	"	"	25
" Babies	"	"	25
English Coltsfoot Rock	" 4	"	40
English Plum Pudding	" 5	"	35
Fig Paste	"	"	20
Fruit Ice	"	"	35
H. H. Rolls	"	"	18
Imperials, all flavors	"	"	25
Jackson Balls	"	"	18
Jelly Gum Drops	"	"	25
Jordan Almonds	"	"	45
Jordan Burnt Almonds	"	"	45
Jujube Sticks	"	"	26
Lady Cream Drops	"	"	18
Lozenges, all flavors	"	"	22
" Conversation	"	"	23
" 1 cent rolls	1 gross in box, per box,		1 00
" 5 cent rolls	32 rolls in box,	"	1 00
Mint Kisses	per pound, 5 lb. boxes,		18
Mixed Candy	"	"	18
" Extra	"	"	22
" French	"	"	25
Molasses Bar	"	"	18
Mint Balls	"	"	18
Mottoes	"	"	18
Peanut Bar	"	"	22
" Stick	"	"	22
" Squares	"	"	22
Rock Candy, all string	"	"	23
Smooth Almonds, No. 1	"	"	30
" " No. 2	"	"	26
Spanish Burnt Almonds	"	"	30
Stick Candy, Assorted, Best	in 3 lb. paper boxes, 36 lbs. in case,		14

CONFECTIONERY—*Continued.*

Strawberry Ice	per pound, 5 lb. boxes,	$	35
Sugar Sands, all colors	"	"	20
Vanilla Ice	"	"	35
Walnut Ice	"	"	35
Wine Drops	"	"	25

GRAINED WORK.

Boston Mints	"	"	18
French Fruit Drops	"	"	23
Grained Sugar Mice	"	"	23
Mint Drops	"	"	18
Washington Stars	"	"	18

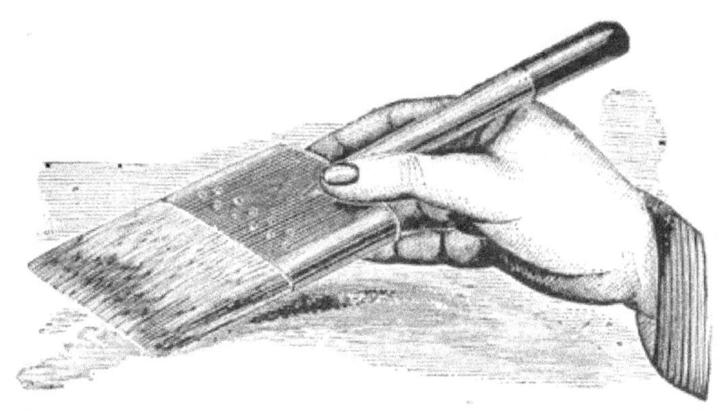

ARTISTS' MATERIALS.

TUBE COLORS, IN PATENT COLLAPSIBLE TUBES.

Ordinary, per dozen, $1.20. Whites, DOUBLE SIZE, per dozen, $2.40.

Am. Vermilion.
Antwerp Blue.
Asphaltum.
Bistre.
Bitumen.
Blue Black.
" Verditer.
Bone Brown.
Brilliant Yellow.
Brown Ochre.
" Pink.
" Red.
Burnt Roman Ochre.
" Sienna.
" Terre Verte.
" Umber.
Caledonian Brown.
Cappah Brown.
Cassel Earth.
Chrome Green, No. 1.
" " No. 2.
" " No. 3.
" Orange.
" Yellow No. 1.
" " No. 2.
" " No. 3.
Chinese Blue.
Cologne Earth.
Cremnitz White.
Crimson Lake.
Emerald Green.
Flake White.
Gamboge.
Gold Ochre.
Indian Lake.
" Red.
Italian Pink.
Ivory Black.
King's Yellow.
Lamp Black.
Light Raw Sienna.
Light Red.
Megilp.
Mummy.
Naples Yellow, No. 1.
" " No. 2.
" " No. 3.
New Blue.
Olive Lake.
Orange Mineral.
Oxford Ochre.
Paris Green.
Permanent Blue.
Persian Red.
Prussian Blue.
Purple Lake.
Raw Sienna.
" Umber.
Roman Ochre.
Rose Pink.
Scarlet Lake.
Silver White.
Sugar of Lead.
Terre Rosa.
" Verte.
Vandyke Brown.
Verdigris.
Venetian Red.
Veronese Green.
Yellow Lake.
" Ochre.
" Orpiment.
Zinc White.
Zinnober Green, light.
" " med.
" " deep.

Per Dozen, $1.80.

Burnt Lake.
Chinese Vermilion.
Citron Yellow.
English Vermilion.
Fine Lake.
French Vermilion.

Per Dozen, $3.00.

Brown Madder.
Cobalt Green.
Madder Lake.
Malachite Green.
Oxide of Chromium.
" Transparent.
Pink Madder.
Rose Madder.

ARTISTS' MATERIALS—*Continued.*

Per Dozen, $3.50.

Cobalt Blue.
French Ultramarine.
Indian Yellow.
Lemon Yellow.
Mars Brown.
Mars Orange.
Mars Red.
Mars Yellow.
Robert's Lake.
Ruben's Madder.
Scarlet Vermilion.

Per Dozen, $4.80.

Antimony Yellow.
Carmine.
Purple Madder.

Per Dozen, $5.40.

Burnt Carmine.
Cadmium, light.
" medium.
Cadmium, deep.
Lemon Cadmium.
Scarlet Madder.
Scarlet.

BEST SHEET WAX For Flowers.

In boxes of 1 gross sheets, 12 sheets in package. Size, 5¼ x 2¾.

White, single thick,	per gross.
Green, all shades,	"
Yellow, "	"
Purple, "	"
Blue, "	"
Light Pink, "	"
Buff, for Tea Rose, all shades,	"
Scarlet,	"
Dark Pink,	"
White, double thick.	"

VARIEGATED WAX.

For small Autumn Leaves.

Size, 5¼ x 2¾.

Plain Tints,	per gross.
Carmine,	"

POND LILY WAX—Thick.

In boxes of half gross each, six sheets in package. Size, 7 x 4¼.

White,	per gross.
Green,	"

Each shade numbered to correspond with sample cards, which will be furnished to dealers on application.

This system will enable parties to obtain with exactness the desired shade without sending samples.

WIRE—22 yards on spools.

Silk Wire, white, on spools, 3 thicknesses, per doz., $1.00.
" green, on spools, 3 thicknesses, per doz., $1.00.
" white, coils, 4 thicknesses, per doz., $1.00.
" green, coils, 4 thicknesses, per doz., $1.00.
Cotton Wire, white, on spools, 2 thicknesses, per doz., 75c.
" green, on spools, 2 thicknesses, per doz., 75c.
Paper Wire, for Stems, in bunches, per doz., $1.20.
Silver " on spools, per doz., $3.50.
Uncovered Wire, on spools, per dozen, $1.20.

VIOLIN STRINGS, ETC.

VIOLIN STRINGS.

Nos.				per bundle of 30 strings.	
61260	E, 2½	lengths,	plain, fair quality		
41271	E, 2½	"	Red Ends, best quality	"	$ 1 65
81274	E, 2½	"	Naples, finest quality	"	2 40
31279	E, 3	"	Red Ends, finest "	"	2 40
41287	E, 3	"	Padua, fine "	"	2 70
21288	E, 3	"	Naples, superfine quality	"	3 00
91300	E, 4	"	Roman, good quality	"	3 70
41301	E, 4	"	Genuine Naples, best quality	"	4 20
71303	E, 4	"	Padua, fine quality	"	3 90
21304	E, 4	"	Naples, superfine quality	"	4 20
31316	E, 6	"	" " "	"	6 00

Note.—Either of the foregoing numbers of E strings are suitable for the Guitar E, (or 1st,) but those with four lengths are usually preferred.

11322	A, 2½	lengths,	good quality	"	1 20
71323	A, 3	"	better "	"	1 50
31324	A, 3	"	Padua, fine quality	"	2 10
61325	A, 3	"	Genuine Naples, best quality	"	4 00
41326	A, 3	"	Roman, good "	"	3 00
81327	A, 3	"	Naples, superfine "	"	4 50

Note.—Either of the foregoing numbers of A strings are suitable for the Guitar B, (or 2nd,) but those of the finest qualities are usually preferred.

31337	D, 2	lengths,	good quality	"	2 10
41338	D, 2½	"	better "	"	2 40
21339	D, 2½	"	Padua, fine quality	"	3 00
71340	D, 2½	"	Genuine Naples, best quality	"	5 00
41341	D, 2½	"	Roman, good quality	"	4 00
61342	D, 2½	"	Naples, superfine quality	"	5 00
11351	G, 1	"	single, wound on gut, ordinary	per doz.	25
71352	G, 1	"	" " good	"	45
21355	G, 1	"	double, " extra fine	"	1 00
31356	G, 1	"	single, " good quality, plain	"	
91357	G, 1	"	" " superior " "	"	60
41358	G, 1	"	double, " extra " "	"	85

VIOLA STRINGS.

41366	A, (or 1st,)	Genuine Naples, best quality	per doz.	2 00
91367	D, (or 2nd,)	" " "	"	3 00
81368	G, (or 3rd,)	wound on gut. "	"	1 30
31369	C, (or 4th,)	" "	"	1 50

VIOLIN STRINGS—*Continued.*

GUITAR (BASS) S GS.

No.	Description	Unit	Price
31375	D, (or 4th,) wound on white silk, extra quality	per doz.	$ 90
21377	A, (or 5th,) " " "	"	1 00
61379	E, (or 6th,) " " "	"	1 10
41383	D, (or 4th,) genuine Spanish, wound on white silk, the finest quality made	"	1 30

VIOLIN NECKS.

No.	Description	Unit	Price
61427	Curly Maple, good quality, neatly carved scroll, each		1 00
41428	" fine " finely " "		1 50

VIOLIN PEGS.

No.	Description	Unit	Price
71430	Maple, ordinary, black	per gross	1 75
21431	" black, with pearl dot inlaid in head	"	4 20
81435	Boxwood, imitation Ebony, good model, with pearl dot inlaid in head	"	8 40
31440	Ebony, oval shape, French, with pearl dot inlaid in head	per doz.	80
91441	" hollow " " " " "	"	90
21451	" oval " with pearl leaf, etc., inlaid on sides	"	1 75

VIOLIN FINGER BOARDS.

No.	Description	Unit	Price
71474	Ebony, plain, best quality	per doz.	4 25

VIOLIN BRIDGES.

No.	Description	Unit	Price
31478	Ordinary, with 2 scrolls, 1st quality	per gross	1 50
41480	" " 3 " well cut	"	3 00
21481	Good model, 3 " " and finished	"	3 60
61486	"Didelot," finely cut scrolls, first quality	per doz.	1 00
41487	"Bausch," " "	"	1 10
91488	"Dresden," " "	"	50

VIOLIN TAIL PIECES.

No.	Description	Unit	Price
71513	Imitation Ebony, with pearl diamond, etc., inlaid	per doz.	1 00
1515	" " flower, "	"	1 50
41518	" " flowers, " fine finish.	"	4 20
81527	Ebony, plain, good model	"	2 40

VIOLIN RESIN.

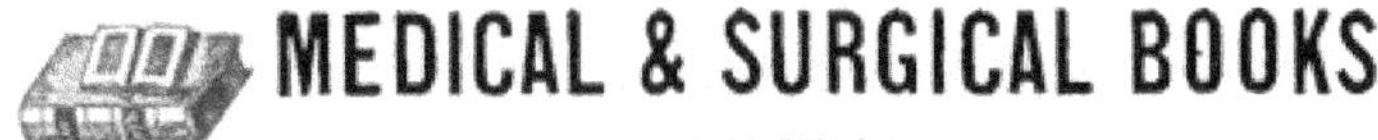

MEDICAL & SURGICAL BOOKS

— SUPPLIED BY —

VAN SCHAACK, STEVENSON & REID,

92 AND 94 LAKE STREET, CHICAGO.

1.	United States Dispensatory, 13th edition	$ 8 00
2.	Pharmacopœia of the United States ....cloth, $1.75; leather	2 25
3.	Griffith's Universal Formulary. New ed. ....cloth, $4.50; leather	5 50
4.	Beasley's Druggists' Receipt Book	3 50
5.	Attfield's Chemistry ....cloth, $2.75; leather	3 25
6.	Bowman's Medical Chemistry	2 25
7.	" Practical "	2 25
8.	Fowne's Chemistry ....cloth, $2.75; leather	3 25
9.	Fresenius' Quantitative Chemistry	6 00
10.	" Qualitative "	4 50
11.	Galloway's Qualitative Chemistry	2 50
12.	Morton and Leed's Chemistry	2 00
13.	Muspratt's Chemistry, 2 vols	25 00
14.	Miller's Elements of Chemistry, 3 vols	18 00
15.	Odling's Chemistry	2 00
16.	Rand's Medical Chemistry ....cloth, $2.00; sheep	2 50
17.	Cleveland's Pronouncing Dictionary	1 25
18.	Dunglison's Medical Dictionary ....cloth, $6.50; leather	7 50
19.	Hoblyn's Dictionary ....cloth, $1.50; sheep	2 00
20.	Thomas' Pronouncing Dictionary ....cloth, $3.25; leather	3 75
21.	Wood's Pocket Lexicon ....75 cts.; tuck	1 00
22.	Aitkin's Practice of Medicine ....cloth, $12.00; leather	14 00
23.	Bennett's Clinical Medicine ....cloth, $6.00; leather	7 00
24.	Cyclopedia of Practical Medicine, 4 vols ....cloth, $11.00; leather	15 00
25.	Flint's Practice of Medicine ....cloth, $6.00; leather	7 00
26.	Hartshorne's Essentials of Medicine ....cloth, $2.38; half morocco	2 62
27.	Tanner's Practice of Medicine ....cloth, $6.00; leather	7 00
28.	Trousseau's Clinical Medicine, 2 vols. complete, cloth, $10.00; leath.	12 00
29.	Watson's Practice, 2 vols ....cloth, $9.00; leather	11 00
30.	Wood's Practice, 2 vols ....cloth, $9.00; leather	10 00
31.	Parrish's Pharmacy ....cloth, $6.50; leather	7 50

We supply, in addition to the above selected list, all Standard Works upon

Anatomy.
Brain Diseases.
Cancer.
Chest.
Heart.
Lungs.
Throat.
Chronic Diseases.
Deformities.
Dislocations.
Dissections.
Eclectic Practice.
Electricity.
Eye and Ear.

Materia Medica.
Insanity.
Microscope.
Neuralgia.
Pharmacy.
Practice of Medicine.
Rectum and Anus.
Rheumatism.
Skin Diseases.
Surgery.
Botany.
Nervous Diseases.
Chemistry.

Children's Diseases.
Dentistry.
Fractures.
Females.
Fevers.
Histology.
Kidneys.
Jurisprudence.
Poisoning.
Obstetrics.
Pathology.
Physiology.
Venereal Diseases, etc.

*** Complete catalogues mailed to any address.

VAN SCHAACK, STEVENSON & REID,

Old Salamander Wholesale Drug Warehouse, 92 and 94 Lake St.,

CHICAGO.

TILDEN'S FLUID EXTRACTS.

We have always in stock a full supply of these Celebrated Standard Fluid Extracts, representing, with few exceptions, for each fluid pint, sixteen troy ounces of the crude material; the exceptions being the compounds prepared according to the Pharmacopœia. Put up in 16oz. bottles, each package labeled with directions and dose. We confidently recommend the goods as now prepared by Messrs. Tilden & Co., as *fluid extracts*, and not *mere tinctures*.

In ordering Fluid Extracts, be particular to specify Tilden's.

The Trade supplied by

VAN SCHAACK, STEVENSON & REID,

92 and 94 Lake Street, cor. Dearborn, Chicago.

PRICE LIST.

☞ *A Deduction of* **50** *per cent. per Pound from List Prices.*

			$
Aconite Leaves	See Special List.	per lb.	$
" Root	" "		
Agrimony		"	1 75
Aloes		"	2 75
Angelica Root		"	1 25
Angustura		"	4 00
Arnica	Arnica Montana	"	1 75
Aromatic Comp.		"	2 25
Avens Root	Geum Rivale	"	1 50
Balmony	Chelone Glabra	"	1 25
Bayberry	Myrica Cerifera	"	1 25
Barberry Bark	Berberis	"	1 25
Belladonna	See Special List.		
Bethroot	Trillium Pendulum	"	1 75
Bitter-Root	Apocynum Andros.	"	2 00
Bitter-Sweet	Solanum Dulcamara	"	1 50
Black Alder	Prinos Verticillatus	"	1 50
Blackberry	Rubus Villosus	"	1 50
Black Cohosh	Cimicifuga	"	2 00
" Comp.	"	"	2 00
Black Hellebore	Helleborus Niger	"	1 75
Black Pepper	Piper Nigrum	"	1 50
Blood Root	Sanguinaria	"	1 75
Blue Cohosh	Leontice Thalic	"	1 50
Blue Flag	Iris Versicolor	"	1 75
Boneset	Eupatorium Perf.	"	1 25
Boxwood	Cornus Florida	"	1 50
Broom Top	Cytisus Scoparius	"	1 75
Buchu	Barosma Crenata	"	2 50
" Comp.	" "	"	2 50
Buchu and Pareira Brava		"	3 50
Buckthorn	Rhamnus Catharticus	"	1 50
Bugle-Weed	Lycopus Virginica	"	1 25
Burdock	Lappa Minor	"	1 50

Butternut	Juglans Cinerea	per lb.	$1 25
Button Snake Root	Liatris	"	1 50
Canella	Canella Alba	"	1 50
Cannabis Indica	See Special List.		
Calabar Bean	Physostigma Venenos	"	6 00
Calamus		"	2 00
Catechu		"	2 00
Cardamom		"	6 00
" Comp.		"	3 00
Cascarilla	Croton Eleuteria	"	1 25
Catnip	Nepeta Cataria	"	1 25
Cayenne Pepper	Capsicum	"	3 00
Centaury, Red	Sabbatia	"	1 50
Chamomile	Anthemis Nobilis	"	1 75
Chestnut Leaves	Castanea Vesca	"	2 00
Cherry Bark	Prunus Virginiana	"	1 60
" Comp.	" "	"	1 50
Cinchona, Pale		"	2 50
" Comp.		"	2 50
" Calisaya		"	4 25
" Red		"	4 25
" Aromatic		"	2 50
Cloves	Caryophyllus	"	1 75
Clover, Red	Trifolium Pratense	"	2 00
Cleavers	Galium	"	1 25
Colchicum Root	See Special List.		
" Seed	" "		
Colocynth	Cucumis Colocynthis	"	2 25
Columbo	Cocculus Palmatus	"	2 50
Coltsfoot	Tussilago	"	1 50
Comfrey	Symphytum	"	1 50
Conium	See Special List.		
" Seed	" "		
Cotton Root Bark	Gossypium Herbaceum	"	3 00
Cramp Bark	Viburnum Opulus	"	1 50
Crane's Bill	Geranium Maculatum	"	1 75
Cubebs	Piper Cubeba	"	2 50
Culver's Root	Leptandra Virginica	"	2 00
Dandelion	Taraxacum	"	2 00
" and Senna		"	1 75
" Comp.		"	2 00
Dwarf Elder	Aralia Hispida	"	1 25
Elder Flowers	Sambucus	"	1 50
Elecampane	Inula	"	1 25
Ergot	See Special List.		
Eucalyptus		"	6 00
Fern Sweet	Comptonia	"	1 25
Fever Few	Pyrethrum	"	1 25
Fever Bush	Benzoin	"	1 25
Fire Weed	Erechthites	"	1 50
Foxglove	See Special List.		
Frostwort	Helianthemum	"	1 50
Garden Celandine	Chelidonium	"	1 50
Gelseminum	See Special List.		
Gentian	Gentiana Lutea	"	1 50
" Comp.	" "	"	1 75
Ginger	Zingiber Officinale	"	2 25
Golden Seal	Hydrastis Canadensis	"	2 00
Golden Rod	Solidago	"	1 25
Gold Thread	Coptis	"	1 75
Gravel Plant	Epigæa Repens	"	1 25
Guarana		"	10 00
Hardhack	Spiræa Tomentosa	"	1 25
Hemlock	Pinus Canadensis	"	1 25
Henbane	See Special List.		
Hop	Humulus Lupulus	"	2 50
Horehound	Marrubium Vulgare	"	1 50

Hydrangea	Hydrangea Arborescens	per lb.	$ 1 75
Ignatia Bean	Strychnos Ignatia	"	3 50
Indian Hemp	Apocynum Cannab.	"	2 00
" Foreign	See Special List.		
" White	Asclepias Incarnata	"	1 75
Indian Physic	Gillenia Trifoliata	"	1 25
Ipecac	Cephælis Ipecacuanha	"	7 50
" and Seneka		"	6 00
Jalap		"	5 00
Jersey Tea	Ceanothus Amer.	"	2 00
Johnswort	Hypericum	"	1 25
Juniper Berries	Juniperus Communis	"	1 25
Lady's Slipper	Cypripedium Pubescens	"	2 25
Lettuce	Lactuca Sativa	"	1 50
Lily White	Nymphœa	"	1 25
Life Root	Senecio Aureus	"	1 50
Liquorice	Glycyrrhiza Glabra	"	1 50
Liverwort	Hepatica Americana	"	1 50
Lobelia	Lobelia Inflata	"	1 75
" Comp.	" "	"	1 75
Logwood	Hæmatoxylon Campech.	"	1 25
Lovage	Ligusticum Levisticum	"	1 25
Lungwort	Pulmonaria	"	1 50
Lupulin, U. S. P.		"	3 50
Male Fern	Aspidium Filix Mas.	"	1 50
Mandrake	Podophyllum Peltatum	"	1 75
" Comp.	" "	"	1 75
Marsh Rosemary	Statice Limonium	"	1 25
Matico	Piper Angustifolium	"	3 00
Mugwort	Artemisia Vulgaris	"	1 25
Nux Vomica	See Special List.		
Opium, Aqueous	" "		
Orange Peel	Aurantium	"	1 50
Orris Root	Iris Florentina	"	1 75
Pareira Brava	Cissampelos Pareira	"	4 00
Parilla Yellow	Menispermum Canadense	"	2 00
Peppermint	Mentha Piperita	"	1 25
Pink Root	Spigelia Marilandica	"	2 50
" and Senna		"	2 00
" Comp.		"	2 50
Pipsissewa	Chimaphila Umbellata	"	1 50
Pitcher Plant	Sarracenia Purpurea	"	2 50
Pleurisy Root	Asclepias Tuberosa	"	2 00
Poke	Phytolacca Decandra	"	1 50
Poplar Bark	Liriodendron Tulipifera	"	1 50
Poppy	Papaver Somniferum	"	1 75
Prickly Ash	Xanthoxylum Fraxineum	"	1 75
Ptelea	Ptelea Trifoliata	"	1 50
Quassia	Simaruba Excelsa	"	1 50
Queen of the Meadow	Eupatorium Purpureum	"	1 50
Rhatany	Krameria Triandra	"	2 00
Rhubarb	Rheum Palmatum	"	5 00
" Aromatic		"	4 75
" and Senna		"	4 00
Rosin Weed		"	3 00
Rue	Ruta Graveolens	"	1 50
Saffron	Crocus Sativus	"	4 00
Sage	Salvia Officinalis	"	1 50
Sarsaparilla, Rio Negro		"	2 50
" and Dandelion		"	2 50
" Comp.		"	2 50
" American		"	1 75
Sassafras	Laurus Sassafras	"	1 50
Savin	Juniperus Sabina	"	1 25
Scullcap	Scutellaria Lateriflora	"	2 25
" Comp.	" "	"	1 75
Seneka	Polygala Senega	"	3 50

Senna, Alex.	Cassia Acutifolia	per lb.	$ 1 50
" Comp.	" "	"	2 00
" and Jalap		"	3 00
" Aqueous		"	1 50
Skunk Cabbage	Symplocarpus Fœtidus	"	1 25
Snake Root	Serpentaria	"	3 00
Solomon's Seal	Convallaria	"	1 50
Southernwood	Artemisia Abrotanum	"	1 50
Spearmint	Mentha Viridis	"	1 25
Spikenard	Aralia Race	"	1 50
Squill	Scilla Maritima	"	1 50
" Comp.	" "	"	3 00
Stillingia		"	2 50
" Comp.		"	2 50
Stone Root	Collinsonia Canadensis	"	1 75
Stramonium	See Special List.		
Sumach	Rhus Glabrum	"	1 25
Sweet Gale	Myrica	"	1 25
Tag Alder	Alnus Rubra	"	1 25
Tansey, Double	Tanacetum Vulgare	"	1 25
Thyme	Thymus Vulgaris	"	1 25
Tobacco		"	2 00
Tonqua	Dipterix Odorata	"	3 00
Turkey Corn	Corydalis Formosa	"	3 00
Turmeric	Curcuma Longa	"	1 25
Unicorn	Aletris Farinosa	"	3 00
" False		"	3 00
Uva Ursi	Arctostaphylos	"	1 50
Valerian	Valeriana Officinalis	"	2 00
Veratrum Viride	See Special List.		
Vervain	Verbena	"	1 25
Wahoo	Euonymus Atropurpureus	"	2 25
Water Pepper	Polygonum Punctatum	"	1 25
White Oak	Quercus Alba	"	1 25
Whitewood Bar	Liriodendron	"	1 25
Wild Indigo	Baptisia Tinctoria	"	1 50
Wild Turnip	Arum Triphyllum	"	1 50
Wild Yam	Dioscorea Villosa	"	1 50
Wintergreen	Gaultheria Procumbens	"	1 25
Witch Hazel	Hamamelis Virginica	"	1 25
Wormseed	Chenopodium	"	1 50
Wormwood	Artemisia Absinthium	"	1 25
Yarrow	Achillea Millefolium	"	1 25
Yellow Dock	Rumex Crispus	"	2 00

SPECIAL LIST—Fluid Extracts.

[10 per cent. discount.]

Aconite Leaves,	(in pound bottles)	per lb.	$ 1 50
" Root,	" "	"	1 50
Belladonna,	" "	"	2 00
Cannabis Indica,	" "	"	2 75
Conium,	" "	"	1 50
" Seed.	" "	"	2 00
Colchicum Root,	" "	"	1 50
" Seed,	" "	"	2 25
Digitalis,	" "	"	1 50
Ergot,	" "	"	3 50
Gelseminum,	" "	"	2 25
Hyoscyamus,	" "	"	2 00
Nux Vomica,	" "	"	1 75
Opium,	" "	"	3 00
Stramonium,	" "	"	1 50
Veratrum,	" "	"	1 75

A Deduction of **50** per cent. per Pound from List Prices.

Alcoholic Solid Extracts.

		In lb. Earth'n Jars.	In ½ lb Earth'n Jars.	In ¼ lb Earth'n Jars.	In 2 oz. Earth'n Jars	In 1 oz. Earth'n Jars.
Aconiti	Aconite	$4 50	$4 75	$5 00	$5 25	$5 50
Aloes		4 50	4 75	5 00	5 25	5 50
Anthemidis	Chamomile	4 00	4 25	4 50	4 75	5 00
Apocyni Andros.	Bitter Root	4 50	4 75	5 00	5 25	5 50
Apocyni Cannab.	Indian Hemp	4 50	4 75	5 00	5 25	5 50
Artemisiæ	Wormwood	4 00	4 25	4 50	4 75	5 00
Asclepiæ Inc.	White Ind. Hemp	4 50	4 75	5 00	5 25	5 50
Belladonnæ		6 00	6 25	6 50	6 75	7 00
Buchu		8 00	8 25	8 50	8 75	9 00
Calumbæ		4 00	4 25	4 50	4 75	5 00
Cannabis Ind.	Ind. Hemp, For.				per oz	2 25
Chimaphilæ	Prince's Pine	4 00	4 25	4 50	4 75	5 00
Cimicifugæ	Black Cohosh	5 00	5 25	5 50	5 75	6 00
Colchici Acetic		6 00	6 25	6 50	6 75	7 00
Colocynthidis Compositum		6 00	6 25	6 50	6 75	7 00
" "	Pow'd, lb. bottles	6 50				
Conii		3 50	3 75	4 00	4 25	4 50
Cornus Floridæ	Boxwood	4 00	4 25	4 50	4 75	5 00
Cubebæ	Cubebs	7 00	7 25	7 50	7 75	8 00
Cypripedii	Lady's Slipper	6 00	6 25	6 50	6 75	7 00
Digitalis	Foxglove	4 00	4 25	4 50	4 75	5 00
Dulcamaræ	Bittersweet	3 50	3 75	4 00	4 25	4 50
Eupatorii	Boneset	3 00	3 25	3 50	3 75	4 00
Filicis Maris	Male Fern	3 00	3 25	3 50	3 75	4 00
Gentianæ		1 50	1 75	2 00	2 25	2 50
Geranii Mac.	Cranesbill	5 00	5 25	5 50	5 75	6 00
Hellebori	Black Hellebore	6 00	6 25	6 50	6 75	7 00
Humuli	Hop	6 00	6 25	6 50	6 75	7 00
Hyoscyami	Henbane	6 00	6 25	6 50	6 75	7 00
Ignatiæ Amaræ	Ignatia Bean				per oz	2 00
Iris Versicol.	Blue Flag	5 00	5 25	5 50	5 75	6 00
Jalapæ	Jalap	8 00	8 25	8 50	8 75	9 00
Juglandis	Butternut	3 00	3 25	3 50	3 75	4 00
Juniperis Com.	Juniper Berries	3 00	3 25	3 50	3 75	4 00
Krameriæ	Rhatany	4 00	4 25	4 50	4 75	5 00
Lactucæ	Lettuce	3 50	3 75	4 00	4 25	4 50
Lappi	Burdock	4 00	4 25	4 50	4 75	5 00
Leontice Thalic.	Blue Cohosh	3 50	3 75	4 00	4 25	4 50
Leptandræ	Culver's Root	6 00	6 25	6 50	6 75	7 00
Marrubii	Hoarhound	4 50	4 75	5 00	5 25	5 50
Nicotianæ Tabac.	Tobacco	3 00	3 25	3 50	3 75	4 00
Nux Vomica		6 00	6 25	6 50	6 75	7 00
Papaveris	Poppy	4 50	4 75	5 00	5 25	5 50
Podophylli	Mandrake	5 00	5 25	5 50	5 75	6 00
Phytolaccæ	Poke	4 00	4 25	4 50	4 75	5 00
Quassiæ		6 00	6 25	6 50	6 75	7 00
Quercus Albæ	White Oak	3 00	3 25	3 50	3 75	4 00
Rhei	Rhubarb	9 00	9 25	9 50	9 75	10 00
Rumicis Crispæ	Yellow Dock	4 50	4 75	5 00	5 25	5 50
Rubi Villosi	Blackberry	3 00	3 25	3 50	3 75	4 00
Rutæ	Rue	3 50	3 75	4 00	4 25	4 50
Sabinæ	Savin	3 00	3 25	3 50	3 75	4 00
Sambuci	Elder	3 00	3 25	3 50	3 75	4 00
Sanguinariæ	Bloodroot	6 00	6 25	6 50	6 75	7 00
Sarsaparillæ, American		5 00	5 25	5 50	5 75	6 00
" "	Compound	5 00	5 25	5 50	5 75	6 00
" Rio Negro		5 00	5 25	5 50	5 75	6 00
" " "	Compound	5 00	5 25	5 50	5 75	6 00
Sennæ, Alexandria		3 50	3 75	4 00	4 25	4 50
Stramonii		4 00	4 25	4 50	4 75	5 00
Taraxaci	Dandelion	2 00	2 25	2 50	2 75	3 00
Trifolii	Red Clover	2 50	2 75	3 00	3 25	3 50
Uvæ Ursi		4 00	4 25	4 50	4 75	5 00
Valerianæ	English	6 00	6 25	6 50	6 75	7 00
Veratri Viridis	White Hellebore	6 00	6 25	6 50	6 75	7 00

CONCENTRATED MEDICINES.

VAN SCHAACK, STEVENSON & REID'S

LIST,

With Doses and Prices Annexed.

As *manufacturers* differ in their *long* prices on the same preparations, we cannot name a *uniform* discount, but always make the largest discount from the kind ordered.

	Av'ge Dose in Grains.	Per oz.
Aconitin	1-24 to 1-12	$5 00
Ampelopsin	2 to 5	1 50
Alnuin	2 to 10	1 00
Apocynin	½ to 1	3 00
Atropin	1-24 to 1-12	5 00
Asclepin	1 to 5	1 50
Baptisin	1 to 8	1 25
Barosmin	1 to 3	3 50
Caulophyllin	2 to 5	1 00
Cerasein	2 to 10	1 00
Chelonin	2 to 5	2 00
Chimaphilin	2 to 5	1 25
Cimicifugin	(See Macrotin.)	
Collinsonin	2 to 5	2 75
Colocynthin	½ to 2	3 00
Copaivin	2 to 5	
Cornin	2 to 5	1 00
Corydalin	1 to 3	3 00
Cypripedin	2 to 4	1 75
Digitalin	⅛ to ½	1 50
Dioscorein	2 to 5	2 00
Euonymin	1 to 4	2 25
Eupurpurin		2 50
Eupatorin (Perfo.)	1 to 4	1 00
" (Purpu.)	1 to 5	2 00
Frazerin	2 to 10	1 25
Gelsemin	½ to 1	3 00
Geraniin	2 to 5	1 00
Gossypiin	3 to 8	2 50
Hamamelin	1 to 3	1 25
Helonin	2 to 5	2 50
Hydrast Sulph.		3 50
Hydrastina, Alk.		$4 75
Hydrastin, Neutral.	1 to 3	2 00
Hydrastin-Muriate		2 75
Hyoscyamin	⅛ to ½	3 50
Irisin	1 to 3	1 25
Jalapin	2 to 5	4 00
Juglandin	2 to 10	1 00
Leontodin	2 to 4	2 50
Leptandrin	2 to 5	1 00
Lobelin		2 50
Lupulin	1 to 4	1 00
Lycopin	1 to 4	1 50
Macrotin	½ to 2	1 00
Menispermin	1 to 5	1 50
Myricin	2 to 5	1 00
Phytolacin	½ to 3	1 50
Podophyllin	½ to 3	1 00
Populin	2 to 5	1 00
Prunin	1 to 3	1 00
Rhein	1 to 4	3 50
Rhusin	1 to 3	1 25
Rumin	1 to 3	1 50
Sanguinarin	½ to 2	1 25
Scutellarin	2 to 5	2 00
Senecin	2 to 5	1 75
Smilacin	2 to 5	
Stillingin	2 to 5	2 25
Trillin	2 to 5	1 00
Veratrin	½ to 1	2 50
Viburnin	2 to 5	2 00
Xanthoxylin	2 to 5	1 75
Aletrin	1 to 3	2 25

A full supply of the above preparations—Tilden's, Merrill's and Keith's manufacture—constantly on hand, by

Van Schaack, Stevenson & Reid,

WHOLESALE DRUGGISTS, CHICAGO.

VAN SCHAACK, STEVENSON & REID,

AGENTS FOR

TILDEN'S SUGAR-COATED

PILLS AND GRANULES

Of the United States Pharmacopœia, and Formulas of Eminent Physicians.

SUGAR-COATED AND ELEGANTLY POLISHED.

RELIABLE, SOLUBLE, PERMANENT.

These prices are subject to a discount of 50 per cent.

FORMULÆ.	Price per 100 Pills in Bottles of 500.	Price per 100 Pills in Bottles of 100.
Aconitine, 1-60 gr.	$ 70	$ 75
Aconite Extract, ¼ gr.	40	45
" " ½ "	45	50
" " 1 "	55	60
Aloetic, U. S. P. { Aloes Soct., 2 grs. Soap, 2 grs. }	35	40
Aloes and Assafœtida, { Aloes, Assafœtida, Soap, } equal parts. 4 grs.	35	40
Aloes and Iron, { Aloes Soct., Extract Conium, *aa* 1 part, Iron Sulphate, Ginger, Jamaica, *aa* 2 parts, } 4 grs.	35	40
Aloes and Mastich. (See Lady Webster's.)		
Aloes and Myrrh, U. S. P. { Aloes Soct., 2 parts, Myrrh, Saffron, *aa* 1 part, } 4 grs.	45	50
Aloes and Ext. Gentian. (See Gentian Compound)	45	50
Aloes and Nux Vomica, { Aloes, 1½ grs. Ext. Nux Vomica, ½ gr. }	45	50
Alterative, { Mass Hydrargyri, 1 gr. Pulv. Opii, ⅓ gr. Pulv. Ipecac, ⅙ gr. }	45	50
Ammonium Bromide, 1 gr.	70	75
Anderson's Soct's, { Aloes Soct., Soap, Colocynth, Oil Annis, } 2 grs.	35	40
Anthelmintic, { Santonin, 1 gr. Calomel, 1 gr. }	95	1 00
Anthemis, Extract, 2 grs.	65	70
Anti-Bilious, { Ext. Colocynth, 2½ grs. Podophyllin, ¼ gr. }	65	70
Anti- Dyspeptic, { Strychnia, 1-40 gr. Ext. Belladonnæ, 1-10 gr. Pulv. Ipecac, 1-10 gr. Mass. Hydrarg., 2 grs. Ext. Coloc. Comp., 2 grs. }	95	1 00

FORMULÆ.	Price per 100 Pills in Bottles of 500.	Price per 100 Pills in Bottles of 100.
Anti-Malarial, (McCaw,) { Quiniæ Sulph. 1 gr. Ferri Sul. Ext., ½ gr. Ol. Res. Pip. Nig., 1-16 gr. Ac. Arsenios., 1-10 gr. Gelseminin, ¼ gr. Podophyllin, 1-8 gr. }	2 20	2 25
Anti-Periodic, { Cinchoniæ Sulph., 1 gr. Ferri Sul. Exsic., 1 gr. Ext. Quass., ½ gr. Ext Rhei, ¼ gr. Pulv. Myrrhæ, ⅔ gr. }	95	1 00
Antimonii Comp., U. S. P. (See Calomel Comp.)	35	40
Apocynum, 2 grs.	65	70
Aperient, { Ext. Nux Vomica, ⅓ gr. " Hyoscyamus, ½ gr. " Colocynth Comp., 2 grs. }	80	85
Arsenious Acid, 1-32 gr.	35	40
Assafœtida, U. S. P., 4 grs.	35	40
" " 2 grs.	30	35
" and Iron, { Assafœtida, 2 grs. Sulphate Iron, 1 gr. }	35	40
" and Rhei, { Assafœtida, Rhei, Iron by Hydrogen, *aa* 1 gr. }	70	75
Atropia, 1-60 gr.	70	75
Ballou, { Ext. Col. Comp., 1 gr. " Jalapæ, 1 gr. Hydr'g. Chlor. Mit., 1 gr. Pulv. Ipecac, ⅛ gr. }	70	75
Belladonna Ext., ¼ and ½ gr. each,	35	40
" " 1 gr.	55	60
Biniodide Mercury, ⅛ gr.	40	45
" " 1-16 gr.	40	45
Bismuth and Nux Vom. { Bismuth Subnit., 5 grs. Ext. Nux Vom., ½ gr. }	1 45	1 50
" Subnitrate, 2 grs.	70	75
" Subcarbonate, 3 grs.	70	75
Blue Pill, U. S. P., 2½ grs	35	40
" " 5 grs.	45	50
" " Compound, { Blue Pill, 1 gr. Opii, ½ gr. Ipecac, ¼ gr. }	70	75
" " and Podophyllin. (See Pod. and Blue Pill.)	45	50
Caffein Citrate, 1 gr.	4 45	4 50
Calomel, ½ gr., 1 gr. and 2 grs., each	35	40
" 3 grs.	40	45
" 5 grs.	45	50
" Compound, (Plummer's), 3 grs.	35	40
" and Comp. Ext. Colocynth. { Calomel, 1 gr. Ext. Coloc. C., 2½ gr. }	70	75
" et Ipecac Com. { Calomel, 1 gr. P. Ipecac Com., 3⅓ gr. Ext. Gent. q. s. }	45	50
" and Opium, { Calomel, 2 grs. Opium, 1 gr. }	80	85
" and Rhei, { Calomel, ⅓ gr. Ext. Rhei, ½ gr. " Coloc. Comp., ½ gr. " Hyoscyamus, ⅙ gr. }	70	75
Camphor and Hyoscyam. { Gum Camphor, 1 gr. Ext. Hyos. Eng., 1 gr. }	45	50
" " and Valerian. { Camphor, 1 gr. Ext. Henbane, 1½ grs. Pulv. Valerian, ½ gr. }	45	50

FORMULÆ.	Price per 100 Pil's in Bottles of 500.	Price per 100 Pills in Bottles of 100.
Camphor and Opium, { Camphor, 2 grs. Opium, 1 gr. }	$ 95	$1 00
" Compound, { Camphor, 1 gr. Kino, 1 gr. Opium, 1 gr. Capsicum, 1-16 gr. } 3 grs.	85	90
Cannabis Indica Extract, ½ gr.	1 00	1 05
" " " 1 gr.	1 85	1 90
Capsicum, 1 gr.	60	65
Cathartic Compound, U. S. P. { Ext. Colocynth Comp. " Jalap, Calomel, Gamboge. } 3 grs.	65	70
" Compound, bottles, 25 each ......doz. net, $ 1 50		
" " " 25 " ......gross, net, 17 00		
" Improved. { Ext. Colocynth Comp. " Jalap, Podophyllin, Leptandrin, Ext. Hyoscyamus, " Gentian, Oil Peppermint. } 3 grs.	65	70
" Improved, bottles, 25 each ......doz. net, $ 1 50		
" " " 25 " ......gross, net, 17 00		
" Comp. (Vegetable). { Ext. Coloc. Simple, ⅓ gr. Podophyllin, ¼ gr. Pulv. Res. Scam. ⅓ gr. " Aloes Socot. 1¼ gr. " Cardamomi, 1-9 gr. " Saponis, ⅙ gr. }	45	50
Caulophyllin 1-10 gr.	35	40
Chapman's Dinner, { Aloes, Mastic, Ipecac, Oil Fennel. } 4 grs.	55	60
Cerii Oxalas, 1 gr.	95	1 00
Chimaphila Ext. 3 grs.	70	75
Chinoidine, 2 grs.	45	50
" 1 gr.	45	50
Chionidine Comp. { Chinoidine, 2 grs. Ferri Sulph. Exsic. 1 gr. Piperini, ½ gr. }	95	1 00
Cimicifugin, 1 gr.	60	65
Cinchonia, Sulphate, 1½ gr.	65	70
" " 3 grs.	95	1 00
Cinchonidia, " 1 gr.	55	60
" " 2 grs.	95	1 00
" " 3 grs.	1 45	1 50
Citrate of Iron, Quinia and Strych. { Cit. Iron and Quinia, 2 grs. Strychnia, 1-60 gr. }	1 70	1 75
Cochia, { Colocynth Comp. Scammony, Aloes, Gamboge, Potass. Sulphate. } 3 grs.	80	85
Codeia, 1-16 gr.	3 00	3 05
Colchicum Extract, ½ gr.	60	65
Colocynth, Compound Extract, U. S. P., 3 grs.	75	80
" " and Blue Pill, { Colocynth Comp. 2½ grs. Blue Pill, ½ gr. }	80	85
" and Calomel, 3 grs.	70	75
" and Hyoscyamus, U. S. P., 3 grs.	80	85
" and Ipecac, 3 grs.	75	80
" Comp. and Jalap, { Ext. Colocynth, Comp. 1⅓ grs. " Jalap, 1 gr. " Henbane, ½ gr. " Gentian, 1 gr. Podophyllin, 1-10 gr. Leptandrin, ¼ gr. Oil Peppermint, q. s. }	70	75
" and Podophyllin	80	85
Cunium Extract, ¼ gr.	40	45
" " ½ gr.	50	55

FORMULÆ.	Price per 100 Pills in Bottles of 500.	Price per 100 Pills in Bottles of 100.
Conium Extract, 1 gr.	$ 60	$ 65
" " and Ipecac, U. S. P., 1 gr.	70	75
Cook's Pill, { Aloes, 1 gr. Calomel, ½ gr. Rhei, 1 gr. Soap, ½ gr. }	45	50
Copaiba, Pure Solidified, 3 grs.	45	50
" " " 4 grs.	55	60
" Compound, { Pil. Copaib: Ferri Cit: Resin Guaiac, Oleo-resin: Cubeb. }	75	80
" and Buchu, { Copaiba, 3 gr. Oleo Resin Buchu, 1 gr. }	45	50
" and Ext. Cubebs, { Pil. Copaib. 2 grs. Oleo-resin: Cubeb. 1 gr. }	45	50
" Ext. Cubebs and Citrate Iron, 3 grs.	45	50
Cornus Florida, Ext., 2 grs.	65	70
Corros. Sublimate, ⅛ gr.	35	40
" " 1-16 gr.	35	40
Cubebs, Extract, 2 grs.	95	1 00
" and Alum, 3 grs.	70	75
" Ext. Rhatany and Iron, { Ext. Cubebs, 1½ grs. " Rhatany, ½ gr. Iron Sulphate, 1 gr. }	75	80
Cypripedium Extract, 2 grs.	70	75
Digitalin, 1-60 gr.	70	75
Digitalis, Compound, { Pulv. Digital. Eng. 1 gr. " Scillæ, 1 gr. Potass. Nit. 2 grs. }	45	50
" Extract, ½ gr.	50	55
Dinner—Lady Webster's, { Aloes Soct., Gum Mastich, Rose Leaves. } 3 grs.	45	50
Diuretic, { Sapon. Hispan. Pulv. 2 grs. Sodæ Carb. Exsic. 2 grs. Ol. Baccæ Junip. 1 drop. }	45	50
Dupuytren, { Pulv. Guaiac, 3 grs. Hydrarg. Chlor. Corros. 1-10 gr. Pulv. Opii, ⅙ gr. }	45	50
Elaterium, Clutterbuck's, ⅛ gr.	95	1 00
Emmenagogue, (Mutter) { Ferri Sulph. 1½ grs. Aloes Socot. Pulv. ½ gr. Terebinth. Alb. 1½ grs. }	35	40
Ferri et Quass. et Nux Vom. { Fer. per Hydrogen, 1½ grs. Ext. Quassiæ, 1 gr. " Nux Vom. ¼ gr. Pulv. Saponis, ¼ gr. }	70	75
" Strychnia. { Strych. 1-60 gr. Ferrum per Hydrogen, (Quevenne's) 2 grs. }	70	75
Ferrocyanide Iron, 3 grs.	45	50
Gamboge Compound, { Gamboge, Soct. Aloes, Ginger, Jamaica, Soap. } 3 gr.	35	40
Gelseminin, 1-16 gr.	35	40
" ⅛ gr.	45	50
Gentian Extract, 2 grs.	50	55
" Compound, { Ext. Gentian, Aloes Soct. Rhei Powdered, Oil Carui. } 4 grs.	45	50
Geraniin, 1 gr.	60	65

FORMULÆ.

	Price per 100 Pills in Bottles of 500.	Price per 100 Pills in Bottles of 100.
Gonorrhœa, { Pulv. Cubebæ, 2 grs. Bals. Copaib. Solid, 1 gr. Ferri Sulph. Exsic, ½ gr. Terebinth. Venet. 1½ grs. }	$ 55	$ 60
Hellebore, Black, Extract, 1 gr.	60	65
Helonin, 1-10 gr.	45	50
Hepatic, { Pil. Hydrarg., 3 grs. Ext. Colocynth Comp. 1 gr. " Hyoscyami, 1 gr. }	85	90
Hooper's (Female Pills). { Aloes Socot. Iron Sulph., Ext. Black Helleb. Myrrh, Soap, Canella, Ginger, } 2½ grs.	35	40
Hydrastin, 1 gr.	90	95
" Alkaloid, 1 gr.	90	95
Hyoscyamus Extract, ¼ gr.	35	40
" " ½ gr.	40	45
" " 1 gr.	50	55
Ignatia Extraxt, ½ gr.	55	60
" " 1 gr.	95	1 00
Iodide of Iron, (*Blancard's Formula*,) 1½ gr	95	1 00
Iodine, ½ gr.	95	1 00
Iodoform, 1 gr.	1 55	1 60
" and Iron, { Iron by Hydrogen, 1 gr., Iodoform, 1 gr. }	1 95	2 00
Ipecac Extract, ½ gr.	55	60
" and Squill, 3 grs.	50	55
" and Opium, { Opium, 1 part, Ipecac, 1 part, Potass. Sulph., 8 parts. } 2½ grs. Dover's powder.	55	60
Ipecac and Opium, { Opium, ½ gr., Ipecac, ½ gr., Potass. Sulph., 1 gr. } equal to 5 grs. Dover's powder.	1 10	1 15
" " { Opium, 1 gr., Ipecac, 1 gr., Potass. Sulph. 2 grs. } equal to 10 grs. Dover's powder.	1 95	2 00
Irisin, ½ gr.	60	65
" 1 gr.	70	75
" Compound, { Irisin, ¼ gr. Podophyllin, 1-10 gr. Strychnia, 1-40 gr. }	45	50
Iron and Aloes. See Aloes and Iron, 4 grs.	35	40
" Citrate, 2 grs.	45	50
" " and Quinine Citrate, 1 gr.	75	80
" " " " 2 grs.	1 35	1 40
" " and Strychnia Cit. { Strychnia Cit. 1-50 gr. Iron Cit. 2 grs. }	70	75
" Compound, U. S. P. { Myrrh, Soda Carbonate, Iron Sulph. } 3 grs.	35	40
Iron, Carbonate, Vallet's, 3 grs.	35	40
" Carbonate of, and Manganese, 3 grs.	70	75
" Hydrocyanate, ¼ gr.	90	95
" and Iodoform. See Iodoform and Iron.		
" Lactate, 1 gr.	45	50
" Phosphate, 2 grs.	60	65
" Pyro Phosphate, 1 gr.	35	40
" Proto Iodide, 1 gr.	70	75
" Quevenne's, by Hydrogen, 1 gr.	45	50
" " " 2 grs.	70	75
" and Strychnia, { Strychnia, 1-60 gr., Iron by Hydrogen, 2 grs. }	70	75
" Sulphate Exsic., 4 grs.	35	40
" Valerianate, 1 gr.	95	1 00

FORMULÆ.	Price per 100 Pills in Bottles of 500.	Price per 100 Pills in Bottles of 100.
Jalap, 1 gr.	$ 95	$1 00
Jalapin, 1 gr.	95	1 00
Kermes, ¼ gr.	95	1 00
Krameria Ext., (*Rhatany*), 2 grs.	60	65
Lactuca Extract, 2 grs.	50	55
Leptandrin, ⅛ gr.	35	40
" ¼ gr.	35	40
" ½ gr.	45	50
" 1 gr.	70	75
" Comp. { Leptandrin, 1 gr. Irisin, ¼ gr. Podophyllin, ⅛ gr. }	95	1 00
Lupulin, 3 grs.	35	40
Macrotin, 1-10 gr.	35	40
Magnesia, Calcined, 2 grs.	35	40
" and Rhei. 1 grain each, 2 grs.	45	50
Mandrake Compound, { Ext. Colocynth Comp. 1 gr. " Mandrake, 1 gr. " Gentian, ½ gr. Gamboge, ⅛ gr. Podophyllin, ¼ gr. Capsicum, 1-16 gr }	70	75
Mercury, Prot. Iodide, ½ gr.	50	55
" Red, " 1-16 gr.	40	45
" Iodide and Opii, Iodide, 1 gr. Opii, ⅓ gr.	85	90
Morphia, Acet., ⅛ gr.	70	75
" Sulphate, ¼ gr.	1 20	1 25
" " ⅛ gr.	70	75
" " 1-16 gr.	60	65
" " 1-32 gr.	55	60
" Valerianate, ½ gr.	1 20	1 25
" Compound, { Morphia Sulph. ¼ gr. Tart. Pot. and Ant. ¼ gr. Calomel, ¼ gr. }	1 45	1 50
Neuralgic, (Dr. Gross,) { Quinia Sulph. 2 grs. Morphia Sulph. 1-20 gr. Strychnia, 1-30 gr. Ac. Arsenious, 1-20 gr. Ext. Aconiti, ½ gr. }	3 20	3 25
" (Dr. Gross,) without Morphia	2 95	3 00
" Idiopath. { Ext. Hyoscyam. ⅔ gr. Ext. Aconiti, ⅓ gr. " Conii, ⅔ gr. Ext. Cannab. Ind. ¼ gr. " Ignat. Am. ½ gr. Ext. Stramon. 1-5 gr " Opii, ½ gr. Ext. Belladonnæ, ⅙ gr. }	1 95	2 00
" (Dr. T. S. Reed,) { Sulph. Quinine, 1 gr. Cit. Iron, 1 gr. Ext. Can. Ind., ¼ gr. " Belladonna, ¼ gr. " Opii, ¼ gr. }	3 20	3 25
" { Ext. Belladonna, 1 gr. Sulph. Morphia, ⅛ gr. Strychnia, 1-10 gr. }	95	1 00
Nitrate Silver, ½ gr.	90	95
Nux Vomica Ext., ⅛ gr.	35	40
" " " ½ gr.	45	50
Opium, 1 gr.	70	75
" and Acetate Lead, { Opium, 1 gr. Acet. Lead, 1 gr. }	75	80
" and Camphor, { Opium, 1 gr. Camphor, 2 grs. }	95	1 00
" " and Tannin, { Opium, ¼ gr. Camphor, 1 gr. Tannin, 2 grs. }	85	90
Phytolaccin, ½ gr.	60	65
Phosphorus, 1-50 gr.	95	1 00
" 1-100 gr.	95	1 00

FORMULÆ.	Price per 100 Pills in Bottles of 500.	Price per 100 Pills in Bottles of 100.
Phosphorus, Comp., { Phosphorus, 1-100 gr. / Ext. Nux Vomica, ¼ gr. }	$1 25	$1 30
" " and Iron, { Phosphorus, 1-100 gr. / Ext. Nux Vomica, ⅛ gr. / Phosphate Iron, ⅜ gr. }	1 25	1 30
" Carb. Iron, Val. and Nux Vomica	1 25	1 30
Podophyllum, Extract, (*Mandrake*), 1 gr.	65	70
Podophyllin, ¼ gr.	35	40
" ½ gr.	45	50
" 1 gr.	70	75
" Comp., { Podophyllin, ⅜ gr. / Ext. Hyos., ⅛ gr. / " Nux Vomica, 1-16 gr. }	75	80
" and Belladonna, { Podophyllin, ¼ gr. / Ext. Bellad. alc., ⅛ gr. / Oleo-resin Capsici, ¼ gr. / Sacchari lactis., 1 gr. }	70	75
" and Blue Pill, { Podophyillin, ½ gr. / Blue Pill, 2½ grs. }	45	50
Poppy Extract, 2 grs.	65	70
Potass., Tartrate of, and Iron, 2 grs.	60	65
" Iodide, 2 grs.	80	85
" Bromide, 1 gr.	70	75
" Permangan, Cryst., ⅙ gr.	45	50
*Quinine Sulph., ¼ gr.	85	90
* " " ½ gr.	1 10	1 15
* " " 1 gr.	1 35	1 40
* " " 2 grs.	2 70	2 75
* " " 3 grs.	3 95	4 00
" and Aloes, { Quinine, ¾ gr. / Aloes, ¼ gr. } 1 gr.	1 25	1 30
" Compound, { Quinine Sulph. 1 gr. / Iron by Hydrogen, 1 gr. / Arsenious Acid, 1-32 gr. }	1 80	1 85
" Comp. and Strychnia, { Quinine, 1 gr. / Iron by Hydrogen, 1 gr. / Arsenious Acid, 1-20 gr. / Strychnia, 1-20 gr. }	1 80	1 85
" and Ext. Belladonna, { Quinine, 1 gr. / Ext. Belladonna, ⅜ gr. }	1 80	1 85
" and Iron, { Quinine, 1 gr. / Iron by Hydrogen, 1 gr. }	1 80	1 85
" Iron and Strychnia, { Quinine, 1 gr. / Iron Carb. Vallet's, 2 grs. / Strychnia Sulph., 1-60 gr. }	1 80	1 85
" et Colocynth Comp. { Quiniæ Sulph. 1 gr. / Ext. Col. Comp. 1 gr. / " Ignat. Amar. ⅜ gr. / Piperina, ⅜ gr. / Morph. Sulph. 1-12 gr. }	2 20	2 25
" et Ferri Carb. { Quiniæ Sulph. 1 gr. / Ferri Carb. (Vallet's,) 2 grs. }	1 70	1 75
" t Ferri Lact. Comp. { Quiniæ Sulph. 1 gr. / Ext. Ignat. Amar. ½ gr. / Ferri Lactat. 2 grs. }	1 95	2 00
" et Ferri Valer., 2 grs.	3 45	3 50
" et Hydrarg. { Quin. Sulph. 1 gr. / Mass. Hydrarg. 2 grs. / Oleo Resin Piper. Nig. ¼ gr. }	1 70	1 75
" et Strychnia, { Quin. Sulph. 1 gr. / Strychnia, 1-60 gr. }	1 70	1 75

* Pills of Bi-Sulphate Quinine of same sizes, furnished at same prices.

FORMULÆ.

	Price per 100 Pills in Bottles of 500.	Price per 100 Pills in Bottles of 100.
Quinine et Zinci. Valer., { Quin. Valer. 1 gr. / Zinci. Valer. 1 gr. }	$3 95	$4 00
" Valerianate, ½ gr.	1 95	2 00
Quassia Extract, 1 gr.	85	90
Rhei Extract, 1 gr.	1 05	1 10
" U. S. P. { Rhei, 3 grs. / Soap, 1 gr. }	70	75
" Compound U. S. P. { Rhei, 2 grs. / Aloes Ext. 1½ grs. / Myrrh " 1 gr. / Oil Peppermint. }	70	75
" and Blue Pill, { Rhei, / Blue Pill, / Soda Carb. } 4 grs.	75	80
" and Iron, 3 grs.	85	90
Rheumatic, { Ext. Colocynth Comp. 1½ grs. / " Colchici Acet. 1 gr. / " Hyoscyamus, ⅓ gr. / Calomel, ⅓ gr. }	85	90
Santonin, ½ gr.	45	50
Sanguinaria Extract, (*Bloodroot*), ½ gr.	60	65
Sanguinarin, ¼ gr.	50	55
" 1 gr.	70	75
Sarsaparilla Ext. 3 grs.	70	75
Savin, Extract, 1 gr.	60	65
Senna, Alex. Extract, 2 grs.	65	70
Soap and Opium, 3 grs.	1 00	1 05
Soda, Bi-Carbonate, 4 grs.	45	50
Squill Compound, U. S. P. 3 grs.	45	50
Stillingin, 1 gr.	70	75
Stramonium Extract, ½ gr.	50	55
" " 1 gr.	60	65
Strychnia, 1-48 gr.	35	40
" 1-32 gr.	35	40
" 1-16 gr.	40	45
" Compound, { Strychnia, 1-20 gr. / Arsenic, 1-20 gr. Quinia, 1 gr. / Iron by Hydrogen, 1 gr. }	1 80	1 85
Syphilitic, { Potass. Iodid. 2½ grs. / Hydrarg. Chlor. Corros. 1-40 gr. }	95	1 00
Tartar Emetic, ⅛ gr.	45	50
Taraxacum Extract, 3 grs.	70	75
Tannin, 1 gr.	75	80
Triplex, { Aloes Ext. 2 parts, / Podophyllin, / Blue Mass, *aa* 1 part. } 3 grs.	70	75
Uva Ursi, Extract, 2 grs.	65	70
Valerian, Extract, 2 grs.	85	90
Valerianate Ammonia, 1 gr.	1 45	1 50
" Iron, 1 gr.	95	1 00
" Morphia, ⅛ gr.	1 20	1 25
" Quinia, ½ gr.	1 95	2 00
" Zinc, 1 gr.	95	1 00
Veratria, 1-32 gr.	65	70
Veratrum Viride, Extract, ¼ gr.	45	50
" " " ½ gr.	60	65

Counter Urn.

French Counter Jar

Ring Jar

(See page 167.

Standard Measures Tin, Copper or Britannia.

Wayne Show Globe, Engraved.

Prismatic Show Mortar.

Silvered Glass Globe, on foot.

Perfume Stand.

"Old Mr." Hercules (about 5 feet high.)

Glass Silver Mortar on foot.

English Counter Urn.

French Counter Jar on foot.

Cut Sponge Jars.

Atomizer.

Counter Urn.

[illegible] Stand.

DRUGGISTS' GLASSWARE.

SHOW GLOBES.

Pine Apple Shape—With Double Stopper.

	Height to centre large Globe,	Height over all.		
1-2 gallon,	6 1-2 inches,	22 inches	each,	$1 75
1 "	7 1-2 "	27 "	"	2 50
2 "	10 "	34 "	"	3 50

PINE APPLE SHOW GLOBES.

Engraved—Double Stoppers.

1-2 gallon	each,	$2 5
1 "	"	3 5
2 "	"	4 5

Pyramid, Union and Pine Apple Show Globes, engraved, various styles, two and three stoppers, from five to twenty gallons.

APOTHECARIES' SHOP FURNITURE.

ROUND SHOULDER, CRYSTAL GLASS—MUSHROOM STOPPERS.

TINCTURE BOTTLES.

				Per Doz.
1 ounce,	height	$3\frac{3}{4}$	inches (includes the stopper),	$ 90
2 "	"	4	" " " "	1 10
4 "	"	6	" " " "	1 30
8 "	"	$6\frac{3}{4}$	" " " "	1 50
Pint,	"	$8\frac{1}{2}$	" " " "	1 70
Quart,	"	10	" " " "	1 95
½ gallon,	"	$12\frac{1}{2}$	" " " "	3 40
1 "	"	$14\frac{1}{4}$	" " " "	5 10

SALT MOUTH BOTTLES.

				Per Doz.
1 ounce,	height	$3\frac{1}{2}$	inches (includes the stopper),	$1 10
2 "	"	4	" " " "	1 30
4 "	"	5	" " " "	1 50
8 "	"	$6\frac{1}{2}$	" " " "	1 95
Pint,	"	$8\frac{1}{4}$	" " " "	2 10
Quart,	"	$9\frac{1}{2}$	" " " "	2 75
½ gallon,	"	12	" " " "	4 25
1 "	"	$13\frac{3}{4}$	" " " "	6 75

TINCT. BELLADON

PULV. CARDAM

GLASS LETTERS

THE INDESTRUCTIBLE GLASS LABEL, IMPERVIOUS TO ACID OR TIME, FOR TINCTURES, SALT-MOUTHS, URNS, JARS, ETC.

Our facilities for supplying Glass Labels, of the *best material*, are unsurpassed. For style, etc., we call attention to the samples annexed. In ordering, please mention the style you wish by the letter above each. We supply them either fitted to the bottle, or separate.

C

H

I

SAMPLES OF GLASS LABELS SUPPLIED BY

Van Schaack, Stevenson & Reid.

J

K

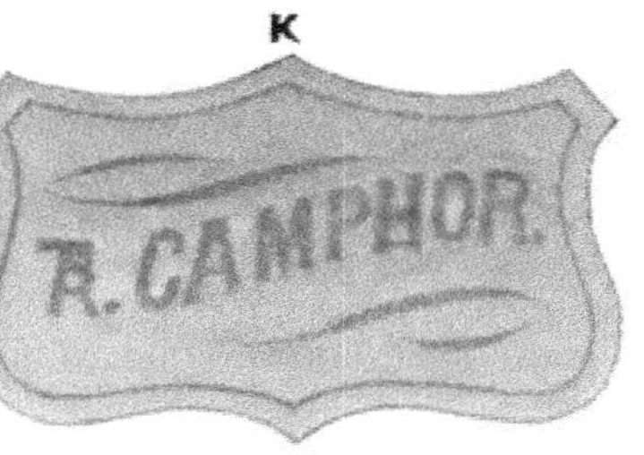

L

The Labels are here represented in one-third their original sizes.

A

B

C

SAMPLES OF GLASS LABELS SUPPLIED BY

Van Schaack, Stevenson & Reid.

D

E

F

The Labels are here represented in one-third their original sizes.

GLASS DRAWER LABELS.

Straight, with Gold or White Ground, Gold Borders and Black Letters.

6 x 1	inch	per doz.
7½ x 1¾	"	"

Straight, with white, pink, green or blue ground, and gold letters, shaded in color.

EXTRA HEAVY BALSAM OR OIL BOTTLES.

These Bottles are preferable for oils to the ordinary ground stoppered article, in that, instead of the liquid running over the outside, it is caught in a cup, and returns through an aperture, into the bottle, preventing all waste, and keeping the bottle and the shelf clean.

With Loose Glass Cover.

				Per Doz
½ pint,	height,	8	inches	$4 50
Pint,	"	9	"	5 00
Quart,	"	10½	"	6 00
½ gallon,	"	12	"	9 00

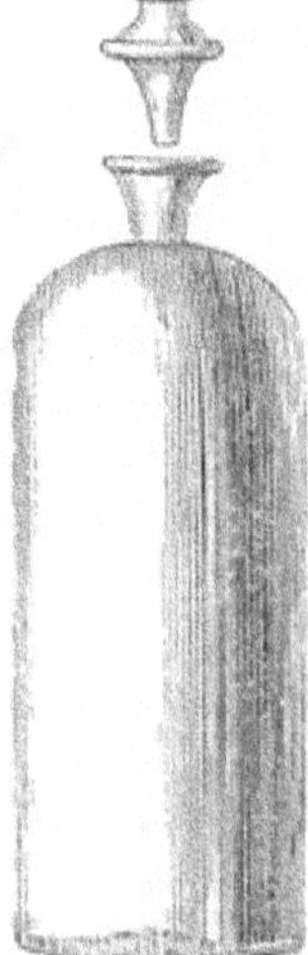

EXTRA HEAVY DISPENSING SYRUP BOTTLES.

Flint or Blue Glass.

Made with loose stoppers, to obviate the difficulty of stoppers sticking in using heavy liquids.

				Per Doz.
Pint,	height,	8½	inches	4 50
Quart,	"	10	"	6 00
½ gallon,	"	12	"	8 00

FLINT PRESCRIPTION VIALS.

PLAIN ROUND SHOULDER PRESCRIPTIONS—NARROW MOUTHS.

Net Prices—Packages Charged at Cost. *By Case.*

Size		Packing				Price
1-2	ounce,	5	gross boxes	....	per gross,	$ 2 70
1	"	5	"	....	"	2 70
2	"	3	"	....	"	3 30
3	"	2	"	....	"	3 90
4	"	2	"	....	"	4 50
5	"	1	"	....	"	5 40
6	"	1	"	....	"	6 00
8	"	1	"	....	"	7 20
10	"	1	"	....	"	9 00
12	"	1-2	"	....	"	9 60
16	"	1-2	"	....	"	12 00
20	"	1-2	"	....	"	13 20
24	ounce,	1-2	gross boxes	....	"	14 40
32	"	1-2	"	....	"	16 80
Assorted, from 1-2 to 8 ounce				....	"	5 40

FRENCH SQUARE PRESCRIPTIONS.

Net Prices—New and Neat. *By Case.*

Size		Packing				Price
1-2	ounce,	5	gross boxes	....	per gross,	$ 2 85
1	"	5	"	....	"	2 85
2	"	3	"	....	"	3 60
3	"	2	"	....	"	4 20
4	"	2	"	....	"	4 80
6	"	1	"	....	"	6 30
8	"	1	"	....	"	7 50
12	"	1-2	"	....	"	10 20
16	"	1-2	"	....	"	12 60
Assorted, from 1-2 to 8 ounce				.....	"	6 00

PLAIN OVALS.

By Case.

Size		Packing				Price
1	ounce,	5	gross boxes	....	per gross,	$ 2 85
2	"	3	"	....	"	3 60
3	"	2	"	....	"	4 20
4	"	2	"	....	"	4 80
6	"	1	"	....	"	6 30
8	"	1	"	....	"	7 50
12	"	1-2	"	....	"	10 20
16	"	1-2	"	....	"	12 60

UNION OVALS—NEW STYLE.

By Case.

Size		Packing				Price
1-2	ounce,	5	gross boxes	....	per gross,	$ 2 85
1	"	5	"	....	"	2 85
2	"	3	"	....	"	3 60
3	"	2	"	....	"	4 20

UNION OVALS—Continued.

			By Case.
4 ounce,	2 gross boxes	per gross,	$ 4 80
6 "	1 "	"	6 30
8 "	1 "	"	7 50
12 "	1-2 "	"	10 20
16 "	1-2 "	"	12 00
Assorted, 1-2 to 8 ounce		"	6 00

PLAIN SQUARE, OR OBLONG PRESCRIPTIONS.

			By Case.
1-2 ounce,	5 gross boxes	per gross,	$ 3 30
1 "	5 "	"	3 30
2 "	3 "	"	3 90
3 "	2 "	"	4 50
4 "	2 "	"	5 10
6 "	1 "	"	6 60
8 "	1 "	"	7 80
12 "	1-2 "	"	11 70
16 "	1-2 "	"	13 20

BALL NECK PANELS.

				By Case.
1 ounce,	holding 3-4 ounce,	5 gross boxes,	per gross,	$3 60
2 "	" 1 1-4 "	3 "	"	3 90
3 "	" 2 "	2 "	"	4 50
4 "	" 2 1-2 "	2 "	"	4 80
5 "	" 3 "	2 "	"	5 10
6 "	" 4 "	1 "	"	7 20
8 "	" 6 "	1 "	"	9 00

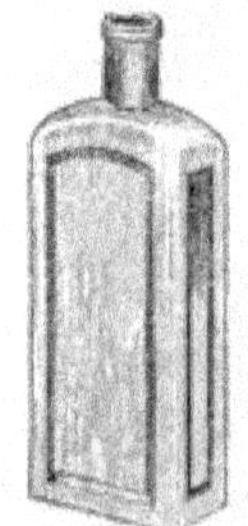

ARCH BALL NECK PANEL—NEW STYLE.

				By Case.
1 ounce,	holding 3-4 ounce,	5 gross boxes,	per gross,	$3 60
2 "	" 1 1-4 "	3 "	"	3 90
4 "	" 1 3-4 "	2 "	"	4 80
5 "	" 2 1-4 "	2 "	"	5 10
6 "	" 3 1-2 "	1 "	"	7 20
8 "	" 6 "	1 "	"	9 00

HOMŒOPATHIC VIALS.

½ drachm Short Vials	per gross,	$ 1 00
1 " "	"	1 10
2 " "	"	1 35
3 " "	"	1 75
4 " "	"	2 50
½ drachm Long Vials	"	1 00
1 " "	"	1 10
2 " "	"	1 25
3 " "	"	1 75
4 " "	"	2 25

HOMŒOPATHIC CASE VIALS.

½ drachm, no brim, extra heavy	per gross,	$1 10
1 " " "	"	1 35
2 " " "	"	1 60
3 " " "	"	2 20
4 " " "	"	3 00
6 " " "	"	3 95
8 " " "	"	5 25

MUCILAGE BOTTLES. (see page 175.)
With Caps and Brushes.

2 oz.	per doz.	$ 60
4 oz.	"	75

DRUGGISTS' GREEN GLASSWARE.

PRESCRIPTIONS.—HEAVY MOULDED, ROUND SHOULDER, NARROW MOUTHS.

At Net Prices. (No Charge for Boxes in Original Packages.) By Case.

1-2, 1 and 2 drachms,	5 gross boxes	per gross,	$ 2 00
1-2, and 1 ounce,	5 "	"	2 05
2 ounce,	3 "	"	2 25
3 "	2 "	"	2 70
4 "	2 "	"	3 25
6 "	1 "	"	4 05
8 "	1 "	"	4 95
12 "	1 "	"	6 30
16 "	1 "	"	7 90
Assorted, 1 to 8 oz.	1 "	"	3 60

MOULDED CASTOR OIL BOTTLES, ROUND.

By Case.

Castor Oils, 6s	to gallon,	1 gross boxes	per gross,	$ 9 00
" 8s	"	1 "	"	8 10
" 10s and 12s	"	1 "	"	6 30
" 16s	"	1 "	"	5 40
" 20s	"	1 "	"	4 50
" 24s and 30s	"	1 "	"	3 85
" 40s and 50s	"	2 "	"	3 15
Assorted		1 "	"	4 50

DRUGGISTS' PACKING BOTTLES.

Narrow Mouths. *By Case.*

1-4 pint,	1 gross boxes	per gross,	$ 3 60
1-2 "	1 "	"	5 40
1 2 ounce,	1 "	"	6 75
1 pint,	1-2 "	"	8 10
1 1 2 pint,	1-2 "	"	9 45
1 quart,	1-2 "	"	11 25
3 pint,	1-2 "	"	14 85
1-2 gallon,	1 4 "	"	18 00
3 quart,	2 doz. boxes	"	22 50
1 gallon,	1 "	"	29 70

Wide Mouths, 50 cents per gross advance.

SHORT NECK OVALS.

			By Case
Ovals, 1 ounce, 3 gross boxes	....	per gross,	$2 05
" 2 " 3 "	....	"	2 25
" 3 " 2 "	....	"	2 95
" 4 " 2 "	....	"	3 40
" 6 " 1 "	....	"	4 05
" 8 " 1 "	....	"	5 40
12 " 1 "	....	"	6 75
" 16 " 1 "	....	"	9 00
" 32 " 1 "	....	"	12 60

PANELS.

			By Case.
1 ounce panels, 5 gross boxes	....	per gross,	$ 2 50
2 " 3 "	....	"	2 80
3 " 2 "	....	"	3 50
4 " 2 "	....	"	4 05
5 " 1 "	....	"	4 62
6 " 1 "	....	"	5 40
8 " 1 "	....	"	6 75
12 " 1 "	....	"	9 00
16 " 1 "	....	"	10 80
20 " 1 "	....	"	12 15
24 " 1 "	....	"	13 50
32 " 1 "	....	"	15 75

ACID BOTTLES, GROUND STOPPERS.

			By Case.
1 2 pint Acids, 1-2 gross boxes	....	per doz.	$ 1 55
1 " " 1-2 "	....	"	1 80
1 quart " 1-2 "	....	"	2 25
1-2 gallon " 1-4 "	....	"	3 38
1 " " 1 doz. boxes	....	"	4 50

CITRATE MAGNESIA BOTTLES.

12 oz. Flint	....	per gross,	$11 35
" Green	....	"	10 13
" Blue	....	"	10 35

FLASKS—PLAIN STYLE.

½ pint flasks, plain style, 6 doz. boxes	....	per gross,	$ 5 40
1 " " " 6 "	....	"	7 20
Quart " " 6 "	....	"	10 80

BLACK BOTTLES, DEMIJOHNS, Etc.

WINE OR BRANDY BOTTLES.

(Superior in style and finish to any Imported.)

Net Prices—Red or Amber Color.			*By Case.*
5's to gallon,	bulb neck, patent, 1 gross box	per gross,	$12 00
6's "	plain and patent shoulder	"	10 50
7's "	plain and patent	"	9 75
8's "	plain and patent	"	9 38
10's "	bulb neck	"	9 38
12's "	"	"	9 38
15's "	"	"	9 38

SCHNAPP BOTTLES.

Quarts	1 gross boxes	per gross,	$12 38
5's to gallon	1 "	"	12 38
6's "	1 "	"	12 38
Pints	1 "	"	9 00

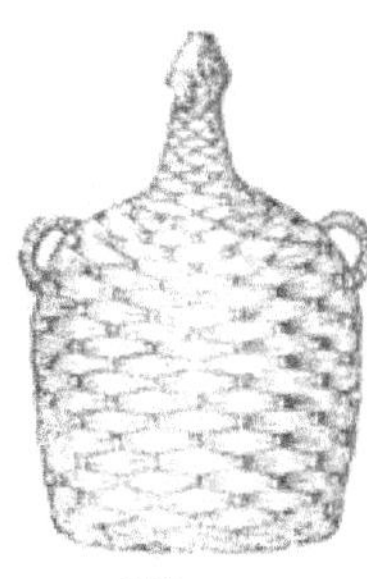

DEMIJOHNS, HEAVY AND UNIFORM, HANDLED AND COVERED.

	Per Doz.
Quarts, holding the quantity	$ 4 00
½ Gallon	6 00
Gallon	8 00
2 Gallons	10 00
3 Gallons	12 00
5 Gallons	14 40

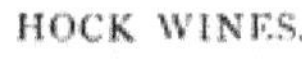

HOCK WINES.

Ruby Color.

			Per Gross.
Hocks,	5's or 6's to gallon,	Clay Mould,	$15 00
"	half size,	" "	12 38
"	5's or 6's to gallon,	Iron "	13 88
"	half size,	" "	11 62

CLARET BOTTLES.

Equal to French.

		Per Gross.
C. M.,	5's and 6's, 1 gross boxes	$15 00
I. M.,	5's and 6's, 1 "	13 88

Lettered Moulds.

We have made arrangements for supplying Moulds for the insertion of names of our customers in their Flint Glassware, which is becoming very general in the larger towns. The price of the labeled ware is not increased, after the first expense of the plate and mould.

Parties sending out special preparations, will find the moulded ware very desirable. We get up monograms, designs, &c., when requested. We annex prices of moulds for a few styles of bottles. No order taken for less than one case, and a special discount made when six sizes moulds are ordered at once.

PRESCRIPTION VIALS.

For insertion of name, etc., on	1 oz. moulds	$5 00
" " " "	2 "	5 00
" " " "	3 "	5 00
" " " "	4 "	6 00
" " " "	6 "	7 00
" " " "	8 "	8 00

UNION OVALS.

For insertion of name, etc., on	1 oz. moulds	$5 00
" " " "	2 "	5 00
" " " "	3 "	5 00
" " " "	4 "	6 00
" " " "	6 "	7 00
" " " "	8 "	8 00
" " " "	16 "	12 00

FRENCH SQUARE PRESCRIPTION BOTTLES.

For insertion of name, etc., on	½ oz. moulds	$6 00
" " " "	1 "	6 00
" " " "	2 "	6 00
" " " "	3 "	6 00
" " " "	4 "	7 00
" " " "	5 "	8 00
" " " "	6 "	8 00
" " " "	8 "	9 00
" " " "	10 "	10 00
" " " "	12 "	1 00
" " " "	14 "	12 00
" " " "	16 "	13 00

THE BEST AND LATEST STYLE OF

FRENCH OR ITALIAN BRONZE DRAWER PULL, (seen in parts,)

With Sliding Glass Label.

New, Chaste, Elegant and Improved Style. Is made of the best bronzed metal, finished in superior style, and can be had in light, dark or antique bronze, combining strength with elegance. The label being placed at a greater angle than in any other Pull, allows it to be read on the lowest drawers, while in a standing position, without the slightest difficulty. There being room enough to use all the fingers in pulling out the drawer, makes it much easier to handle in case the drawer sticks or is heavy.

PORCELAIN DRAWER PULL, PLAIN.

PORCELAIN DRAWER PULL, EMBOSSED.

PORCELAIN DRAWER KNOB.

Fig. 1.

Fig. 1. With Glass Label Lettered to Order.

		Brz. Iron.	Nick. Plated
3¼ by 1¼, for small drawers,	per dozen,	$3 60	$5 50
4¼ by 1⅜, for medium drawers,	"	3 60	5 50
5½ by 1⅝, for large drawers,	"	4 50	6 50

The size given is that of the frame, exclusive of the Pull.

We can also furnish a Paper Label, black letter on white ground. This being protected by a glass in front, forms an equally durable article, is very neat and attractive in appearance, and at the same time the price is very moderate. We recommend them to those to whom economy is an object.

Fig. 1. With Paper Label behind Glass Front.

	Brz. Iron.	Nick. Plated.		Brz. Iron.	Nick. Plated
3¼ by 1¼, per doz.	$2 50	$4 50	5½ by 1⅝, per doz.	$3 25	$5 50
4¼ by 1⅜, "	2 50	4 50			

The smallest size will suit our No. 1 Drawers, the medium size will be required for the No. 1⅜, and the large Pulls are for the large drawers of Wholesale Druggists.

Fig. 2.

Fig. 2 represents a new Pull with frame for a curved label.

The ends of the scroll underneath the frame project sufficiently to allow the fingers to be inserted to pull out the drawer. This has all the advantages of Fig. 1, and is preferred by many as being neater in form, while it has the advantage of great compactness. The extreme length is 4 inches, and the width 2 inches. It allows a label 3⅛ inches in length, giving ample space to allow the lettering to be distinct and legible. We have applied for a patent for this design.

Fig. 2. With Glass Label Lettered to Order.

Bronzed Iron, - - per doz. $3 60 | Nickel Plated, - per doz. $5 50

Fig. 2. With Paper Label behind Glass Front.

Bronzed Iron, - - per doz. $2 50 | Nickel Plated, - per doz. $4 50

Paper Labels can be furnished with either style of Pull at once. We will furnish for any of the Pulls, the Glass Label, either black letters on white or gold ground, gold shaded black letters on white ground, gold letters on any colored ground, and with or without gold border. Generally the more simple the label, the better the effect; when very ornate the letters must be small and less legible.

DRUGGISTS' FURNITURE,

UTENSILS, ETC.

ALCOHOL LAMPS—(see page 175.)

With Ground Glass Caps, to Prevent the Evaporation of Alcohol.

2 oz. with wick tube	each,	$	40
4 " " "	"		50

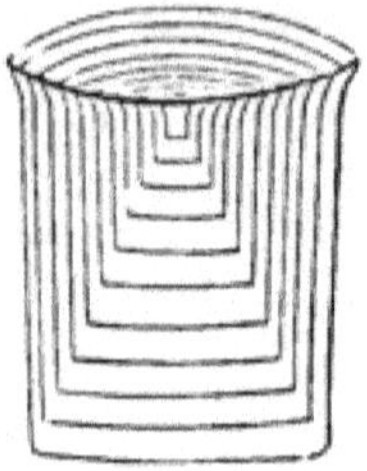

BEAKER GLASSES.

Best Bohemian, of Uniform Thickness.

5 in a Nest,

2 to 10 oz.	per nest,	$1 50

BED-PANS. (See pages 152 and 187.)

French-shaped—Earthen Ware.

No. 1, white	per doz.	$10 50
1, yellow	"	9 00

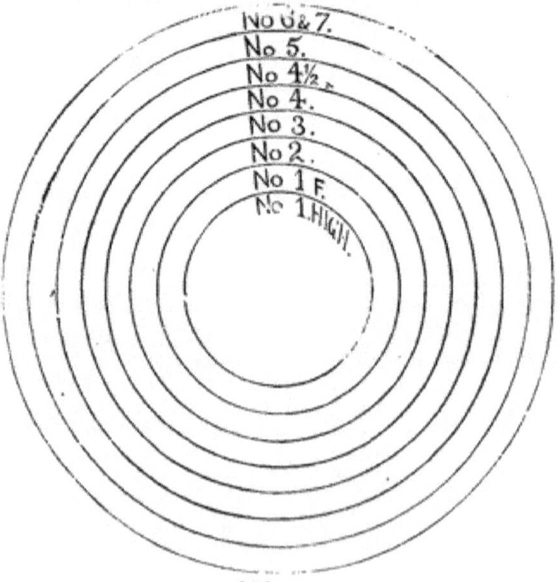

No 6 & 7.
No 5.
No 4½.
No 4.
No 3.
No 2.
No 1 F.
No 1 HIGH.

BOXES, TURNED WOOD.

For prices, etc., see page 15.

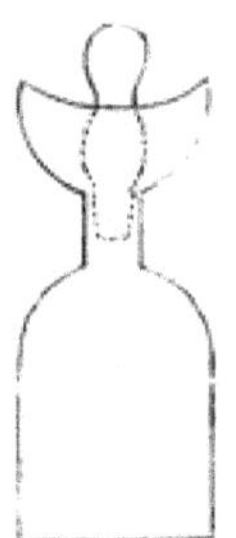

Eye Bot., Stopped

(See page 157.)

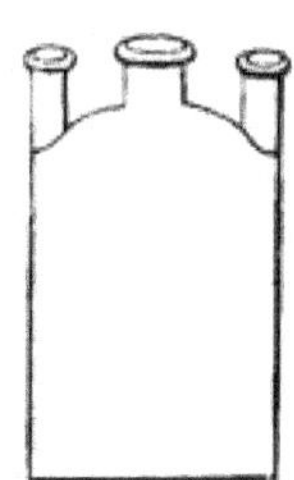

Woulf Bottle.

(See page 186.)

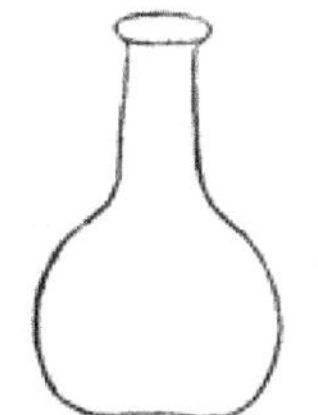

Chemical Flask.

(See page 155.)

RECEIVERS—(See page 170

Plain.

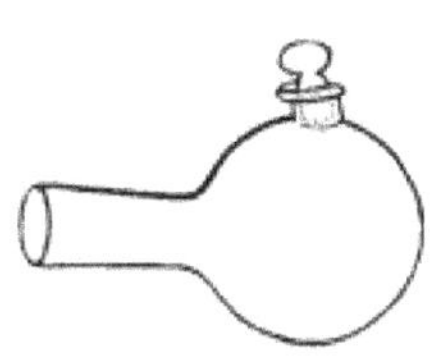

Keyed.

German Saltmouths.

BED PANS.

White and Yellow.
French Style and Common.

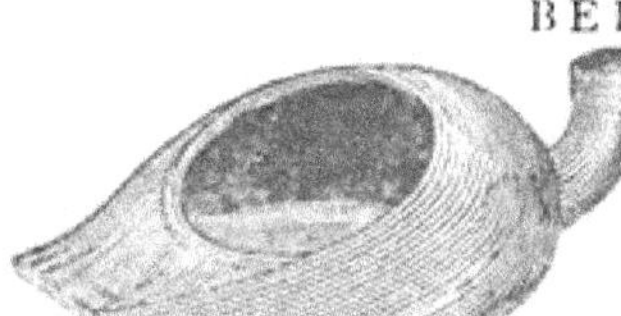

See page 151 and 187.

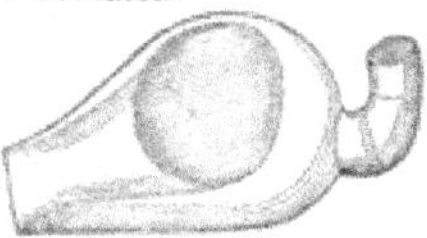

EVAPORATING BASIN.

Porcelain, with Spout for Wooden Handle.

EVAPORATING DISH.

Iron, enameled inside, with handles.

(See page 160)

White Gallipots, Nested.

Crucibles, German, Covered.

Crucibles, German, not covered.

Porcelain Galli-pot, covered

BRACKETS—WINDOW.

Italian Bronzed, a New Article, and very Handsome.

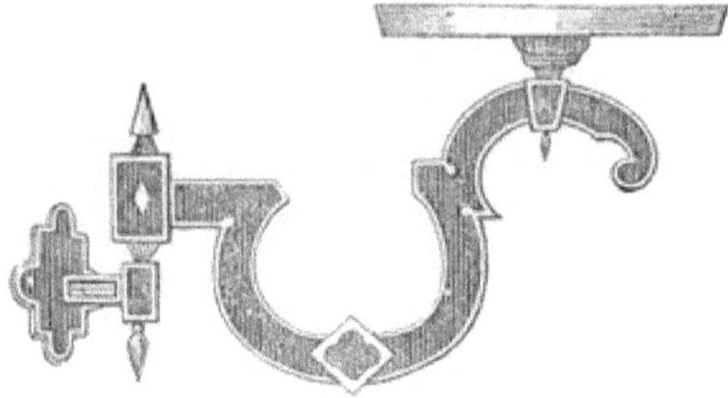

Small, 6 inch Plate	each,	$1 25
Large, 7 " "	"	1 50

BREAST PIPES.

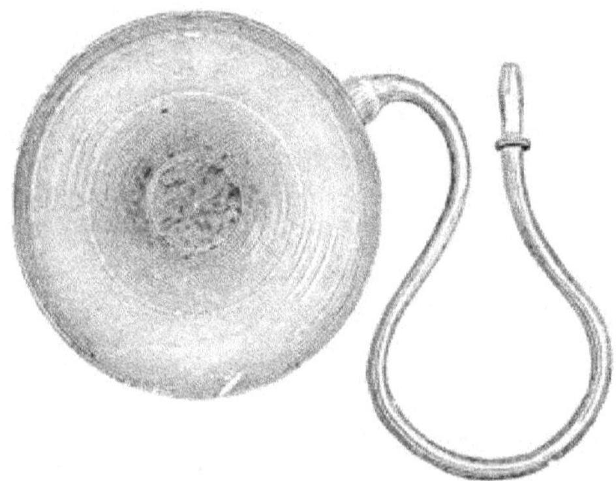

Improved Flexible Tube, No. 1	per doz.	$ 3 50
" " " No. 2	"	3 75
Old Style (all glass)	"	3 00
Allen's Improved Flexible Tube, No. 1	"	3 50

CORK PRESSES—Every Variety.

(See page 176.)

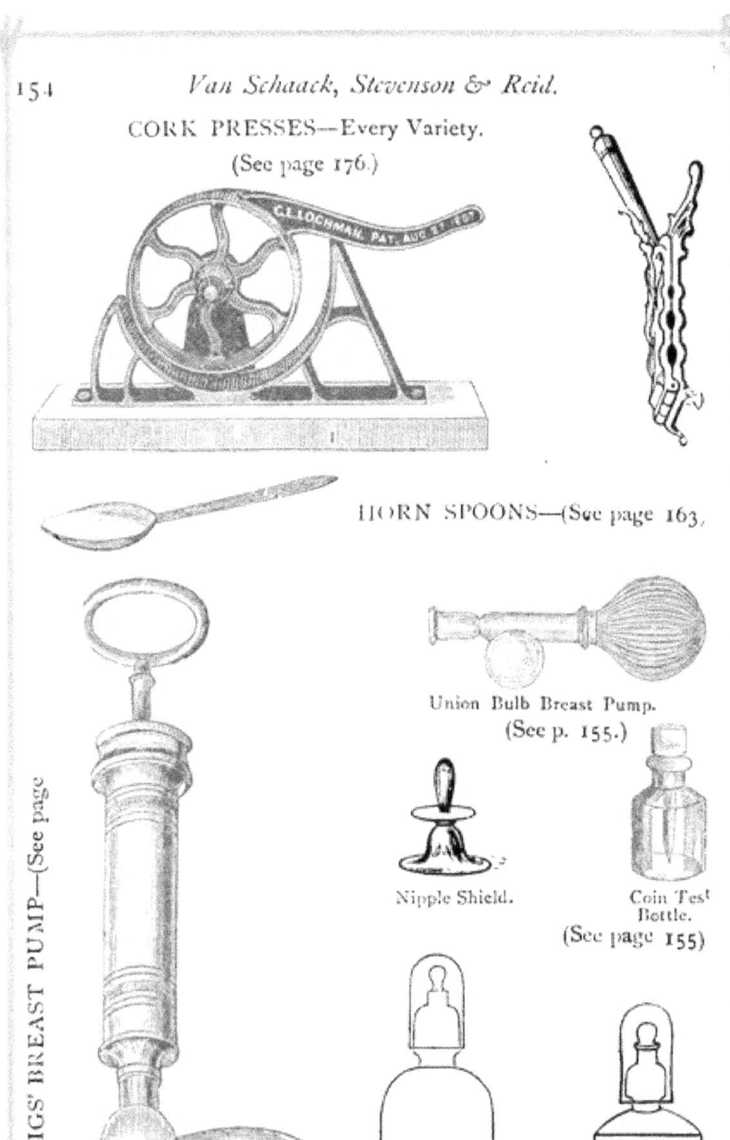

HORN SPOONS—(See page 163.

Union Bulb Breast Pump.

(See p. 155.)

Nipple Shield.

Coin Test Bottle.

(See page 155)

MEIGS' BREAST PUMP—(See page

Ether Bottles.

No. 2 Breast Pipe.

BREAST PUMPS.

	Per Doz.
Alexandra	$14 00
English Bulb	7 50
Hagerty's Bulb, Metal	9 00
Hard Rubber	15 00
Meigs's (see page 154)	6 00
Matson's Improved	15 00
Needham's	14 00
Union, (Bag (see p. 156)	9 00

CAUSTIC HOLDERS.

No. 1, Hard Rubber	each,	$	35
" 2, " "	"		40
" 3, " "	"		45
" 4, " "	"		5[illegible]

CHEMICAL FLASKS. (see page 152.)

Flat or Round Bottom, with Ring at Mouth. Narrow or Wide Mouth.

½ pint	each,	$	25
1 "	"		35
1 quart	"		50
½ gallon	"		75

COIN TEST BOTTLES. (see page 154.)

With Long Stopper by which to apply a drop of Acid to Metals.

1 oz.	each,	$	40
2 "	"		50

CUPPING GLASSES. (see page 172.)

Bell-shape, assorted	per doz.	$1	25
National Rubber Bulb, with Glass Cup	each,		75

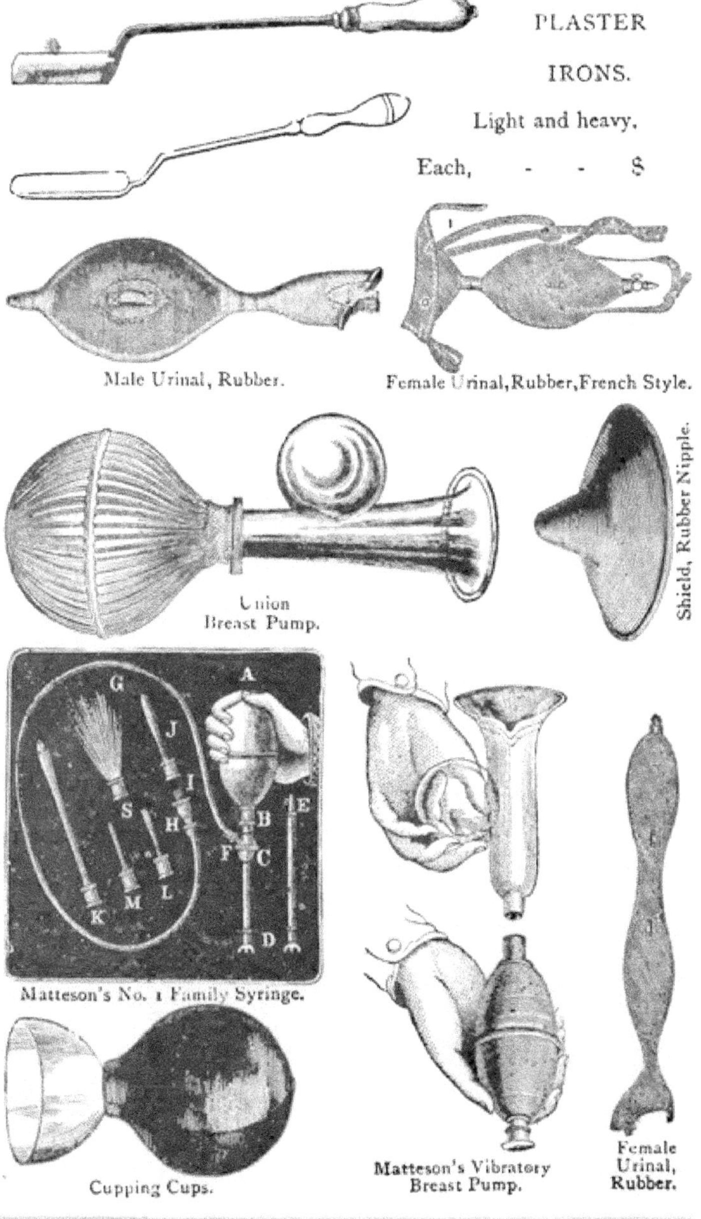

PLASTER IRONS.

Light and heavy.

Each, - - $

Male Urinal, Rubber.

Female Urinal, Rubber, French Style.

Union Breast Pump.

Shield, Rubber Nipple.

Matteson's No. 1 Family Syringe.

Cupping Cups.

Matteson's Vibratory Breast Pump.

Female Urinal, Rubber.

DRUG MILLS.

See page 158 & 159.

This cut represents Henry Troemner's celebrated Mills, so highly recommended by the trade.

(See next page.)

(Also see Instrument List.)

EARTHEN COLD CREAM POTS.

Labels burnt in.

½ oz.	per doz.	$1 00
1 "	"	1 15
2 "	"	1 35

ELECTRO-MAGNETIC MACHINES.

Davis & Kidder's	each,	$ 8 00
Jerome Kidder's, six current	"	17 00

EYE BATHS.

Plain	per doz.	$2 25
" cut out	"	2 75
Moulded, Fluted,	"	2 25
Porcelain	"	2 75
Vapor Eye Baths, stoppered	"	6 00

FILTERS--French.

Round, one hundred sheets.

No. 15,	6 inches diameter	per pack,	$ 25
19,	8 "	"	30
25,	10 "	"	40
33,	13 "	"	50
40,	15 "	"	65
45,	18 "	"	80
50,	20 "	"	90

SWIFT'S DRUG MILLS.

The "NEW STYLE"

Is Fitted with the

Heaviest Fly Wheel

And can easily be taken apart for cleaning.

Price, $12.00.

The "OLD STYLE"

Is well known throughout the country.

Price, $10.50.

We supply new grinding plates for $2.50 per set, for either mill—no extra charge for boxing mills.

The Trade supplied by

VAN SCHAACK, STEVENSON & REID,

Old Salamander Drug House, 92 & 94 Lake St., cor. Dearborn.

AMERICAN DRUG MILLS

MANUFACTURED BY

ENTERPRISE MF'G CO., OF PA.

S. W. Cor. American and Dauphin Sts., Philadelphia.

The Best Drug Mills in the World.

TWENTY DIFFERENT SIZES, FROM $2 TO $100 EACH.

IMPROVED CORK PRESSERS.

THE BEST ARTICLE OF THE KIND IN THE MARKET.

No. 1, Small Size, $1.00; No. 2, Large Size, $1.25.

ENTERPRISE HERB & ROOT CUTTER

PRICE, $3.50.

SEND FOR ILLUSTRATED CATALOGUE.

FOR SALE BY

Van Schaack, Stevenson & Reid,

WHOLESALE DRUGGISTS, 92 & 94 LAKE ST., CHICAGO.

EVAPORATING DISHES.

Best German Porcelain, glazed inside, with heavy rim around the top.

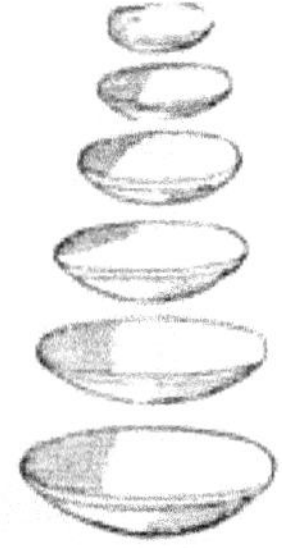

No.	00,	diameter	16 inches,	contents	3 gall.,	each	$5 00
"	0,	"	15 "	"	2 "	"	3 50
"	1,	"	13 "	"	1 "	"	2 10
"	2,	"	12 "	"	3 quarts,	"	1 75
"	3,	"	11 "	"	½ gall.,	"	1 30
"	4,	"	10 "	"	1½ quarts,	"	1 00
"	5,	"	9 "	"	1 "	"	85
"	6,	"	8 "	"	1½ pints,	"	75
"	7,	"	7 "	"	20 oz.,	"	65
"	8,	"	6 "	"	1 pint,	"	55
"	9,	"	5½ "	"	½ "	"	45

EVAPORATING DISHES—IRON. (see page 152.)

1 gallon	each	$1 75	½ gallon	each	$1 25
¾ "	"	1 25	¼ "	"	75

FUNNELS—GLASS.

1 oz.	each,	$ 10
2 "	"	10
4 "	"	12
8 "	"	15
Pints	"	20
Quart	"	25
½ gallon	"	35
1 "	"	50

FUNNELS—HARD RUBBER.

No. 0, gill	each,	$ 25
" 1, ½ pint	"	35
" 2, 1 pint	"	50
" 3, 1 quart	"	65

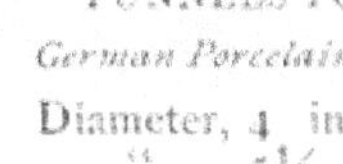

FUNNELS FOR FILTERING. (see page 161.)

German Porcelain, with Staves inside.

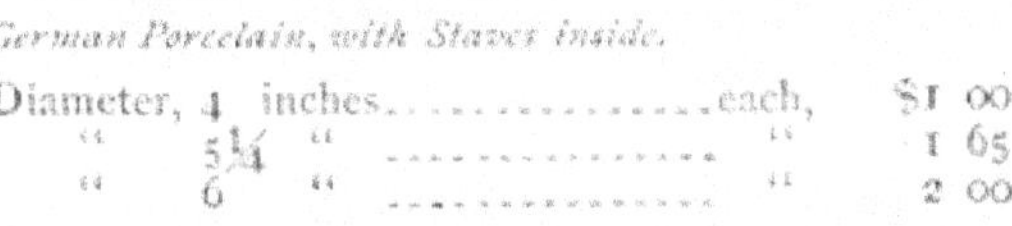

Diameter, 4 inches	each,	$1 00
" 5¼ "	"	1 65
" 6 "	"	2 00

FUNNELS—PORCELAIN, PERFORATED.

All sizes, from $1 25 to $3 00 each.

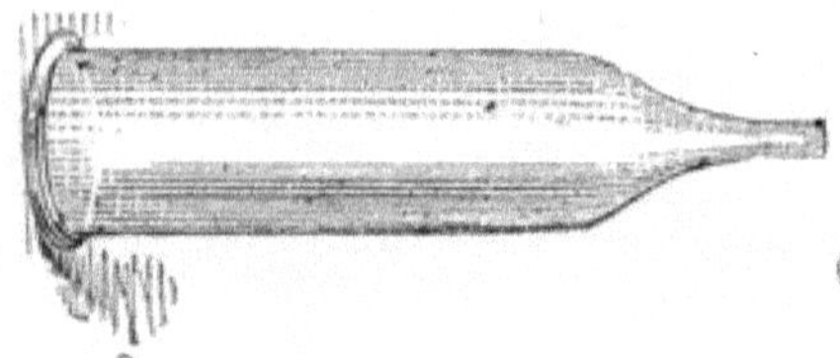

PERCOLATORS.

Glass, Heavy.

(See page 171.)

Glass, for Volatile Liquids,

Pints, Quarts, and Half Gallons.

NURSING BOTTLES—(See page 171 & 162.)

Narrow Mouth.

Wide Mouth.

PORCELAIN FUNNELS—(See page 160.)

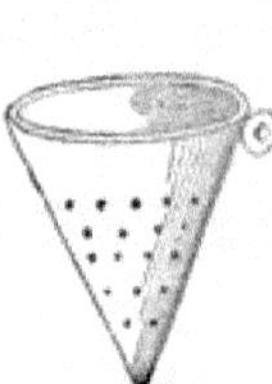

GLASS NIPPLE SHELL.

(See page 171.)

PORCELAIN MORTARS—(See page 169.)

Shallow.

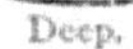

Deep.

THERMOMETERS, all styles.
(See page 183.)

Maw's Alexandra Feeder

Maw's Export Flint Feeder.
(See list, page 171.)

Maw's Trade Mark (or Export) Feeders *Green*, (wood top.)

(See page 171.)

Morocco Covered, with Cup.

Britannia, (White Metal Pocket Flasks.

Van Schaack, Stevenson & Reid's Feeders, same style as Maw's, only Flint Glass, with trimmings complete, at $2.00 per doz.

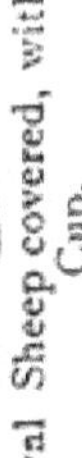

Oval Sheep covered, with Cup.

Champagne Taps, for drawing effervescent liquids, without removing the cork.

GALLIPOTS—GERMAN PORCELAIN.

With Wood Tops, and Labeled.

½ oz. capacity	per doz.	$ 70
1 " "	"	80
2 " "	"	1 00
3 " "	"	1 10
4 " "	"	1 25

GALLIPOTS—WHITE, NESTED. (see page 152.)

1 to 4 oz. Straight	per gross,	$4 50
1 to 4 oz. "	per doz.	50

GALLIPOTS—YELLOW.

¼ oz.	per 100,	$ 1 50
½ "	"	1 50
1	"	1 75
1½	"	2 25
2	"	2 75
3	"	3 25
4 oz.	per 100,	$ 5 25
6 "	"	6 00
8 "	"	7 25
12 "	"	10 00
16 "	"	12 00

GLASS TUBING.

Heavy	per lb.	$ 50
Light	"	50
Solid Rods	"	75

GLASS BOXES.

For Tooth Powders. (see page 175.)

½ oz. Opaque	per doz.	$1 00
1 " "	"	1 25
2 " "	"	1 50
2 " Sandwich	"	1 75

HORN SCOOPS. (see page 164.)

No. 0	per doz.	$1 75
½	"	65
1	"	75
2	"	1 00
3	"	1 25
4	"	1 40
5	"	1 50

HORN SPOONS.—Long Handle. (see page 164.)

No. 1	per doz.	$ 90
2	"	1 25
3	"	1 38
4	"	1 50
5	"	1 75

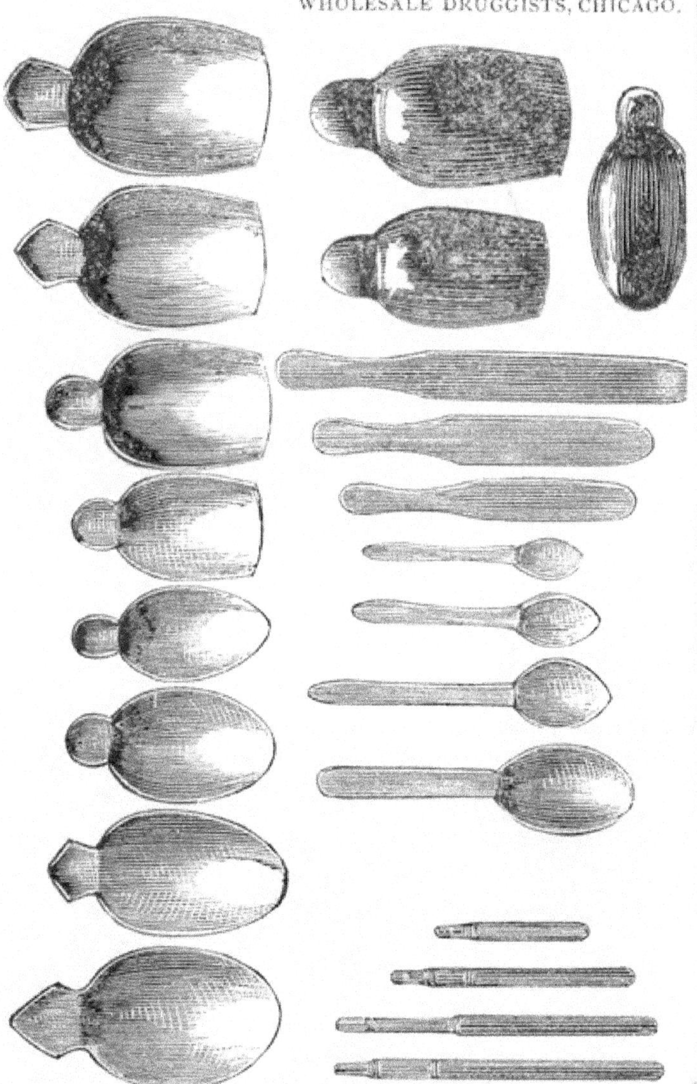

GRADUATES, GLASS.

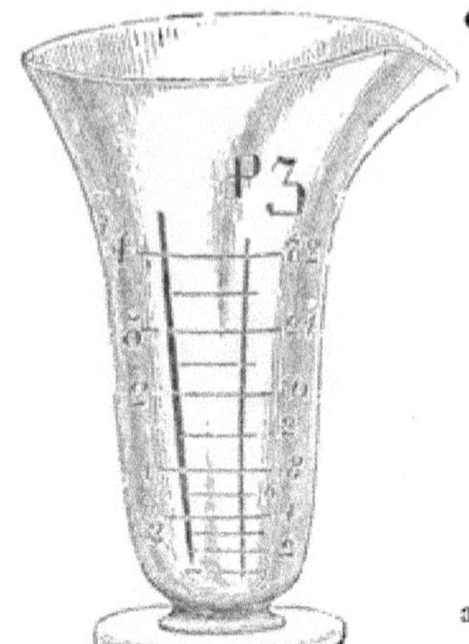

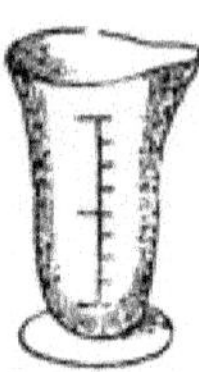

The utmost care is taken to secure accuracy in measure.

½ oz.	each,	$ 25
1 "	"	25
2 "	"	30
3 "	"	35
4 "	"	40
6 "	"	50
8 "	"	60
10 "	"	70
12 "	"	75
16 "	"	1 00
24 "	"	1 25
32 oz.	each,	$1 75
Minim, tall shape	"	35
" short "	"	30
Medicine Glass, graduated for the Medical Tea and Table Spoon:		
On foot	each,	
" Goblet shape	"	50
In Morocco Box	"	60
" " with Minim	"	75
Drop, in cases	"	50

HYDROMETERS. (see page 172.)

For Acid, Ammonia or Alkali	each,	$ 75
" Lyes, Lime or Syrup	"	75
" Beer, Bark or Vinegar	"	75
" Cider, Malt or Wine	"	75
" Spirits—one scale	"	75
" Alcohol—one scale	"	75
" Coal Oil—80@90—shot	"	1 00
" Milk	"	75
Urinometers, with Test Glass, in paper case	"	1 00
U. S. Custom House Spirit Hydrometer, two scales, with four-scaled Thermometer	"	3 00
Coal Oil Hydrometer, one scale, with Thermometer,	"	3 00
8 inch Laboratory Thermometer, all glass	"	1 50

HYDROMETER JARS.

On Foot.

8	inches high,	1½	inches diameter	each,	$ 50
10	"	"	"	"	50
12	"	"	"	"	50
8	"	2	"	"	50
10	"	2	"	"	50

Richly cut, with names.

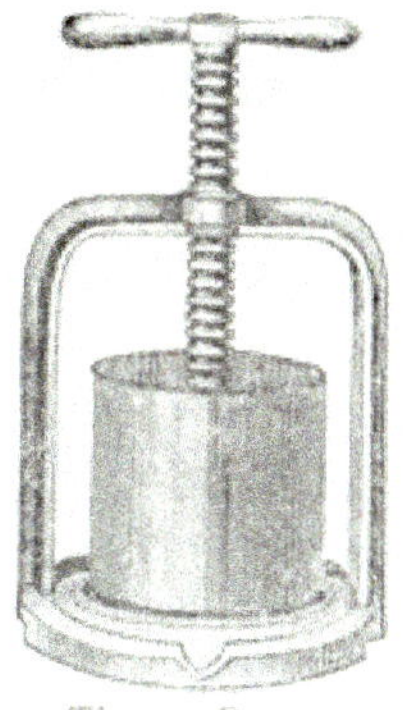

Tincture Presses
Enameled inside.

Quarts, -	$—
1-2 Gallon, - -	—
1 "	—
2 "	—

(See page 186.)

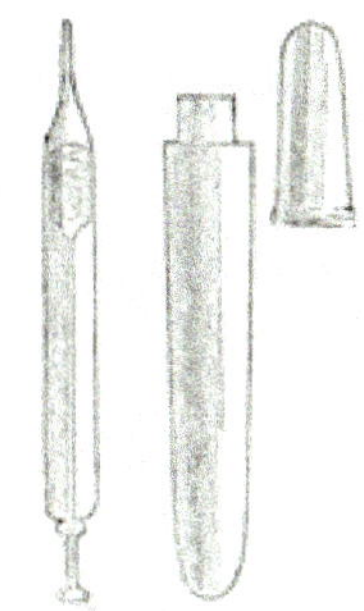

Male Glass Syringes
Metal Cap, in
wood cases.

Per dozen, $—.

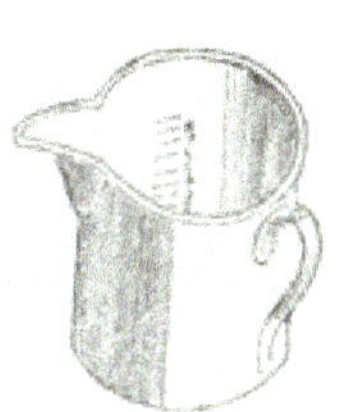

Berlin Porcelain Measures, Graduated.
Handled, for hot liquids or acids.

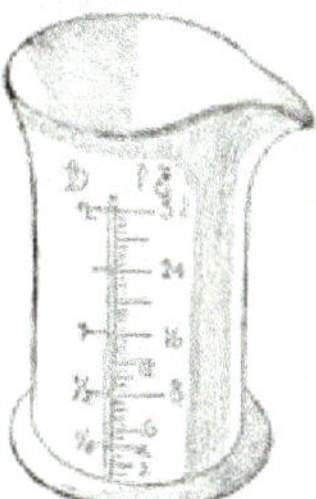

Glass, Graduated
Tumbler shape.

(See page 165.)

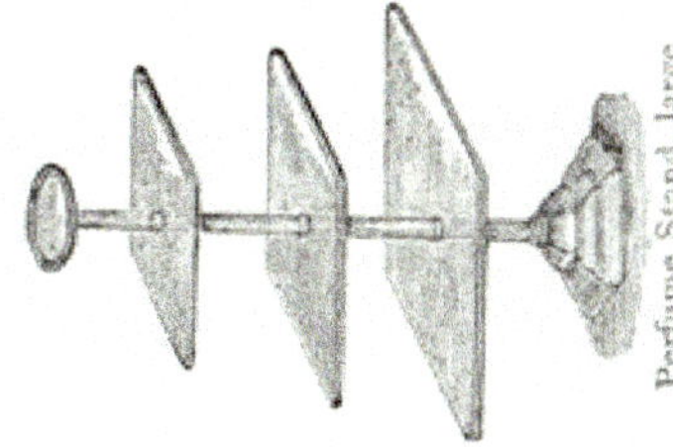

Perfume Stand, large square.

INHALING BOTTLES. [see page 175.]

Dr. Hunter's	per doz.	$12 00
Olive Tar	"	6 00
Pocket Inhaler	"	3 50

JARS—PORCELAIN LEECH.

Labeled in Black Letters.

VASE SHAPE.

No. 1	each,	$5 00
" 4	"	3 75
" 6	"	3 00

CANDY JARS.

With Japanned Tin Covers—Low shape.

Quart, height 6 inches	per doz.	$ 1 88
½ gallon, " 7 "	"	2 25
1 " " 9 "	"	3 75

CONFECTIONERY JARS, RINGED.

¼ gallon	per doz.	$5 50
½ "	"	7 50
1 "	"	10 50

(Not including Glass Labels.)

INFUSION JARS.

With Strainers.

1 pint	each,	$1 00
1 quart	"	1 50

ENGLISH STEEPLE TOP JARS.

White.

8	ounces	per doz.	$2 25
16	"	"	3 00
32	"	"	5 00

Blue.

8	ounces	per doz.	$2 75
16	"	"	3 50
32	"	"	7 00

SPECIE JARS.

With Japanned Tin Covers—Tall shape.

1 pint,	height	7 inches	per doz.	$1 50
1 quart,	"	8 "	"	2 00
½ gallon,	"	10 "	"	2 50
¾ "	"	11 "	"	3 50
1 "	"	12½ "	"	4 00

WHITE FLAT TOP JARS.

½ oz.	per doz.	1 00
1 "	"	1 00
2 "	"	1 10
3 "	"	1 15
4 "	"	1 15
8 "	"	1 25
16 "	"	1 50
2 lbs.	"	2 50

MORTARS, WEDGEWOOD.

No. 0000,	3	inches across the top	each,	$ 35
" 000,	3¼	" " "	"	45
" 00,	3½	" " "	"	50

MORTARS, WEDGEWOOD—(continued.)

No. 0,	4 inches across the top	each,	$ 55
" 1,	4½ " " "	"	65
" 2,	5 " " "	"	85
" 3,	6 " " "	"	1 00
" 4,	6½ " " "	"	1 25
" 5,	7 " " "	"	1 60
" 6,	8 " " "	"	2 00
" 7,	8½ " " "	"	2 50
" 8,	9½ " " "	"	3 00
" 9,	10½ " " "	"	3 75
" 10,	12 " " "	"	4 50
" 11,	13 " " "	"	5 25
" 12,	14 " " "	"	6 00

MORTARS, GLASS.

All sizes..per pound, 30

MORTARS, PORCELAIN. [see page 161.]

No. 3,	5 inches across the top,	16 oz.	each,	$1 25
" 4,	4½ " " "	8 "	"	1 00
" 5,	3¼ " " "	4 "	"	80
" 6,	3 " " "	2 "	"	65

MORTARS, IRON.

½ pint	each,	$ 45
1 "	"	65
1 quart	"	75
½ gallon	"	1 25
1 "	"	2 25
2 "	"	2 75

NASAL DOUCHES.

	Per Doz.
Hagerty's	$12 00
Nichols's	12 00
Quinlan's	12 00
Pierce's	
Allen's	

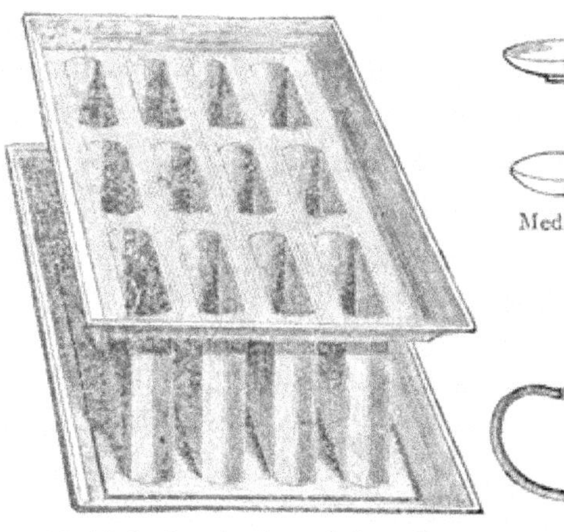

Moulds for Camphor Ice and Cosmetics.

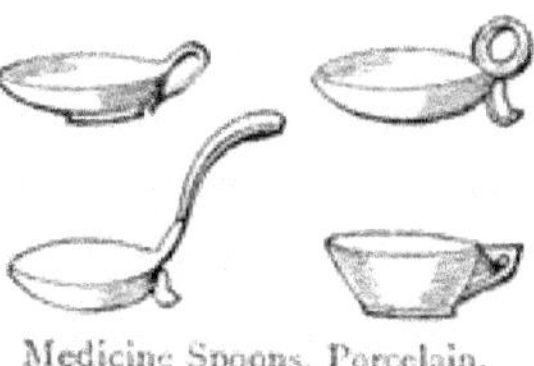

Medicine Spoons, Porcelain.

Speculum, Vaginal.

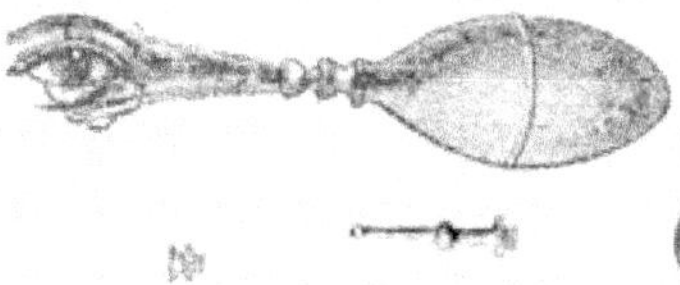

Eye and Ear Syringe.

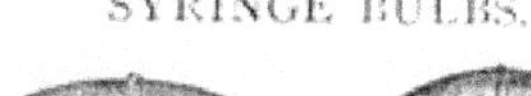

SYRINGE BULBS.

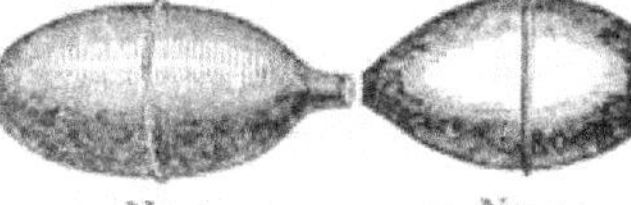

No. 1. No. 2.

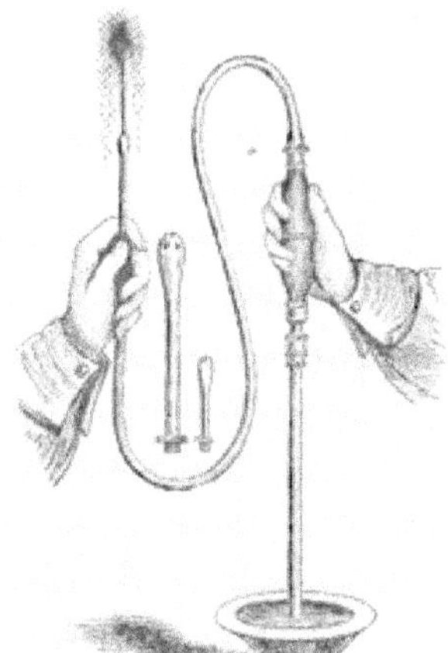

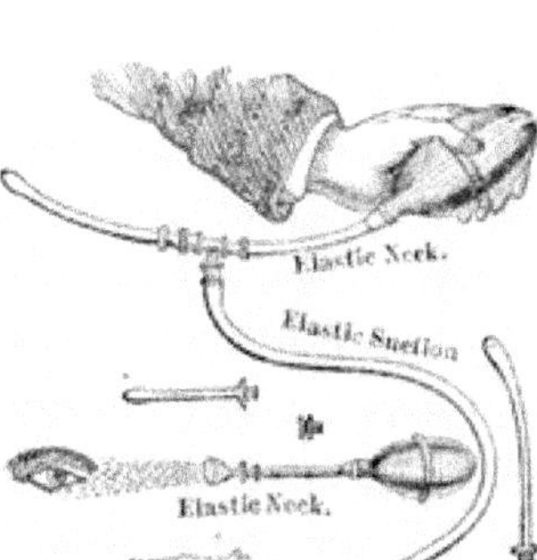

Essex Family Syringe.

NIPPLE SHIELDS.

Dr. Needham's Patent	per doz.	$3 75
White Rubber	"	1 00
Glass and Rubber	"	2 00
" " Hose	"	3 50
Wood, untrimmed	"	60
Nipple Shells, Glass (see page 161)	"	75

NURSING BOTTLES. [see pages 161 and 162.]

Maw's Alexandra, Flint	$5 25
" " Green	3 50
" Export, Flint	3 50
" Trademark, Green	1 75
Plain, 10 oz. N. M., Flint	75

Van Schaack, Stevenson & Reid's Nursing Bottles are same pattern as "Maw's Trade Mark," but are of the *best flint* ware, and very desirable and cheap 2 00

PATCH-BOXES, OR OINTMENT POTS.

With Covers.

1/8 oz.	per gross,	$ 8 00	1½ oz.	per gross,	$12 00
1/4 "	"	8 00	2 "	"	12 50
1/2 "	"	9 00	3 "	"	15 00
3/4 "	"	9 50	4 "	"	16 00
1 "	"	10 00			

PERCOLATORS, GLASS, HEAVY. [see page 161.]

Pints	each,	$ 38
Quarts	"	50
½ gallon	"	75
1 gallon	"	1 00
2 gallons	"	1 75

PERCOLATORS, TIN.

Quart	each,	$ 85
½ gallon	"	1 00
1 gallon	"	1 50
2 gallons	"	
Keyed, quart	"	1 85
" ½ gallon	"	2 25
" 1 gallon	"	3 00
" 2 gallons	"	

PILL TILES, PORCELAIN, GRADUATED.

7 inches	each,	$ 65
8 "	"	75
9 "	"	90
8 x 10 "	"	1 00
10 x 12 "	"	1 50

Maw's.

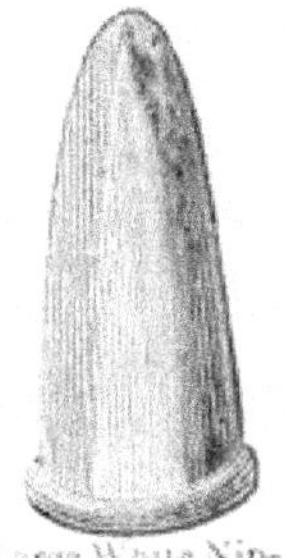
Large White Nipple No. 2.

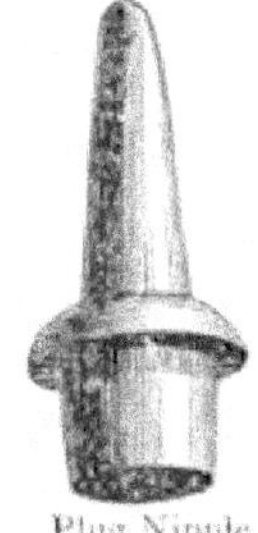
Plug Nipple.

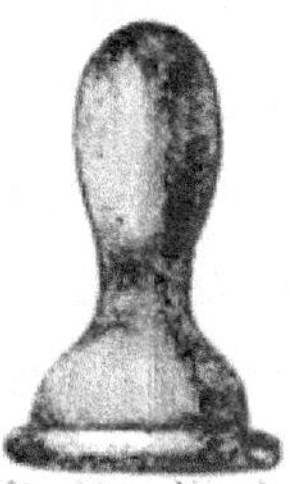
Davidson Nipple No. 4.

SICK FEEDERS.

Porcelain.

(See page 177.)

SUPPOSITORY MOULDS.

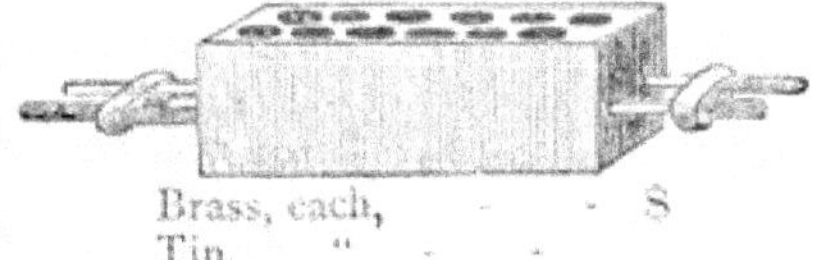

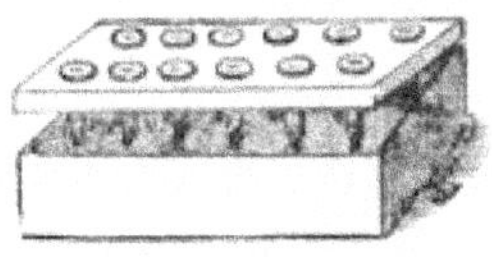

Brass, each, - - - $
Tin, " - - -

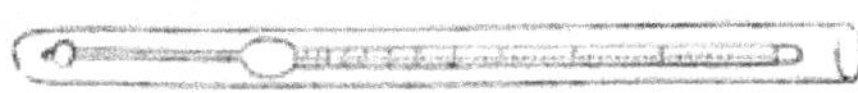

HYDROMETER.

(See page 165.)

SPATULAS—(See page 177.)

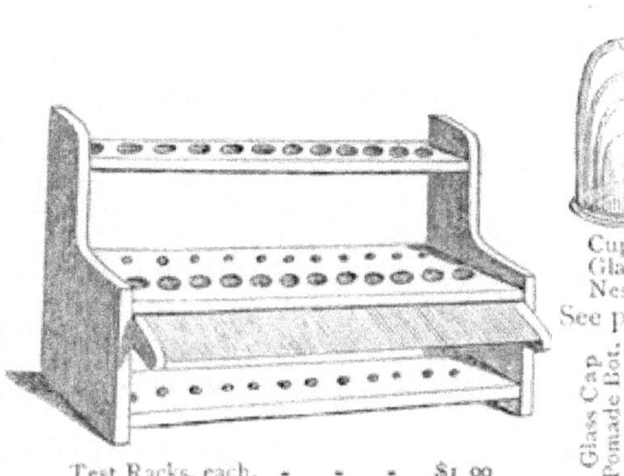
Test Racks, each, - - - - $1 00

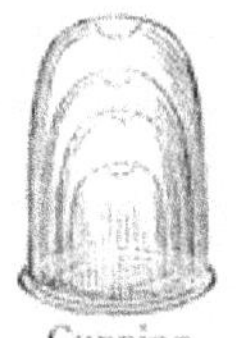
Cupping Glasses, Nested.
See p. 155.

Metal Cap, Pomade Bottle.
See p. 173.

Glass Cap Pomade Bot.
See p. 173.

PILL MACHINES.

Walnut Frames, with Side Rollers.

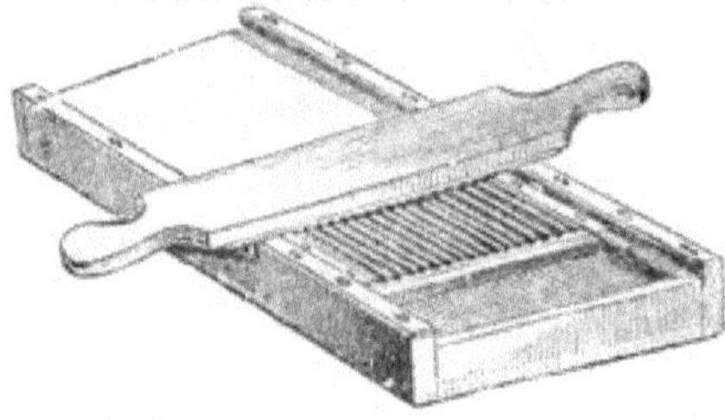

12 Pil.., 3. 4. or 5 grain	each,	$5 00
18 " 3. 4. or 5 "	"	6 00
24 " 3. 4. or 5 "	"	7 00

POMADE BOTTLES. [see page 172.]

1 oz. Phillicombs, no Cap	per doz.	35
2 " " "	"	50
3 " " "	"	60
4 " " "	"	75

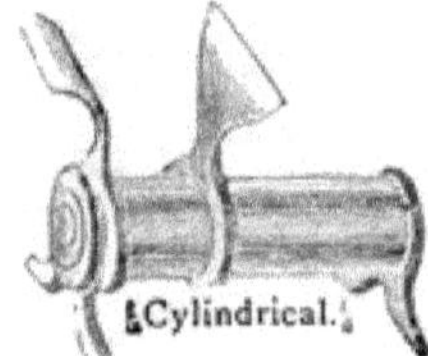

POWDER FOLDERS.

Various Styles.

$2 50 to $3 00 each.

PULLS, DRAWER, PORCELAIN.

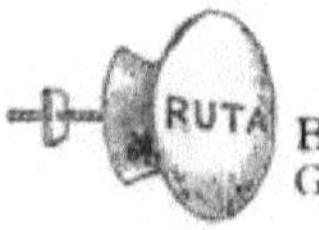

Plain or Embossed Pattern, Decorated.

For Druggists' Drawers, per doz., $3 50 to $5 00

KNOBS, DRAWER, PORCELAIN.

Size, one to two inches.

Black Letters	per doz.,	$2 50
Gold Band and Black Letters	"	3 25

PESSARIES. [See page 175.]

Glass Globular, assorted	per doz.	$1 25
" Concave, "	"	1 25
Hard Rubber Concave, assorted	each,	35
" Bow Pattern, assorted	"	50
" Horse-shoe Pattern, assorted	"	50
Long Stem Rubber, white	"	75
" " " black	"	75
Meigs's Ring	"	50
Wadsworth's Uterine Stem, large, medium and small,	"	6 00
White French Rubber, inflated	"	50
" " " short tube	"	50

THE NEW PATENT NURSERY

Pocket Spirit Stove.

— THE —

FRIEND

OF THE

SICK ROOM

You can use any vessel for the top that will stand heat. None goes with the Stove.

The best selling article in the line yet brought before the public

The NEW STOVE, by the aid of a few drops of Alcohol, will

Cook a Chop or Steak, Boil Water, Eggs, &c.

2 Tablespoonfuls of Alcohol will burn for Half an Hour, and will Boil 1 Pint of Water in from 5 to 7 Minutes.

As a HEATING LAMP for the nursery, or other purposes, it is the best ever invented. No traveler should be without one.

Men who take their meals at the shop, will find this just the thing to get up a hot dinner with.

The inside of the stove being filled with the indestructible substance, *Asbestos*, (which absorbs the spirit), can neither explode, boil over, nor get out of order.

It can be folded up and put into a box that goes with it.

PRICE, $1.00 EACH. BY MAIL, $1.10.

Address orders to the Wholesale Agents,

VAN SCHAACK, STEVENSON & REID,

Old Salamander Wholesale Drug Warehouse, Chicago.

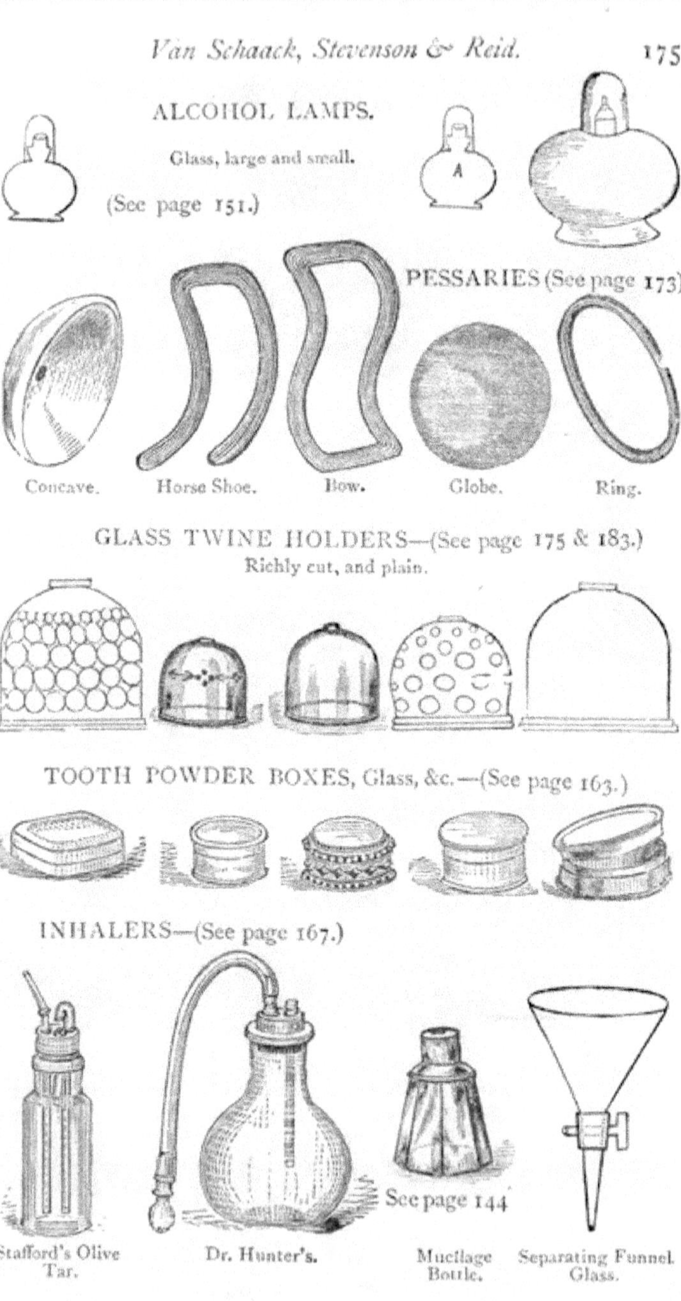
ALCOHOL LAMPS.
Glass, large and small.
(See page 151.)
A
PESSARIES (See page 173)
Concave.
Horse Shoe.
Bow.
Globe.
Ring.
GLASS TWINE HOLDERS—(See page 175 & 183.)
Richly cut, and plain.
TOOTH POWDER BOXES, Glass, &c.—(See page 163.)
INHALERS—(See page 167.)
See page 144
Stafford's Olive Tar.
Dr. Hunter's.
Mucilage Bottle.
Separating Funnel Glass.

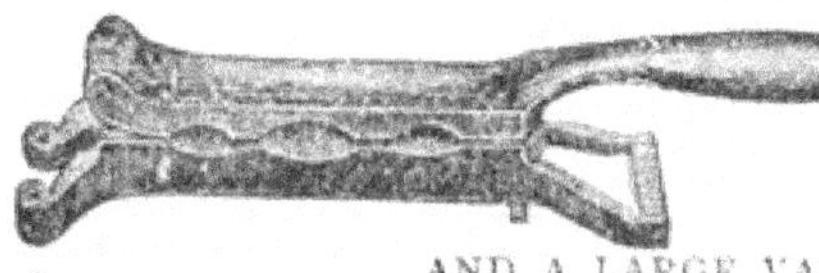

PRESSES, CORK.

40 to 60 cents each.

AND A LARGE VARIETY OF BRONZED CORK PRESSES.

$1 00 to $1 50 each.

Lochman's (see page 154) . . . . each, $ 1 50

PUNGENTS, PRESTON SALTS, Etc.

Pungents, Assorted colors, in box . . . . per doz.,
" Crystal (see page) . . . . "

Also a fine variety of Fine Cut and Silver Capped.

RETORTS. (See page 152.)

Tubulated and Stoppered.

½ pint	each,	$ 55
1 "	"	75
1 quart	"	90
½ gallon	"	1 25

Plain.

½ pint	each,	$ 35
1 "	"	60
1 quart	"	70
½ gallon	"	1 00
1 "	"	1 25

RETORT STANDS, IRON, No. 1.

With three rings for supporting Retorts, Flasks, Funnels, etc., each, $1 25 to $3 00.

Enameled Base Style, 4 Rings, etc. (see Instrument List.)

SADDLE-BAG BOTTLES—FLAT SQUARE.

Stoppered.

½ oz.	per doz.	$1 00
1 "	"	1 25
1½ "	"	1 40

SADDLE-BAG BOTTLES—ROUND.

Stoppered.

½ oz.	per doz.	$1 00
1 "	"	1 25
2 "	"	1 40

SADDLE-BAG BOTTLES—ROUND.

Not Stoppered.

½ oz.	per doz.	$ 50
1 "	"	60
2 "	"	75

SIEVES.

Nested, 4 in nest, diameter, 5½, 4¼, 3½, and 2½ inches,	per nest,	$1 25
Bolting Cloth, " 8 inches	each,	1 50
" " " 12 "	"	1 75
Drum, " 12 "	"	5 00

SICK FEEDERS, &c. [see page 172.]

Earthenware	per doz.	$3 00
" Spoons, dessert size	"	2 50
" " table "	"	3 50
" Spitting Mugs	"	6 00
Sick Tubes, Glass, light, bent	"	25
" " " straight	"	25

SPATULAS, STEEL.

With Riveted Handle.

3 inch blade	each,	$ 20
4 "	"	25
5 "	"	30
6 "	"	35
7 "	"	40
8 "	"	50
9 "	"	65
10 "	"	80
11 "	"	1 00
12 "	"	1 25

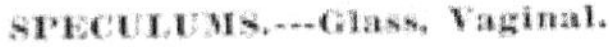

SPECULUMS.---Glass, Vaginal.

Per Doz.

Assorted Sizes, Silvered and Coated..... $7 50

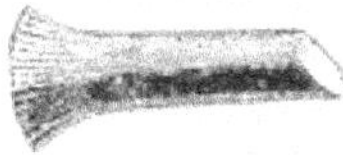

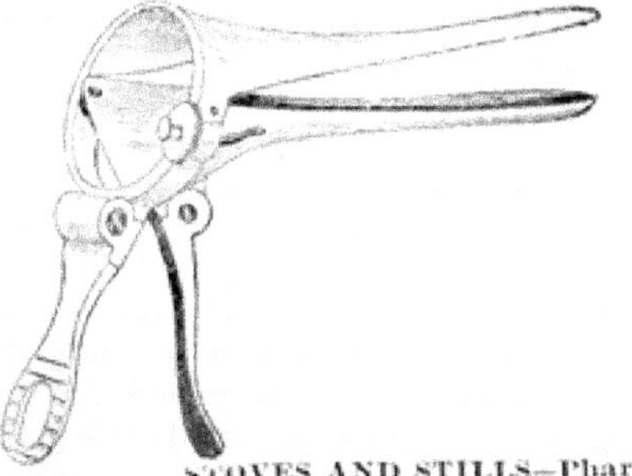

Handled Metal Vaginal Speculums, a great variety.

STOVES AND STILLS—Pharmaceutists'.

Mitchell's, see page ——.

SYRINGES, GLASS.

			per doz.,	$
No. 00	Male	Syringes, Metal Cap...............	per doz.,	$ 45
" 0	"	" " "	"	50
" 1	"	" " "	"	60
" 2	"	" " "	"	65
" 3	"	" " "	"	75
" 4	"	" " "	"	85
" 5	"	" " "	"	1 10
" 6	"	" " "	"	1 35
2 oz.	Injection	" " "	"	1 50
3 "	"	" " "	"	2 25
4 "	"	" " "	"	2 50
No. 1	Female	" " "	"	60
" 2	"	" " "	"	75
" 3	"	" " "	"	85
" 4	"	" " "	"	90
" 5	"	" " "	"	1 15
" 6	"	" " "	"	1 40
2 oz.	"	" " "	"	1 60
3 "	"	" " "	"	2 10
4 "	"	" " "	"	2 45

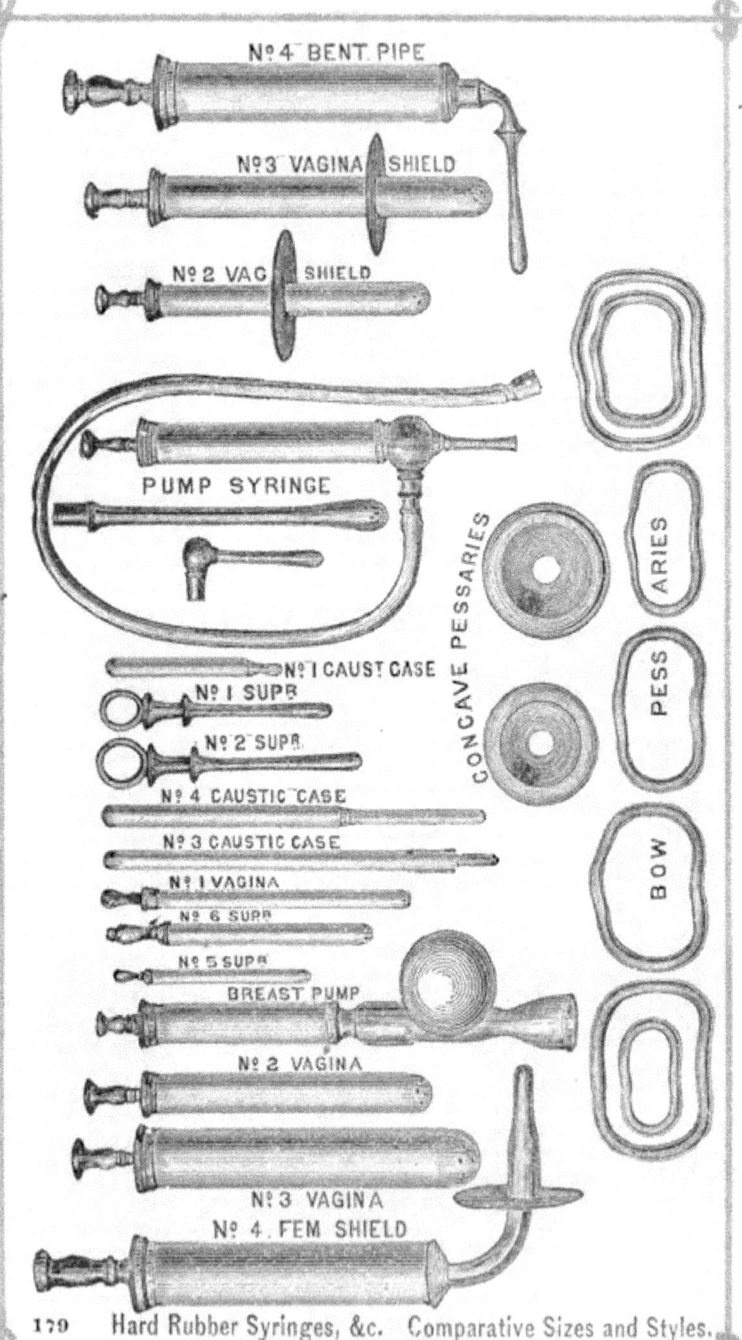

179 Hard Rubber Syringes, &c. Comparative Sizes and Styles.

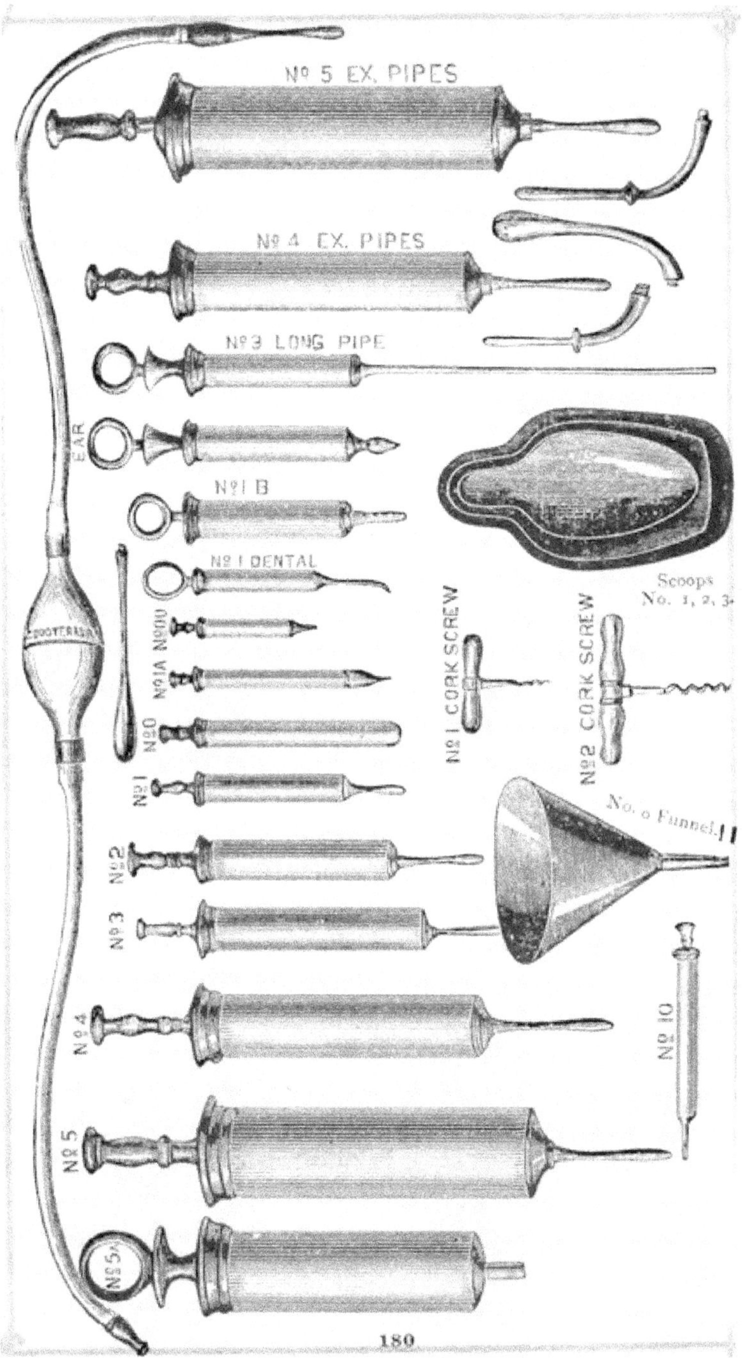
Nº 5 EX. PIPES
Nº 4 EX. PIPES
Nº3 LONG PIPE
EAR
Nº1 B
Nº 1 DENTAL
Nº00
Nº1A
Nº0
Nº1
Nº2
Nº3
Nº4
Nº5
Nº5A
Scoops
No. 1, 2, 3
Nº1 CORK SCREW
Nº2 CORK SCREW
No. o Funnel
Nº10

2 oz. Womb Syringes, with Bent Pipe	per doz.,	$1 95
3 " " " " "	"	2 45
4 " " " " "	"	3 15
Dentist "	"	88
Ear Syringes, bent or straight	"	88
Eye "	"	88

WHITE METAL SYRINGES.

8 oz., Self, in Wood Case, 2 Pipes	per doz.,	$ 9 63
6 " " " " "	"	7 88
4 " " " " "	"	6 12
24 " Single, in Boxes	"	15 00
16 " " "	"	10 00
12 " " "	"	9 00
10 " " "	"	7 88
8 " " "	"	6 00
6 " " "	"	4 25
4 " " "	"	3 50
2 " " "	"	2 50
1 " " "	"	2 20
P. P.	"	88
1 oz., Female	"	1 25
2 " "	"	1 75
4 " "	"	2 62
24 " Horse, in Wood Case	"	30 00
36 " " " "	"	36 00

HARD RUBBER SYRINGES.

No. 00, ⅛ oz.	per doz.,	$ 4 00
" 0, ⅜ "	"	5 50
" 1, ⅜ "	"	5 25
" 1a, ⅜ "	"	5 50
" 1b, ½ "	"	6 50
" 2, ¾ "	"	6 50
" 3, 1 "	"	7 50
" 4, 3 "	"	11 00
" 5, 6 "	"	15 00
" 6, 12 "	"	20 00
" 10, ⅜ "	"	3 00
" 30, 1 "	"	6 25
" 50, 6 "	"	12 00
" 4, Female, 3 oz.	"	13 00
" 5, " 6 "	"	17 00
" 4, " with Shield, 3 oz.	"	17 00
" 4, Male, Bent Pipe, 3 oz.	"	
" 5, " " " 6 "	"	
" 4, with extra Pipes, (Male and Female,) 3 oz.	"	16 00
" 5, 6 oz.	"	20 00
" 6, 12 "	"	24 00

DR. MOLESWORTH'S

Vaginal Injecting and Suction Syringe

Is Indorsed and Recommended by more than Nine-Tenths of all the Physicians who have examined it.

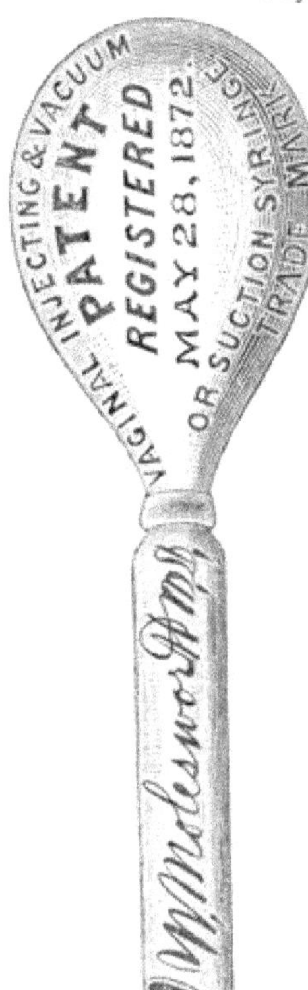

The Vaginal Injecting and Suction Syringe is the ONLY instrument that will carry a cleansing or medicated wash to EVERY and ALL parts of the vagina and mouth of the womb.

This is the ONLY instrument that dilates and detaches from the walls, and removes by suction, without injury, all impurities or foreign matter that may be upon the walls and in the folds of the mucous membrane of the vagina, or within the mouth of the womb, if open.

This is the ONLY instrument with which you can make a thorough application to the mouth of the womb and upper part of the vagina, without INVERTING your patient—this instrument making a perfect application either in a standing, sitting, or horizontal position.

DR. MOLESWORTH'S

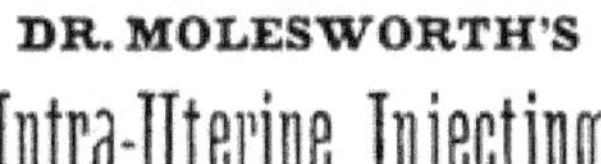

Intra-Uterine Injecting

—AND—

VACUUM SYRINGE.

Tube 10 Inches long.

As the tube of the Intra-Uterine Injecting and Vacuum Syringe is a double Canula, the operator, by closing the mouth of the escape tube, can retain the fluid within the cavity as long as the pressure on the bulb is continued, when by removing the pressure from the bulb, a suction is caused, which will draw all the fluid back into the bulb, or the escape pipe may be opened and allow it to escape through that, or that and the suction may both be used.

The tube being susceptible of different curvatures, this Syringe can be used as well and with equal facility for injecting the BLADDER and Urethra, both in the Male and Female, as for the uterine cavity.

As the bulb is readily detached from the tube, the bulb may be charged as often or as much fluid used as desired without removing the tube, or when a large quantity is desired, the ordinary Pump Syringe can be attached with ease.

For prices see page " Syringes."

The Trade supplied by

VAN SCHAACK, STEVENSON & REID,

92 and 94 Lake Street, cor. Dearborn, Chicago.

No. 1, Vaginal, ¼ oz.	per doz.,	$ 5 25
" 2, " 1 "	"	6 00
" 3, " 1½ "	"	8 00
" 2, " with Shield, 1 oz.	"	8 50
" 3, Vaginal, with Shield, 1½ oz.	"	11 00
Catarrh	"	9 00
Dental	"	6 00
Ear	"	8 50
Pump	"	24 00
Long Pipe, No. 3	"	9 25

ELASTIC BULB SYRINGES.

Van Schaack, Stevenson & Reid's Syringes are made of the best rubber, and are superior to any other low-priced Syringe in market.

American, 2 pipes	per doz.	$ 5 50
Davidson's, No. 1	"	17 00
" " 2	"	14 00
" " 3	"	16 50
" " 4, (Goodyear Patent), H. R. T.	"	22 50
Fountain Syringe, small, No. 1	"	17 00
" " medium, No. 2	"	19 00
" " large, No. 3	"	21 00
Mattson's Family, No. 1	"	16.50
" Diamond Package (do not make now.)		
" Original		8 50
" Vaginal Irrigator	"	18 00
New York, 2 pipes	"	[illegible]
Richardson's, No. 2, 3 pipe	"	7 50
" " 2, 2 "	"	7 00
" " 3,	"	13 50
Van Schaack, Stevenson & Reid's, 3 pipe	"	7 50

TEETHING RINGS.

Teething Rings	per doz.	$ 50

TEST TUBES.

Nested, from 3 to 6 inches	per doz.	$ 40
6 inch	"	50

THERMOMETERS.

6 inch, in Japanned Tin Cases, White Tube	per doz.	$3 00
7 " " " "	"	3 25
8 " " " "	"	3 50
10 " " " "	"	4 00
12 " " " "	"	5 00

The above with Ruby Tube, 50 cents per dozen extra.
See Kendall's, page 313

TWINE HOLDERS.---Glass [see p. 175.]

With Heavy Rim at Bottom.

Glass, plain	each,	$ 50 to 1 00
" engraved	"	1 00 to 1 50
" Bohemian	"	1 75 to 2 00

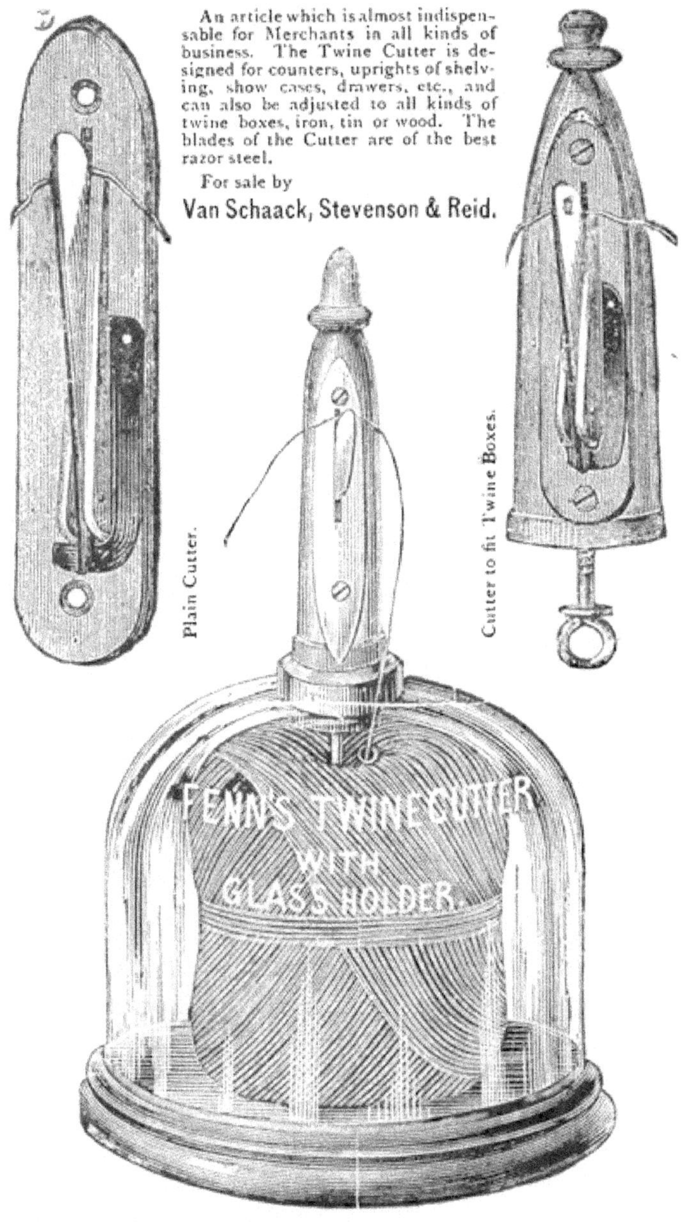
An article which is almost indispensable for Merchants in all kinds of business. The Twine Cutter is designed for counters, uprights of shelving, show cases, drawers, etc., and can also be adjusted to all kinds of twine boxes, iron, tin or wood. The blades of the Cutter are of the best razor steel.
For sale by
Van Schaack, Stevenson & Reid.
Plain Cutter.
Cutter to fit Twine Boxes.
FENN'S TWINE CUTTER
WITH
GLASS HOLDER.

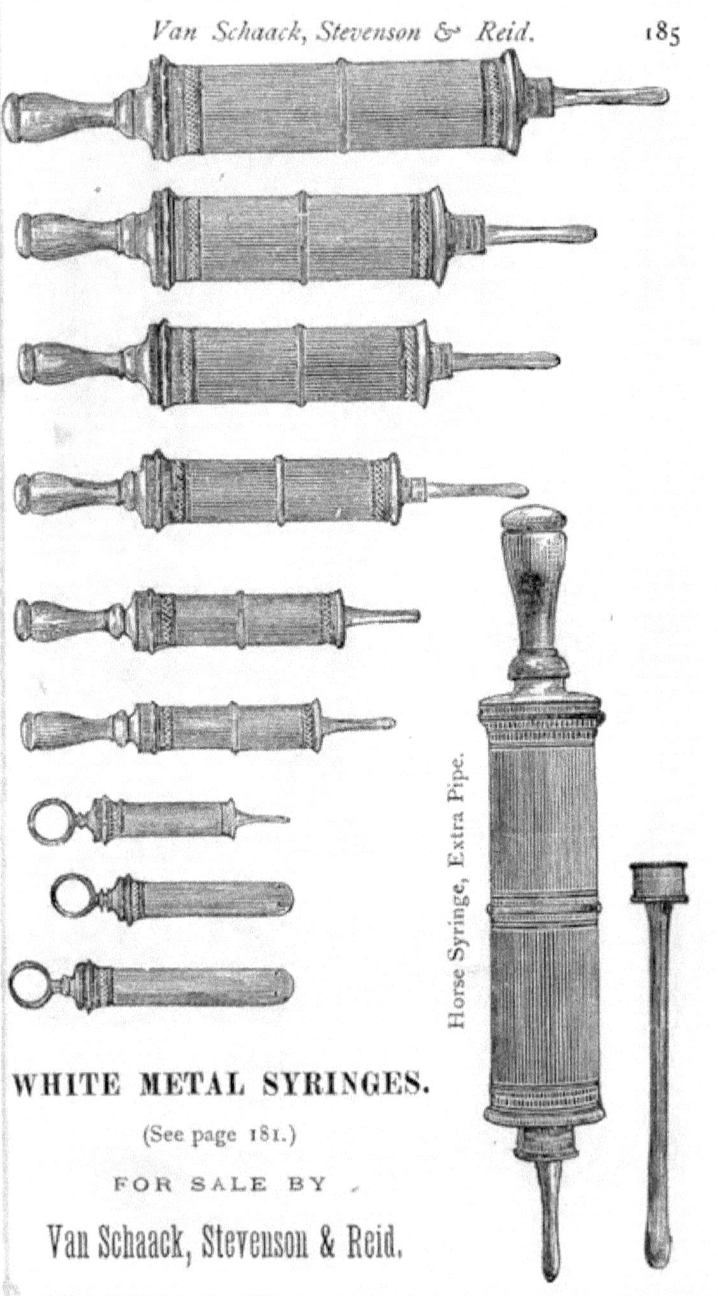

WHITE METAL SYRINGES.

(See page 181.)

FOR SALE BY

Van Schaack, Stevenson & Reid.

TINCTURE PRESSES.---Enameled Inside.

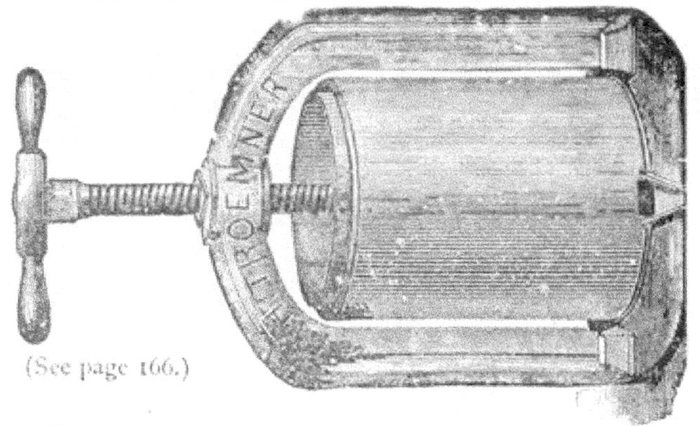

(See page 166.)

¼ gallon, $4 50 ; ½ gallon, $6 50 ; 1 gallon, $9 00.

URNS.---Counter, Plain.

½ gallon..........each, $1 00	2 gallon.... ......each,	$2 50
1 " .... " 1 50		

PAINTED AND LETTERED.

½ gallon..........each, $2 25	2 gallon ....each,	$4 00
1 " .... " 3 00		

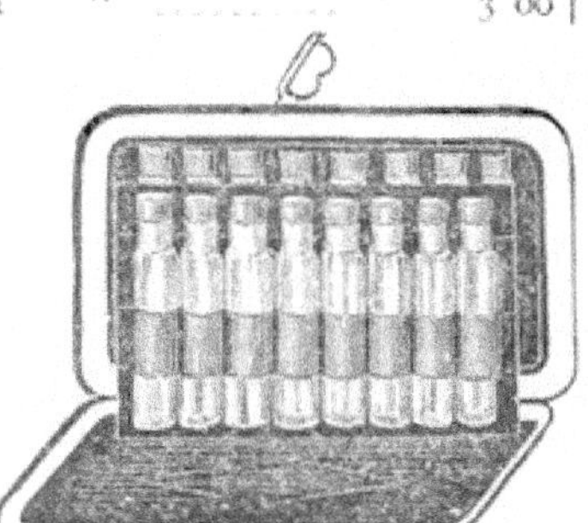

VIAL CASES.

FOR PHYSICIANS.

A great variety.

Per Doz.

Morocco, with Tuck,....
Russet Leather, with Elastic Bands.... ....
Calfskin, with Strap Bands
Turkey Morocco, Gilded, with Strap Bands.....
Morocco, Gilded, with Spring Tucks.........

WOULFE BOTTLES. [see page 152.]

With Three Necks.

1 pint.............each, $1 00	½ gallon........ each,	$2 50
1 quart. .... " 1 25	1 " .... "	3 00

WHITE EARTHEN URINALS.

Male.

Female.

(Bed Pans—" Slipper Shape "—new and desirable style—see page 151.

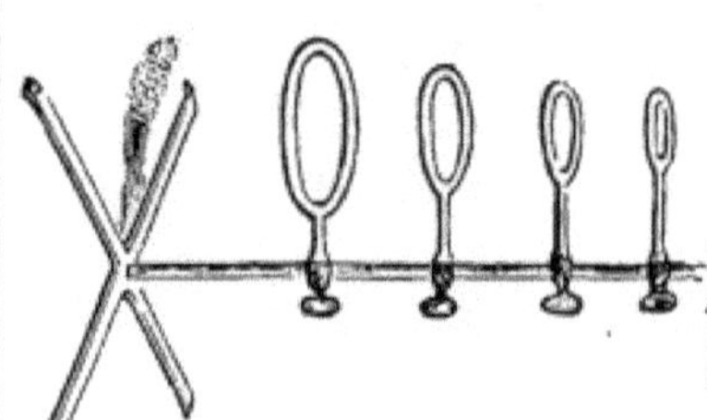

RETORT STAND.

New pattern. Not readily overturned. The arms of the base projecting to prevent it, *even* if loaded all on one side. All four rings can be used at same time.

MAW'S TINCTURE PRESS—(See page 166 & 186)

COUNTER SCALES.

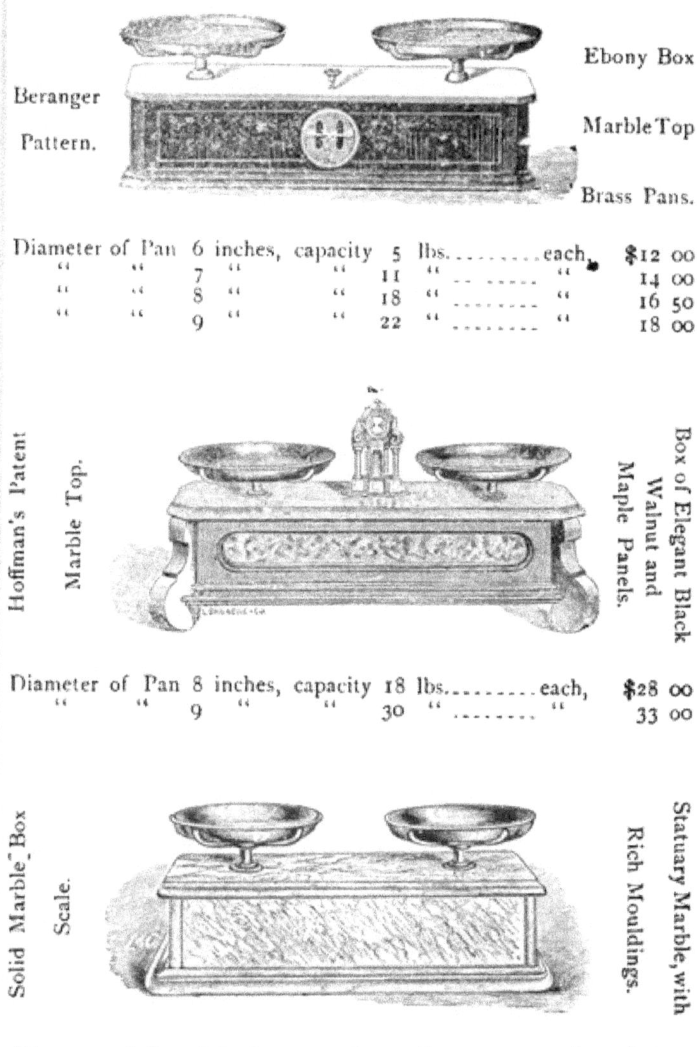

Beranger Pattern.

Ebony Box Marble Top Brass Pans.

Diameter of Pan 6 inches,	capacity 5 lbs.	..........each,	$12 00
" " 7 "	" 11 "	.... "	14 00
" " 8 "	" 18 "	.... "	16 50
" " 9 "	" 22 "	.... "	18 00

Hoffman's Patent Marble Top.

Box of Elegant Black Walnut and Maple Panels.

Diameter of Pan 8 inches,	capacity 18 lbs.	..........each,	$28 00
" " 9 "	" 30 "	.... "	33 00

Solid Marble Box Scale.

Statuary Marble, with Rich Mouldings.

Diameter of Pan 8 inches,	capacity 12 lbs.	..........each,	$37 00
" " 9 "	" 15 "	.... "	40 00

Robervahl Pattern.

Brass Pans.

Diameter of Pan	6	inches,	capacity	5	lbs	........each,	$	6 00
" "	8	"	"	10	"	.... "		8 00
" "	10	"	"	15	"	.... "		10 00

HOWE'S IMPROVED UNION SCALES.

With one Square and one Round Platform; for Bulky Goods, Putty, Pigments, etc. Capacity, ½ oz. to 240 lbs.

With Tin Scoop, single beam..............................	$14 00
" " double "	15 00
" Brass Scoop, single beam..............................	15 00
" " double "	16 00

PRESCRIPTION SCALE CASES.

No. 2.

No. 4.

No. 2 Oiled Walnut, Glass sides and front............each,	$10 00
No. 4 Oiled Walnut, Sliding Door.................... "	12 00

When boxed, 75 cents additional will be charged.

PRESCRIPTION SCALES.

FRENCH PATTERN MARBLE BASE.

No. 1, all Brass......each,	$10 00
" 1, all Silver...... "	14 00
" 1, all Brass ...... "	12 00
" 2, all Silver...... "	16 00

MARBLE BASE ON BOX, FRENCH PATTERN.

No. 1, all Brass.....each,	$12 00
" 1, all Silver Plated "	16 00
" 2, all Brass ...... "	14 00
" 2, all Silver Plated "	18 00
" 3, all Brass ...... "	16 00
" 3, all Silver Plated "	20 00

MAHOGANY BOX WITH WEIGHTS, ARMY PATTERN.

No, 1.....................each,	$5 50
No. 2..................... "	4 50

HAND SCALES.

Brass Pans, Morocco Box,...each,	$1 00
" " Tin "..... "	1 00
" " Wood " ... "	1 00
Glass " Morocco " ... "	1 50
Horn " " " ... "	1 25

SCALE PANS, ADJUSTED.

Glass..per pair, $ 75 | Horn..per pair, $ 1 00

WEIGHTS.

Block, Brass, 1 lb. to ¼ oz.	$3 50
" " 2 lbs. to ¼ oz.	5 00
" " 4 lbs. to ¼ oz.	8 00

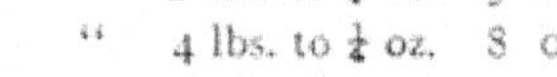

Cups. (Troy.) Brass. 16 ounces down	each.	$4 00
" " " 8 " "	"	3 00
" " " 4 " "	"	1 50
Nest, Brass, 4 lbs. to 1-2 oz.	each,	4 50
" " 2 " to 1-2 "	"	2 50
" " 1 " to 1-2 "	"	1 75
" Iron, 4 " to 1-2 "	"	1 50
" " 2 " to 1-2 "	"	1 00
Nest, Iron, 1 lb. to 1-2 oz.	"	75
" Zinc, 4 " to 1-2 "	"	3 00
" " 2 " to 1-2 "	"	2 50
" " 1 " to 1-2 "	"	1 50

PRESCRIPTION WEIGHTS.

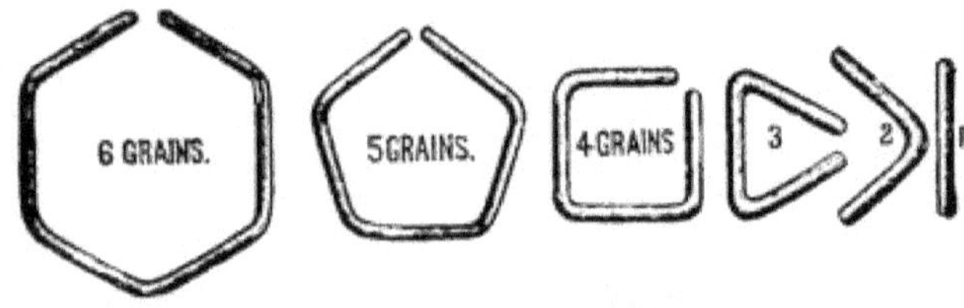

Aluminum Wire, Grains only	per set,	$ 50
Brass, Coin Drachms	"	35
" Square Drachms and Grains	"	25
" Grain Weights	"	15

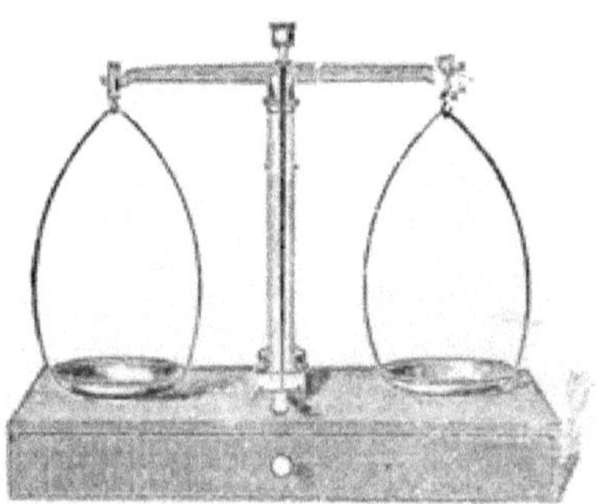

ANALYTICAL SCALES.

For weighing Ores, Gold and Silver, Chemicals, etc. On Polished Box. Improved Drop Lever Ivory Indicator, and sensitive to 1-10 grain.

No. 3, 3 inch Pan, 8 oz. capacity.......................... $15 00

NINE PRIZE MEDALS TAKEN.

HENRY TROEMNER'S

(710 Market Street, Philadelphia.)

Standard Scales and Weights.

ANALYTICAL BALANCES.

No. 1 Analytical Balance, capacity 200 grammes in each pan, in fine polished glass case; beam divided in 1-10 milligramme; sensitive to 1-20 milligramme; all agate bearings, with improved arrest for pans, and apparatus for specific gravity, etc., etc. 3 inch pans. Beam 14 inch. Price.....................$105 00

No. 2 Analytical Balance, in fine polished glass case, capacity 100 grammes in each pan; beam divided in half parts of milligrammes; sensitive to 1-10 milligramme; with apparatus for specific gravity, all bearings agate. 2¾ inch pans 12 inch beam. Price...$86 00

No. 3 Analytical Balance, in fine polished glass case, capacity 2000 grains; sensitive to 1-20 grain; fine steel bearings; movable 3⅛ inch pans. Beam 10 inch. Price $40 00

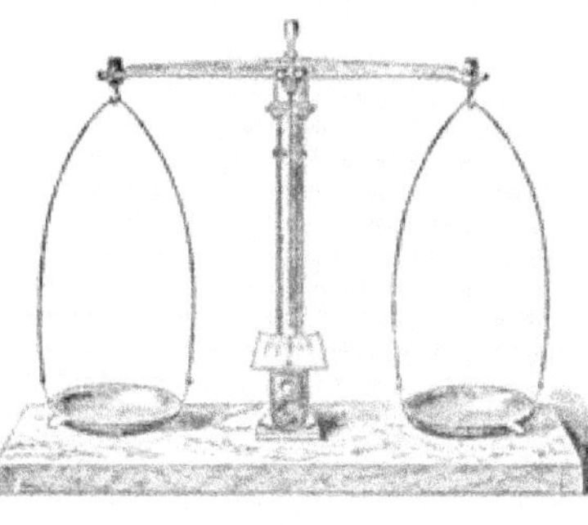

Fine Etched Beam, Brass lackered, and mounted on a Rich Marble Platform. Sensitive to 1-20 grain.

No. 1, 9 in. Beam, 3 in. Pans.

Price $20 00

GLASS BOX SCALE.

Diameter of Pan 9 inch, capacity 20 lbs. $115 00

BELGIAN BLACK MARBLE SCALE.

Very ELEGANT, GOLD LINES Gilt Ornaments.

Diameter of Pan 9 inch, capacity 15 lbs. $50 00

FRENCH WALNUT BOX SCALE.

No. 1, 8 in. Pan, capacity 15 lbs. $28 00

No. 2, 9 inch Pan, capacity 30 lbs. 33 00

EBONY BOX SCALE.

Fine Gilt Ornaments. Marble Top. Very Rich.

8 inch Pan, capacity 15 lbs. $40 00

The Trade supplied by **Van Schaack, Stevenson & Reid,**
Chicago, Ill.

ACKNOWLEDGMENT.

We avail ourselves of this opportunity—which the confusion arising from the *great fire* rendered it impossible for us to do in many instances—to perform the pleasing duty of returning our *most sincere* thanks to those kind friends, both at home and abroad, who upon the first announcement of the destruction of our warehouse, tendered their sincere sympathy, many of them in a most substantial manner—the promptness of our customers in their remittances enabling us to meet all our obligations 100 *cents on the dollar, and promptly at maturity.*

Truly yours,

Van Schaack, Stevenson & Reid.

VAN SCHAACK, STEVENSON & REID,

Submit to those about to embark in the Drug trade, or to refit their Stores, the annexed list, as a well adapted

List of Glass Labels suitable for a first-class Drug Store.

DRAWER LABELS.

Acid, Tart. P.
Acacia, Sec.
Allium.
Aloe, Capen.
Aloe, Socot.
Althæ.
Alumen.
Ammoniac.
Ammo. Mur.
Amylum.
Anisum.
Anthemis.
Ant. Sulph. Nig.
Arnica.
Assafœtida.
Aven. Far.
Benzoinum.
Buchu.
Calumba.
Carum.
Caryophyllus.
Cascarilla.
Catechu.
Cera Alba.
Cera Flava.
Cera Myrica.
Cetaceum.
Chondrus.
Cincho. Flav.
Cincho. Rub.
Curcuma, P.
Cinnamonum.
Colchici, Rad.
Colocynthis.
Coptis.
Coriandrum.
Cupri, Sulph.
Creta, Prepared.
Calamus.
Cypripedium.
Cetraria.
Cort. Auran.
Emp. Adhes.
Emp. Hydrag.
Emp. Plumbi.
Emp. Resina.
Emp. Diach. Comp.
Emp. Ichthy.
Euonymus.
Ext. Glycyr. Cal.
Ex. Glycyr. Sic.
Fœniculum.
Ferri Sulph.
Galla.
Gambogia.
Gentiana.
Glycyrrhiza.
Guaiaci, Lig.
Guaiaci, Res.
Hydras. Can.
Hordeum.
Humulus.
Iris, Flor.

VAN SCHAACK, STEVENSON & REID,

Ichthyocol.
Juniperus.
Krameria.
Lavendula.
Lini Farina.
Linum.
Linteum.
Magnes. Carb.
Magnes. Sulph.
Manna, L. F.
Manna, S. F.
Maranta, Am.
Maranta, Ber.
Myristica.
Myrrha.
Matricaria.
Opium.
Ossa Sepia.
Potass. Bi Tart.
Pareira.
Pimenta.
Piper, Nig.
Pix, Burgund.
Pix, Canaden.
Plumbi, Acet.
Panax.
Potass. Nit.
Prun. Virg.
Quassia.
Rad. Anchus.
Rad. Aconite.
Resina, Alb.
Resina, Com.
Rheum.
Sodæ, Sulph.
Sodæ, Carb.
Sodæ, Bi Carb.
Sodæ, Boras.
Sago.
Sanguinaria.
Santalum.
Sapo. Cast.
Sapo. Cast. Alb.
Sarsaparilla.
Sarsap. Cont.
Sassaf. Rad. C.
Scillæ.
Senega.
Senna, Alex.
Senna, Tin.
Senna, Ind.
Serpentaria.
Sinapis, Alb.
Sinapis, Nig.
Spigelia.
Sulph. in rolls.
Sulphur, Sub.
Sem. Fœnigræc, P.
Terebinthina.
Tapioca.
Tragacanth, Opt.
Tragacanth, Sec.
Ulmus.
Ulmus, Con.

WHOLESALE DRUGGISTS, CHICAGO.

Ulmus, Pul.
Uva Ursi.
Valeriana.
Zingiber, Jam.
Zingiber, Rad.
Zingiber, Rad. Pul.

MISCELLANEOUS.

Annato.
Argols.
Asphaltum.
Black, Ivory.
Black Drop.
Black Lead.
Brown, Spanish.
Blue, Prussian.
Blue, Ultra-Marine.
Blue, Celestial.
Chalk, White.
Chalk, Red.
Cubbear.
Corks.
Corks.
Corks.
Composition.

DYE WOODS.

Brazil Wood.
Camwood.
Emery Flour.
Emery.
Fustic.
Glue, White.
Glue, Common.
Gelatine.
Green, Paris.
Hypernic.
Indigo, Manilla.
Indigo, Madras.
Lac Dye.
Litharge.
Logwood.
Madder.
Mineral Paint.
Nicewood
Ochre, Yellow.
Peachwood.
Red Lead.
Red, India.
Red, Venetian.
Redwood.
Rose Pink.
Seidlitz Mixture.
Shellac.
Smalts, Blue.
Sienna, Raw, Pow'd.
Sienna, Bt.
Seeds, Hemp.
Seeds, Canary.
Stone, Pumice.
Stone, Pumice, Pow'd
Stone, Rotten.
Umber, Raw, Pow'd.
Umber, Bt.
Vermillion, Am.
Vermillion, Trieste.

DYE WOODS.

Verdigris.
Whiting.
White, Ext. Paris.
White, Cremnitz.
White Lead.
Yellow, Chrome.
Zinc, White.

ONE GALL. TINCTURES.

Alcohol.
Aqua Calcis.
Bay Rum.
Benzine, Pure.
Holl. Gin.
Spts. Frumenti.
Spts. Vini Gallici, Am.
Spts. Gallici, Fr.
Vin. Oporto.
Vin. Xeric.

HALF GALLON TINCTURES.

Aq. Camphoræ.
Aq. Menth. P.
Aq. Menth. V.
Aq. Ammo. F.
Aq. Ammo. F. F. F.
Aq. Rosa.
Aq. Dist.
Alcohol.
Alcohol, Dilut.
Æth. Sulph.
Æth. Chlo.
Liq. Calcis.
Lini. Sapo.
Spts. Eth. Nit.
Spts. Camph.
Spts. Laven. Co.
Spts. Limons.
Spts. Menth. Pip.
Spts. Myriæ.
Spts. Vini. Gal.
Syr. Simp.
Syr. Limonis.
Syr. Rhei Aromat.
Syr. Sarsap. Co.
Syr. Scillæ.
Syr. Zingib.
Syr. Rhei.
Syr. Senna.
Tr. Arnica.
Tr. Cincho.
Tr. Cincho. Co.
Tr. Colombæ.
Tr. Aurantii.
Tr. Genth. Comp.
Tr. Opii.
Tr. Frumenti.
Tr. Opii Comp.
Tr. Rhei.
Vin. Album.
Vin. Port. Bur.
Vin. Xeric.
Vin. Port. P. J.

QUART TINCTURES.

Ac. Acetic.
Ac. Pyrolig.
Alcohol.
Alc. Dil.
Æth. Sul.
Aq. Ammon. F.
Aq. Camph.
Aq. Cinnam.
Aq. Destil.
Aq. Menth. P.
Aq. Rosæ.
Bateman, D.
Cologn. Opt.
Cologn. Com.
Copaiba.
El. Calisay.
Ess. Menth. P.
Ex. Bayber. Fl.
Ex. Tarax. Fl.
Ex. Vanil. Fl.
Godfrey's Cordial.
Hair Tonic.
Lin. Sapo.
Lq. Plum. S. A. D.
Ol. Olive, Com.
Ol. Ricini.
Ol. Spicæ.
Ol. Terebin.
Mist. Glyc. Co.
Petrol. Bar.
Sp. Vi. Ga. Ar.
Sp. Vi. Ga. Lav.
Sp. Æth. Nit.
Sp. Æth. S. Co.
Sp. Fomen.
Sp. Lavend. C.
Syr. Ac. Cit.
Syr. Aurant.
Syr. Ipecac.
Syr. Prun. Virg.
Syr. Rhei. Ar.
Syr. Scill.
Syr. Scil. Co.
Syr. Senega.
Syr. Simp.
Tr. Hellebor.
Tr. Lobelia.
Tr. Verat. Vir.
Tr. Arnica.
Tr. Assafœt.
Tr. Camph.
Tr. Cap. Myr.
Tr. Cinch. Co.
Tr. Myrrhæ.
Tr. Opii.
Tr. Opii. Cam.
Tr. Valeri.
Tr. Zingib.
Vin. Antim.
Vin. Colch. R.

PINT TINCTURES.

Acet. Opii.
Ac. Nit. Dil.
Ac. Sulph. Ar.
Æth. Chlor.
Ant. Chl. Sol.
Aq. Ammon.
Bals. Sulph.
Canth. Col.
Cher. Pect.
Collodion.
Collyrium.
Chlorofor.
Creasotum.
Dal. Carmin.
Dew. Carmin.
El. Valer. Am.
El. Val. Stryc.
Ess. Bergam.
Ess. Cinnam.
Ess. Gaulther.
Ess. Limon.
Ess. Picis Lig.
Ess. Spruce.
Ex. Amygdal, F.
Ex. Buchu, Fl.
Ex. Cinch. Fl.
Ex. Chinoid, F.
Ex. Cubeb. Fl.
Ex. Ergot, Fl.
Ex. Parier, F.
Ex. Rosæ, Fl.
Ex. Sarsap. F.
Ex. Senna, F.
Ex. Valer. Fl.
F. E. Podophyl.
Glycerin. C.
Glycerin. O.
Lig. Blue.
Lin. Camph.
Lin. Canthar.
Lin. Chlorof.
Lq. Fer. Subsul.
Lq. Plumb. S. A.
Lq. Potassa.
Lq. Pot. Arsen.
Lq. Sod. Chl.
Mist. Creta.
Mist. Cret. Co.
Ol. Absinth.
Ol. Anisi.
Ol. Amygdal.
Ol. Bergam.
Ol. Camph.
Ol. Caryoph.
Ol. Cedar.
Ol. Chenopod.
Ol. Cinnam.
Ol. Conii.
Ol. Gaulther.
Ol. Hedonia.
Ol. Jessam.
Ol. Juniper.
Ol. Lavend.
Ol. Limon.
Ol. Menth. P.
Ol. Monard.
Ol. Olive, Opt.
Ol. Origan.
Ol. Piper, Nig.
Ol. Rosemar.
Ol. Sabinæ.
Ol. Sassaf.
Ol. Stone.
Ol. Spruce.
Ol. Tanacet.
Ol. Tiglii.
Sp. Ammon. Ar.
Syr. Fer. Iod.
Syr. Rhei.
Terebin. Can.

PINT TINCTURES.

Terebin. Ven.
Toothache D.
Tr. Aconit. F.
Tr. Bayber.
Tr. Belladon.
Tr. Benz. Co.
Tr. Canthar.
Tr. Capsici.
Tr. Car. Co.
Tr. Catechu.
Tr. Cimicif.
Tr. Cinnam.
Tr. Cinch. Fer.
Tr. Colomba.
Tr. Conii.
Tr. Cubeba.
Tr. Curcuma.
Tr. Digital.
Tr. Ergotæ.
Tr. Ferri Chl.
Tr. Gallæ.
Tr. Guiac. Am.
Tr. Hyosciam.
Tr. Iodine.
Tr. Ioden. Co.
Tr. Kameria.
Tr. Nuc. Vom.
Tr. Opii. Acet.
Tr. Quassia.
Tr. Rhei.
Tr. Rhe. et Sen.
Tr. Sanguinar.
Tr. Scillæ.
Tr. Scutellar.
Tr. Serpentar.
Tr. Stramonii.
Tr. Tolutan.
Tr. Valer. Am.
Vin. Cloch. S.
Vin. Ergotæ.
Vin. Spruce.

ONE GALL. SALTMOUTHS.

Anthemis.
Avenæ, Farina.
Camphora.
Carbo Ligni.
Caryophylus.
Cinnamon, P.
P. Zingiber.
P. Zingiber, J.
Pimenta.
Piper, Pulv.
Sassafras, R. C.
Sinapis, P.
Sago, Perlat.
Tapioca.

HALF GALL. SALTMOUTHS.

Acacia, Opt.
Ammonia, Carb.
Acid, Citric.
Acid, Tart.
Alum, Pulv.
Ammon. Mur. P.
Anisum.
Capsicum, Afr.
Capsicum, Am.
Carum.
Caroph. P.
Coriandrum.
Diosma.
Fœnugreek, P.
Ferri, Sub. Carb.
Fœniculum.
Gentiana.
Juniperus.
Lavendula.
Macis.
Magnesia, Calc.
Marrubium.
P. Pimenta.
P. Pot. Nit.
Pruni. Virg.
Quassia.
Sal. Rochel.
Sod. et Pot. Tart.
Sulph. Præx.
Soda, Bi Carb.
Tragacanth.
Uva Ursi.
Valeriana.
Cantharides.
Cardamon.
Cinnamon, P.
Rhei. P.
Canella, P.

QUART SALTMOUTHS.

Absinthium.
Acacia, Pulv.
Ac. Benzoic.
Ac. Oxalic.
Aloes, Pulv.
Anchus. Tinc.
Antim. Sul.
Ant. Sul. Præc.
Aralia. Nud.
Arctium Lap.
Bism. Subcar.
Calc. Carb. Præc.
Canth. Pulv.
Cataria.
Century.
Cetaceum.
Cinch. Rub. P.
Chenopod.
Coccus.
Coloc. Pulp. P.
Colomba.
Coptis.
Cornus, Flor.
Creta. Prep.
Cubeb. Pulv.
Curcuma. P.
Digitalis.
Dentifrice.
Dracontium.
Dulcamara.
Eupatorium.
El. Coloc. Co.
El. Glycyr.
Fer. Ferroc.
Gentianæ, P.
Ground Ivy.
Guaiac, Res.

VAN SCHAACK, STEVENSON & REID,

WHOLESALE DRUGGISTS, CHICAGO.

Hedeoma.
Hepatica.
Hydrastis.
Hyssopus.
Inula.
Inula. Pulv.
Ipecac. Pulv.
Iris Flor. Pulv.
Jalap, Pulv.
Melissa.
Plumb. Acet.
Popul. Trem.
Prinos.
Rhei, Pulv.
Rumex, Cris.
Ruta.
Salvia.
Sambucus.
Santalum.
Sapo. C. Pulv.
Sod. Boras. P.
Sulph. Subl.
Sum. Savory.
Swt. Marjor.

PINT SALTMOUTHS.

Ac. Citric.
Alv. et Can. P.
Ammon. Carb.
Ant. et Pot. Tart.
Arum.
Benzoinum.
Bism. Subnit.
Bole, Armen.
Canel. Pubo.
Cardamon.
Cascarill.
Cabaltum.
Catechu, P.
Colomba, P.
Comp. Pulv.
Comfrey.
Crocus.
Cubeba.
Cupre. Sul. P.
Cydonium.
Dau Carota.
Dextrine.
Ergota, Con.
Gallar Pulv.
Glycyrrh. P.
Gr. Paradis.
Guaiac, Res.
Hyd. Chil. Mit.
Hydrastis.
Lobel. Fol. P.
Myrrhæ.
Myrrhæ, P.
Opii Pulv.
Potas. Acet.
Pot. Bicarb.
Pot. Bromide.
Potas. Chl.
Potas. Cit.
Potas. Ferroc.
Potas. Iod.
Potas. Nit. P.
P. Ipec. et Op.

PINT SALTMOUTHS.

Sacharum, P.
Sal. Prunel.
Sanguinar.
Sanguinar, P.
Sassaf. Med.
Scammon. P.
Senna, P.
Serpentar. P.
Spigelia,
Valer. Pulv.
Sod. Bisul.
Xanthoxyl.
Zedoary, P.
Zinci Acet. P.
Zinci. Oxid.
Zinzib. Con.
Ac. Gallic.
Alum, Exsic.
Ammoniac.
Argen. Nit.
Calc. Phos.
Ergotæ.
Gallac.
Gambogia.
Hydrarg. Am.
Hyd. Chl. Cor.
Kino.
Nucis Vom.
Pot. Bichro.
Pot. Caust.
Sabinæ, P.
Zinci. Carb.

2 LB. OINTMENT JARS.

Cerat. Resina.
Cerat. Resina, C.
Cerat. Cantharides.
Cerat. Cetacei.
Cerat. Plumb. S. Acet.
Cer. Simplex.
Ung. Aq. Rosa.
Ung. Hydrarg.
Ung. Hydrarg. A.
Ung. Hydrarg. Rub.
Ung. Hydrarg. Nit.
Ung. Picis Liquida.

1 LB. OINTMENT JARS.

Cerat. Saponis.
Cerat. Zinci. Carb.
Cerat. Sabina.
Cerat. Adipis.
Ung. Belladonna.
Ung. Benzoin.
Ung. Gallæ.
Ung. Plumb. Carb.
Ung. Pot. Iodid.
Ung. Stramoni.
Ung. Sulphuris.
Ung. Veratria.

FOUR OZ. EXTRACT JARS.

Ext. Belladonna.
Ext. Juglandis.
Ext. Taraxaci.
Ext. Gentian.
Ext. Jalapa.
Ext. Hæmatox.
Ext. Canabis Pur.
Ext. Cinchona.
Ext. Colchici.
Ext. Colocynth.
Ext. Colocynth. Comp.
Ext. Conii.
Ext. Digitalis.
Ext. Dulcamara.
Ext. Hellebori.
Ext. Hyosciam.
Ext. Ignat. Am.
Ext. Krameria.
Ext. Nucis Vomica.
Ext. Opii.
Ext. Podophyl.
Ext. Quassia.
Ext. Rhei.
Ext. Senega.
Ext. Stramonii.
Ext. Valerian.

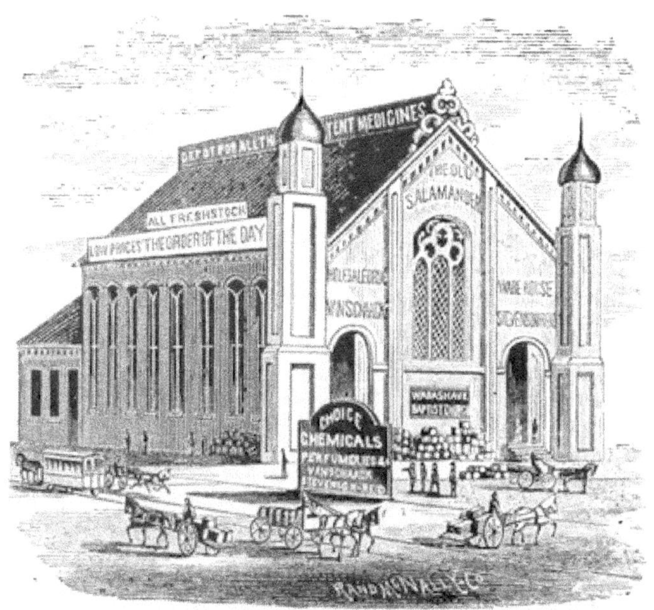

Van S., S. & R.'s Headquarters after the Great Fire.

EDITORIAL FROM

SCHOOLEY'S CHICAGO DRUG REPORTER.

THE OLD SALAMANDER

Has become "familiar as a household word" to the Drug Merchants of the Northwest. This title is most appropriately attached to the Wholesale Drug House of Messrs. VAN SCHAACK, STEVENSON & REID, who were the leading Drug House burned out in our memorable fire, and were the

FIRST WHOLESALE DRUG HOUSE

RE-ESTABLISHED ON THEIR OLD SITE IN THE BURNT DISTRICT.

This is of itself no unmeaning evidence of pluck and enterprise, and when we consider the extent and elegance of their present stores, the credit due for the enterprise is greatly enhanced. Less than eight months after "the great fire," (having occupied the Wabash Avenue Baptist Church in the meantime) we find this firm fully re-established *on the old site*, with a larger stock and more commodious building than before.

We chanced to visit Messrs. Van S., S. & R.'s capacious warehouse a few days since, and our call was both pleasant and instructive. The building, *erected expressly for the business*, has all the conveniences that long experience can suggest—heated by steam, and handling goods by steam power. The stores, 92 & 94 Lake Street, corner of Dearborn Street, are of stone, fifty feet wide, and including the basement, occupy five floors admirably arranged, the two stores being thrown into one, supported by iron columns, and present, altogether, the most spacious and elegant appearance of any Drug Store we have seen in this country or Europe. No business seems to require a longer apprenticeship or more extensive information, than that absolutely essential to the *successful* Jobber of Medicines. The whole habitable globe seems to have paid tribute in making up the variety of this Drug Stock, including most of the medicinal combinations known to the Pharmacopœia.

Each department is under the supervision of practical heads, which no doubt largely accounts for the order and perfect working of all. We rejoiced to look over the *foreign* invoices, showing that Messrs. Van Schaack, Stevenson & Reid were fully alive to the benefits to be derived by Chicago Merchants importing goods direct in bond to this Port, and thus saving at least *two profits*, which they can divide with their customers. We saw their invoices from London, Paris, Messina, Germany, etc., for goods lately arrived and in transit.

Messrs. Van S., S. & R. have for a number of years issued a wondrously complete Druggist Illustrated Price Current, in book form, they being the originators of thus posting up the Drug Trade of the Northwest, and we are glad to hear through numbers of the trade throughout the country, that they consider the Book issued by Messrs. Van S., S. & R., *the most complete, the most easy of reference, the most fully illustrated, and the most reliable of any published.*

A peculiarity of the management of the business of this firm, is that they employ no *drummers* or "tourists," but rely upon close prices and first-class goods. We can easily see how advantageous this plan can be made to the Merchant, both buyer and seller, and surely in the case of Messrs. Van S., S. & R., it has been a great success for years, and we commend an investigation to all our Merchants of

"New Chicago, the glory of the Northwest."

The trade of Messrs. Van Schaack, Stevenson & Reid is immense, and a *growing* one. Handling goods only of the most unquestioned genuineness, has insured the confidence of the trade from Ohio to Nevada, and the pre-eminent success to which they have attained is attributable to the confidence of the trade that their goods are always as represented, and the natural result of inflexible integrity, energy, ability, and a thorough knowledge of their business.

For the Convenience of Dealers

REQUIRING

Show Colors for Window Globes,

Van Schaack, Stevenson & Reid

SUBMIT THE FOLLOWING:

BLUE.

1.—Take of Sulphate of Copper, 1 ounce; Solution of Ammonia, half pint; Water, a gallon. Dissolve the Copper salt in a pint of water, filter, add the Ammonia, add afterwards the rest of the water. This Solution is of a beautiful dark shade. A Light Blue Solution is obtained as follows:

2.—Take of Sulphate of Copper, 8 ounces; Water, a gallon; dissolve and filter. The addition of a pint of Solution of Acetate of Ammonia renders it more decidedly blue, and similar to the next.

3.—Take of Carbonate of Copper, 6 ounces; Nitric Acid, 12 fluid ounces; Water, a gallon. Dissolve the Copper salt in the Acid, and add the water.

GREEN.

1.—By mixing Solutions of *Tersulphate of Iron* and *Sulphate of Copper*, various shades of Green can be obtained.

2.—A beautiful Emerald Green is obtained by adding to a Solution of *Nitrate of Copper*, a little of a Solution of *Bichromate of Potassa*: (too much of the latter renders it brownish green.)

3.—A similar color of a much darker shade is made as follows:—Take of Sulphate of Copper, 10 drachms; Bichromate of Potassa, 6 drachms; Solution of Ammonia, half pint; Water, a gallon. Dissolve the salts separately in half a pint of water, mix the filtered Solutions, add the Ammonia, and when the precipitate has been dissolved, add the water.

4.—Take of Carbonate of Copper, 8 ounces; Muriatic Acid, 2 pints; Water, a gallon. Dissolve the Copper salt in the Acid, and add the water. This Solution has a splendid light green color.

5.—Take of Sulphate of Nickel, 8 ounces; Water, a gallon. Dissolve and filter. This Solution is scarcely superior to No. 4, and is more costly.

VAN SCHAACK, STEVENSON & REID.

PURPLE.

The cheapest and best is made by dissolving pure Crystalized Permanganate of Potassa in distilled Water, in the proportion of 1-4 to 1-2 grain to the fluid ounce. Avoid dust and all organic substances while preparing it.

RED.

1.—Take of Bichromate of Potassa, 4 ounces; Water, a gallon. Dissolve and filter. This Solution has a yellow tinge; the following is more of a brownish red color:

2.—Take of Iodide of Potassium, ½ ounce; Iodine, 2 drachms; Water, a gallon. Dissolve the Iodide and Iodine in 1 ounce of water, then mix the Solution with the water.

3.—Take of Cochineal, 1 ounce; diluted Sulphuric Acid, 1 drachm; Water, a gallon. Powder the Cochineal, triturate it successively with the water in several portions, filter and add the Acid.

4.—Take of pure Carbonate of Cobalt, 8 ounces; pure Muriatic Acid, 1 pint; distilled Water, 1 pint. Mix the Acid with the water, dissolve it in the Cobalt, heat to boiling, and filter. This is of a very fine rose red color, and may be considerably diluted for large globes.

YELLOW.

1.—Take of Bichromate of Potassa, 3 ounces; Carbonate of Soda, 3 ounces; Water, a gallon. Dissolve the Bichromate of Potassa in half a gallon, and the Soda in a pint of water; pour the latter Solution slowly into the former, and when the evolution of Carbonic Acid ceases, add the rest of the water.

2.—Tincture of Chloride of Iron, 1 pint; Water, 7 pints. Mix them.

3.—Solution of Tersulphate of Iron, 8 fluid ounces; Water, 7½ pints. Mix them. This Solution has a brownish red tinge; by the addition of a little Sulphuric Acid, the color becomes of a purer yellow. Too much Acid renders it pale and nearly colorless.

For the convenience of our friends and customers intending to re-label their shop furniture with glass labels, we submit the following

Directions for Making the Cement:

Take one-third Beeswax to two-thirds Rosin, heat in an iron kettle until it is well dissolved.

DIRECTIONS FOR ATTACHING GLASS LABELS.

Melt the cement in an open pan, with a gentle heat, to the consistency of syrup; pour it on the back of the label with a spoon, then apply the label to the bottle with a gentle pressure. The bottle must be kept in a horizontal position, and laid away in the same position for about three hours, in order to allow the cement ample time to harden. After the cement is hard, cut around edges of the label with a square-pointed putty knife, scraping off as much as possible with the knife, then clear off the residue with a cloth saturated in benzine or turpentine. Care must be taken not to bend the label out of its original shape; the ends must not be forced down, but an equal pressure applied to all parts of the label. If the labels are bent out of their original shape, they will be liable to crack when the cement gets hard. In order to ensure uniformity in putting on the label, a gauge must be made of wood, for each size of bottle, to measure from the bottom of the bottle to the bottom of the label. Parties ordering labels without bottles, must, in all cases, give circumference of a medium size for each side of the bottle to be labeled; this can be done by putting a strip of paper around the parts of the bottle where the label is to be attached. The circumference of a medium bottle, of each size, will enable us to supply the labels of the proper bend.

Care should be taken to have the bottle dry and free from grease.

A piece of thin board, or wooden paddle, is the safest thing to use in putting on the Cement.

LABEL BOOKS, SPONGES AND CHAMOIS.

Philadelphia College of Pharmacy Labels for Shop Furniture, revised edition, executed in bronze	$15 00
Same as above. Yellow letters, black ground work	5 00
Physicians' Yellow Label, small size	1 00
Abridged edition of Latin Labels, executed in bronze, containing about 600 labels	7 50
Smaller edition	3 00
Physicians' edition	2 50

SPONGES.

Bath, Mediterranean.		
Large, from 3 to 8 pieces to lb. extra	per lb.	3 00
Small, " 5 to 12 " " ordinary	"	3 00
Small string of 25 pieces, all forms	per string,	3 50
Florida Sheep's Wool.		
Large, from 1 to 3 pieces to lb.	per lb.	2 00
Medium, from 5 to 8 pieces to lb.	"	2 00
Small, from 10 to 16 pieces to lb.	"	1 75
Reef, or Slate Sponges, (finest quality).		
Large, from 30 to 40 pieces to lb.	"	1 50
Small, from 50 to 100 pieces to lb.	"	1 50
Surgeons' Sponges, fine.		
Loose, large, 5 to 15 pieces to lb	"	6 00
" small, 15 to 30 pieces to lb.	"	3 50
Strings, small, about 45 pieces each	per string,	3 00
Zimoca Sponges, (for cleaning cloth, &c.)		
Large, 15 to 20 pieces to lb.	per lb.	2 50
Small, 20 to 30 " "	"	2 50
Thin, flat, for Potters' use	"	5 50

CHAMOIS SKINS.

No. 1. per kip of 30 pieces	per doz.	10 00
A. " " "	"	7 00
B. " " "	"	6 50
C. " " "	"	6 00
D. " " "	"	5 00
E. " " "	"	4 50
F. " " "	"	4 00
G. " " "	"	3 50
Plaster Skins	each,	75
Split " for capping bottles	"	75

ANGUSTURA.
VINUM ERGOTÆ
EMP. ADHÆS.
ACIDUM HYDROCYAN: D:
AMMONIÆ NIT:
ZINCI CARB. PRÆ.
ANISUM.
3d Series.
ACIDUM TANNICUM.
PULVIS POTAS: SULPH:
SPIRIT: LAVAN: COMP:
OL: ORIGANI.
TINCT: FERRI CHLOR:

LABELS

of every description kept in stock & supplied at Manufacturers price by

VAN SHAACK, STEVENSON & REID,

WHOLESALE DRUGGISTS, CHICAGO.

A new Book of Druggists furniture Labels, contains 700 imitation Glass Labels

PRICE $ 18.00

PERFUMERY.

HAIR OILS, SOAPS, &c.

DRUGGISTS' SUNDRIES.

No. 5 Atomizer—Spray.

Per Doz.

Amandaline,

(Mann & Co.'s)..........$1 75

Amandine, (Bazin's) 3 38

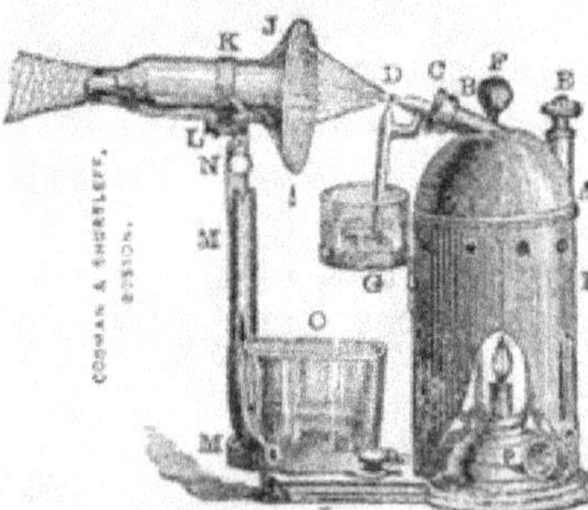

No. 15, the Complete Steam Atomizer for Inhalation, etc.

Codman & Shurtliff's, No. 15, Steam.... 5 00

Aurilaves, Ear Cleaners.

Aurilaves, Lovell's Patent; bone handles	Per Doz.,	$1 75
Improved Ear Cleaner; plated wire handles........	"	75
" " " and Ear Spoon combined..	"	1 50

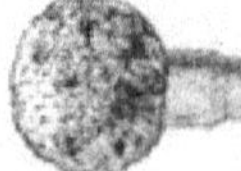

No. 56 Atomizer.

Atomizers, Medicinal.

	Each.
Codman & Shurtliff's, No. 5, Spray	$ 3 50

Atomizers, Perfumers

	Per Doz.
Boston	$12 00
Maw's 6d.	1 75
Silver Spray	12 00
Universal, No. 56	12 00
Essex (see page 232)	13 50

Boston Atomizer.

Bandoline.

Bazin's, Small	Per Doz.,	$1 50
Coudray's, Small	"	1 75
" Large	"	3 00

Sponge Baskets.

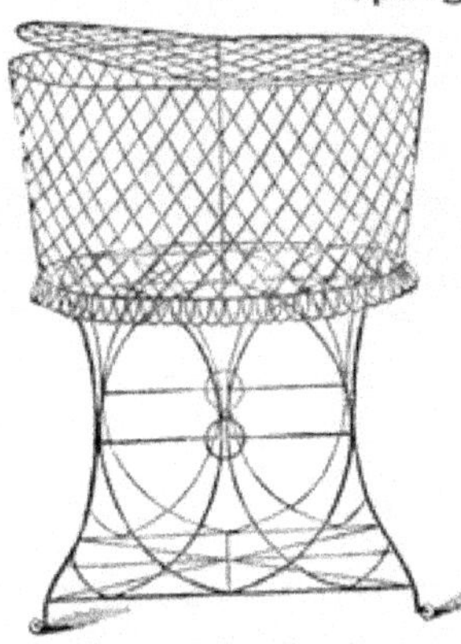

Baskets, Comb and Tooth Brush.

	Per Doz.
No. 1, oval	$2 75
" 2, square	3 00
" 3, "	5 50
" 4, "	8 00
" 5, large fancy, for show case	

Bougies and Catheters. The cut below shows sizes.

(See Instrument List, page ——.)

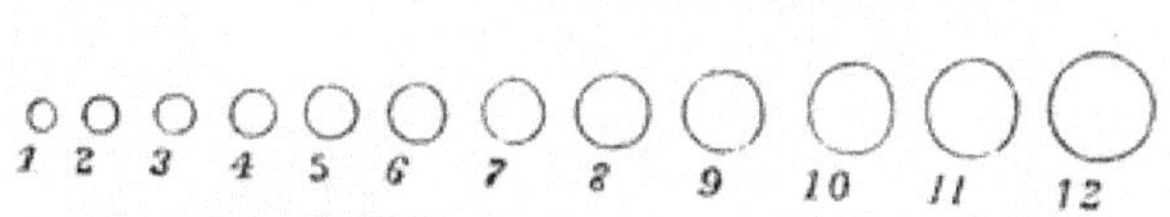

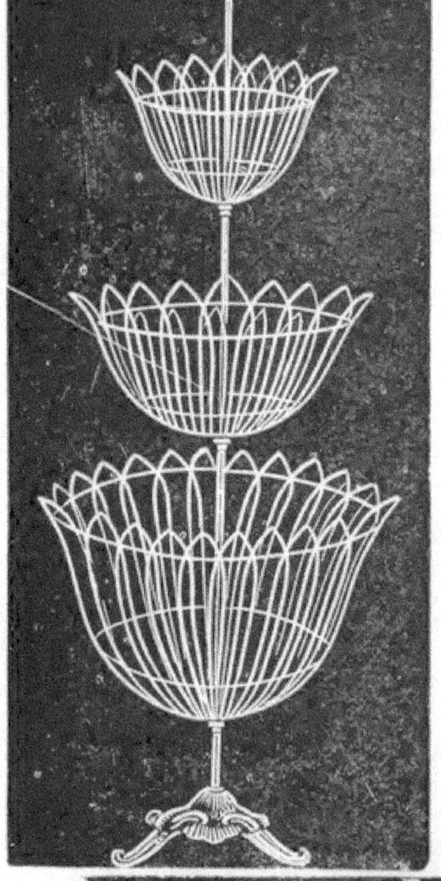

Sponge Baskets.

WIRE.

Green and Bronze.

STYLE No. 1.

"3 Story."

Holding 2 Bushels.

STYLE No. 2.

"3 Story."

Holding 1 1-2 Bushels.

The Trade supplied by

Van Schaack, Stevenson & Reid,

92 and 94 Lake Street, cor. Dearborn, Chicago.

URN STYLE

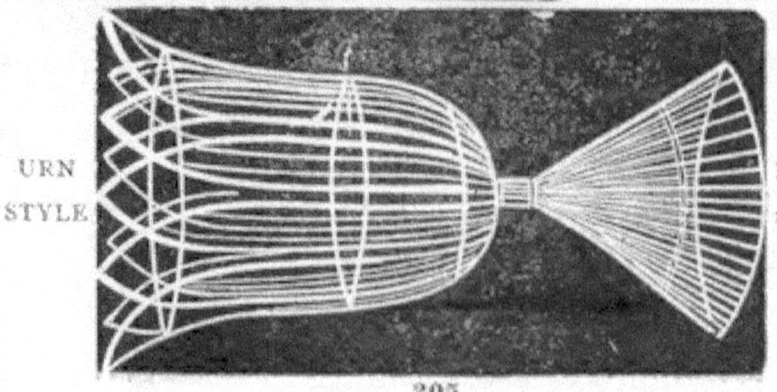

Holding 1 Bush.

Bath Belts.

Horse Hair, Gent's	Per Doz.,	$21 00
" Ladies'	"	21 00

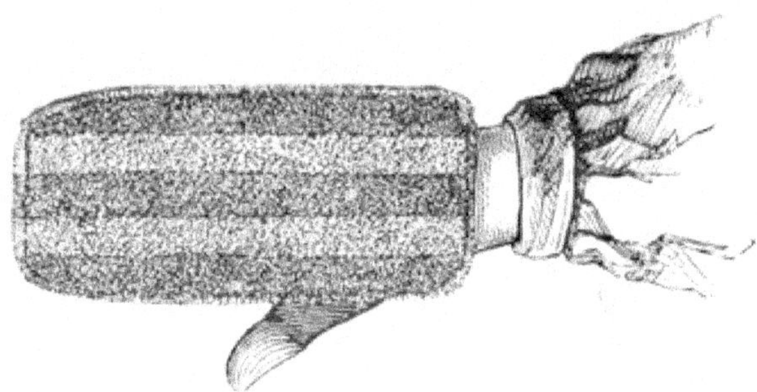

Bath Gloves.

Goat Hair, Friction	Per Doz.,	$13 00
Horse Hair, Friction	"	13 00
" Gent's	"	21 00
" Ladies'	"	21 00
Turkish Brown	"	4 50
" White	"	3 00

Blacking.

	Per Doz.
Army and Navy, Large	$ 75
" Small	45
Brown's French Dressing	1 75
" Liquid Bronze	2 25
Day & Martin's, Large	4 00
" Medium	3 00
" Small	1 75
Mason's (see page 63)	

	Per Doz.
Miller's Harness Oil No. 1,	$4 25
" " " 2,	6 25
" Polish, " 1,	35
" " " 3,	55
" " " 4,	70
" Water-proof, " 2,	1 75
French, Marcerou's	1 50
" Bresson's	1 00

Blanc de Pearl.

	Per Doz.
Coudray's Liquid	$6 00
Dorin's Liquid	3 50
" Tablet, No. 4	2 50

	Per Doz.
Lubin's Liquid	$9 00
" Powder	8 00

Boxes, Powder Puff.

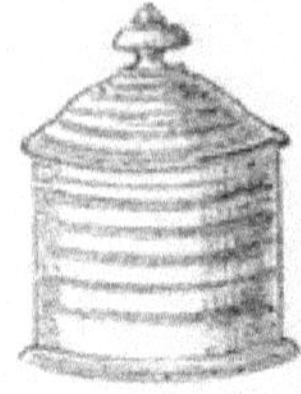

						Per Doz.
Metal, As'ted Styles,	$4 00	$5 00	$6 00	$7 00	$9 00	$12 00
Paper, "			1 00	1 75	2 00	2 50
Relic Wood, "					7 50	10 50
Scotch, " "				7 50	9 00	10 50

Boxes, Soap.

Britannia, Large, Per Doz., $7 50 | Britannia, Small, Per Doz., $6 00
Plated, " 13 50 | For Brown Windsor Soap.

Boxes, Shaving.

Wood, with Mirror, (tin cup) .... Per Doz., $1 50
" " (china cup) .... " 1 75

Brooms, Whisk.

Per Doz.
Ordinary, 1 75

Broom Corn, Large, (Barber's,) $2 50
Broom Corn, Medium Size, Select, $2 00
Broom Corn, Small, Pock't, Select $2 00
Velvet-covered, Large 3 50
" Small 3 00
Horse Hair, (for velvet) 6 50
Tampico, " 3 00

Brushes, Cloth, French Osier.

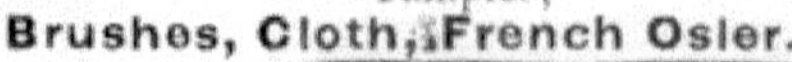

Velvet Back .... Per Doz., $6 00
" handles large .... " 7 00
" " small .... " 5 50
Wood Back .... " 6 00
" fancy stock .... " 7 50
" with handles .... " 7 00

Brushes, Cloth.

No. 1,	small plain Roach,	Dark Backs, White Stock,	Per Doz.,	$ 2 00
" 3,	medium " "		"	3 00
" 5,	large " "		"	3 50
" 154,	small fancy " very stiff	Black Centres	"	4 50
" 155,	large " " "		"	5 00
" 250,	Cedar Back fine Roach, all Black Stock		"	6 50
" 144,	small Black Walnut Back	Quality, White Stock Improved.	"	6 00
" 166,	medium " "		"	7 00
" 188,	large " "		"	8 00
" 122,	small Scalloped Backs		"	5 00
" 101,	medium	Concave Backs, highly polished	"	10 50
" 103,	large	Rosewo'd, fine long white bristle	"	15 00
" 100,	Rosewood Back and White Stock		"	10 50
" 105,	Satinwood " " Black "		"	10 50
" 60,	Florence Manufacturing Co.'s		"	12 50
" 62,	" " "		"	9 00

Brushes, Cloth, with Handles.

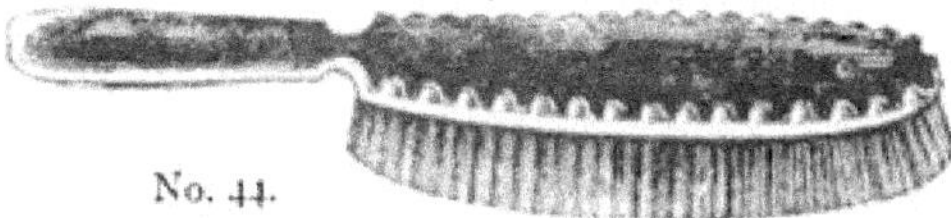

No. 44.

		Price
No. 10, Straight Handle	Per Doz.,	$ 3 50
" 20, Curved Handle	"	4 50
" 44, see Illustration	"	6 00
" 31, Cedar Back, Beveled Handle, for fine cloths or velvets,	"	12 00
Lorenzo's Solid Back	Per Doz., $18.00 and	24 00
No. 24, Cloth and Hat combined	Per Doz.,	12 00

Brushes, Bottle.

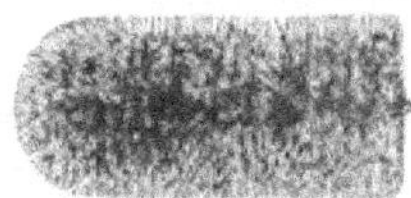

Maw's Bottle Brush.

		Price
Maw's Nursing Bottle Brushes	Per Doz.,	$ 75
" " " Tube Brushes	"	25
" Registered Tube Cleaners	"	40
Druggists' Bottle Brushes		

Brushes, Bath and Flesh.

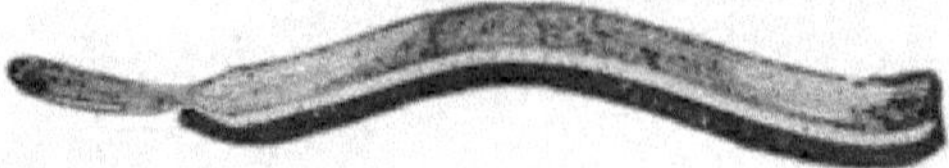

Curved Flesh Brushes, large	Per Doz.,	$12 00
" " " small, for ladies' use	"	9 00

Oval Flesh, (see Illustration)	Per Doz.,	$ 6 00
" " short Handle	"	8 00
Oblong Bath, durable	"	8 00
" " double	"	16 50
New England Bath, No. 1	"	9 00
" " No. 2	"	10 50
Brush and Sponge combined	"	12 00
Turkish Bath, no Handles	"	7 00
Panstrepton, with Handles	"	16 50

Brushes, Blacking.

Common	Per Doz.,	$ 1 50
Medium	"	3 00
" square	"	3 75
Fine Black Stock	Per Doz., $5.00 and	7 50
" Black Center and White Border	Per Doz.,	6 50
" Grey Stock, Holly Handles	"	12 00
Traveling Sets	"	6 00
Hotel Sets	"	15 00

Yankee Shoe Brushes, (no Handles)	Per Doz.,	$ 2 75
" " " (Handles); fine	"	6 00

Maguire's Patent Shoe Brushes....Per doz., $4.00, $5.50 and $8 00

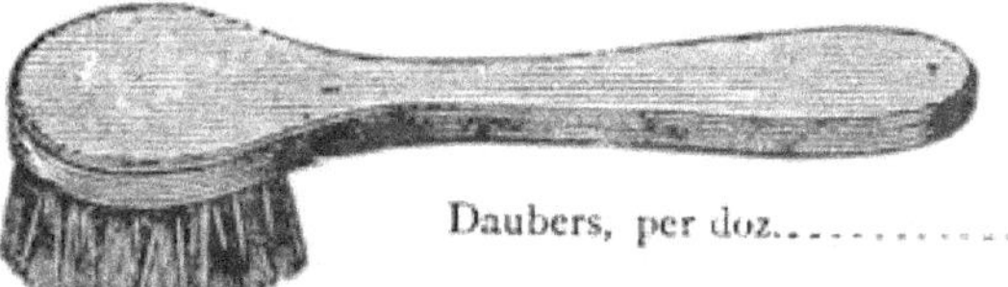

Daubers, per doz.................$ 1 75

Brushes, Crumb.

No. 1, per doz	$ 4 50	No. 3, per doz.	$12 00
" 2, "	8 00		

Travelers' Hair Brushes.

Without Handles, per doz.................................$24 00

Unbleached Hair Brushes.

Per doz.......................................$9 00 to $15 00

Curling Hair Brushes, etc., etc.

Florence Hair Brushes.

No.	Per Doz.	No.		Per Doz.
100	$19 00	935		$ 5 00
115	15 50	680		7 00
225	12 00	495		9 50
445	8 25	522		10 50
446	9 50	358	Monogram or Initial.	10 00
665	9 00	658	Monogram or Initial.	8 50
670	8 00	335	Mirror Backs.	11 00
795	9 00	445	Mirror Backs.	9 75
825	9 00	Infant		5 00

Brushes, Hair.

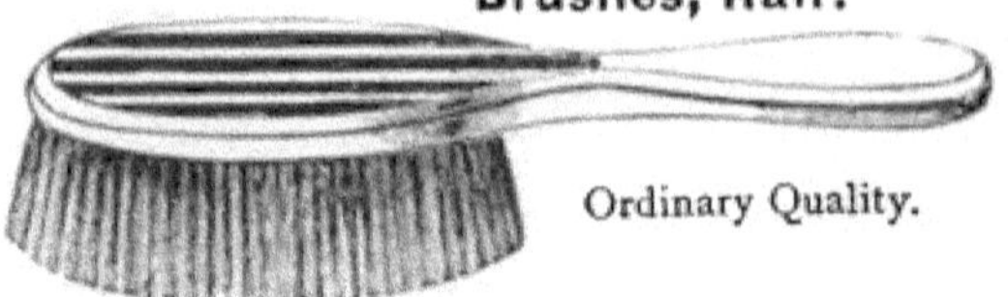

Ordinary Quality.

No.	Per Doz.
6	$ 1 75
28	2 75
69	3 00
71	3 50
878	3 75

Rosewood Backs and St'ck White

No.	Per Doz.
026	$ 4 50
180	4 50
0106	6 00
97	6 00
999	7 25
0266	9 50
0272	10 50
0263	12 00
0345	12 00

Fancy Backs and Stock White.

No.	Per Doz.
37	$ 3 50
58	3 50
182	4 50
092	4 50
0246	5 00
0247	6 00

No.	Per Doz.
048, Backs of six different colors in each box,	$ 7 00
0233	7 00
8768	7 50
0229	9 00
0249	9 00
0115, Rosewood and Pearl	12 00

American Penetrators, per doz., ...$ 6 00, $ 8 00 and 12 00

English " " .... 11 00, 15 00 " 18 00

French " Solid Backs, per doz., 12 00, 15 00, 18 00 and $24 00.

Buffalo, Plain Backs, per doz.........9 row, $12 00 ; 11 row, 15 00

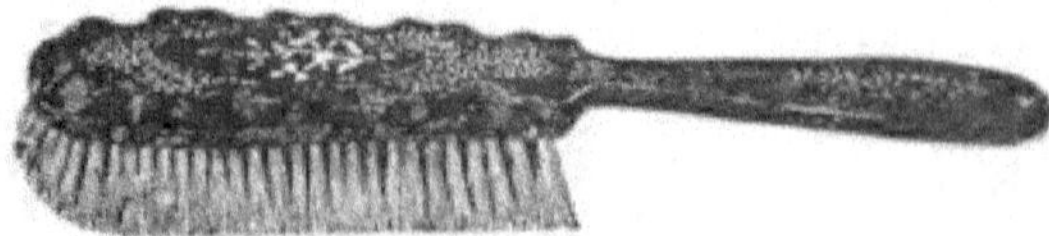

Buffalo Inlaid Backs, per doz.........9 row, $15 00 ; 11 row, 18 00 15 row, $24 00.

14

Brushes, Hat.

Common Wood, $2 00 to $4 00 Per Doz.
Extra, Velvet Backs, $3 50 to $6 00 Per Doz.
Superfine, Assorted Patterns, $7 50 to $12 00 Per Doz.

Brushes, Infant Hair.

	Per Doz.		Per Doz.
Bone Handles, 4 row	$3 00	Buffalo Handles, 5 row	$4 00
" 5 "	3 50	" " 6 "	5 00
" 6 "	4 50	" " 7 "	6 00
" 7 "	5 50	" " 8 "	7 00
" 8 "	6 50	" " 9 "	8 00
" 9 "	8 00	Wood " small	1 75
Florence	5 00	" " large	6 00

Brushes, Nail. In great variety.

Bone Handles, plain and winged,	per doz., from	$1 50 to $3 75
" " fine, plain and winged	" "	3 00 to 6 50
Buffalo " " " " "	" "	4 50 to 9 00

Brushes, Nail, Bath—No Handles.

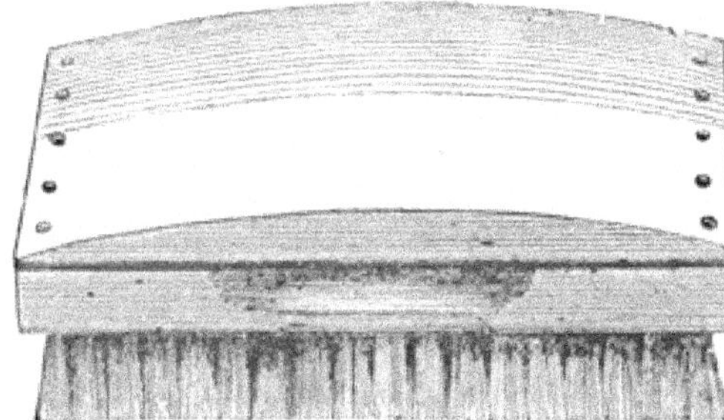
Fibre Nail Brush.

Bone, $2 50 to $7 00 Per Doz.
Wood, $4 00 to $6 00 Per Dox.
Wood, $4 00 to $6 00 Per Doz.
Wood, Fibre, $1 50 Per Doz.

Brushes, Shaving.

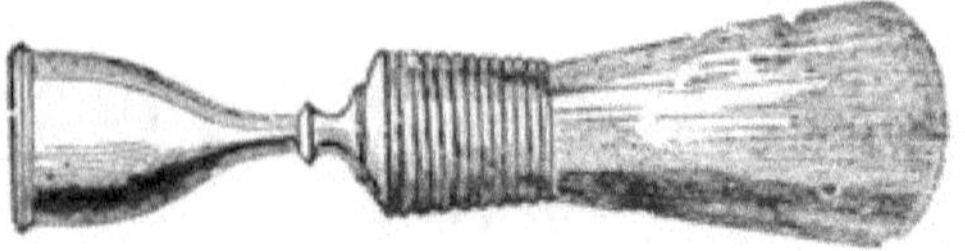

Styles 55, 60, 65.

Styles 123, 125, 126, 127.

No.	Enameled Handles.	Per Doz.
123	These are the very best quality of cheaper goods, and prices reduced.	$ 75
125		1 00
126		1 25
127		1 50

Black Handles, French Bristle.

No.		Per Doz.	No.		Per Doz.
13,	Wire Bound	$1 75	55,	Twine Bound	$1 75
14,	" "	2 00	60,	" "	2 50
15,	" "	2 25	65,	" "	3 50

No.		Per Doz.
9,	Walnut Handles, Barbers'	3 50
400,	Club Pattern, fine Stock	2 00

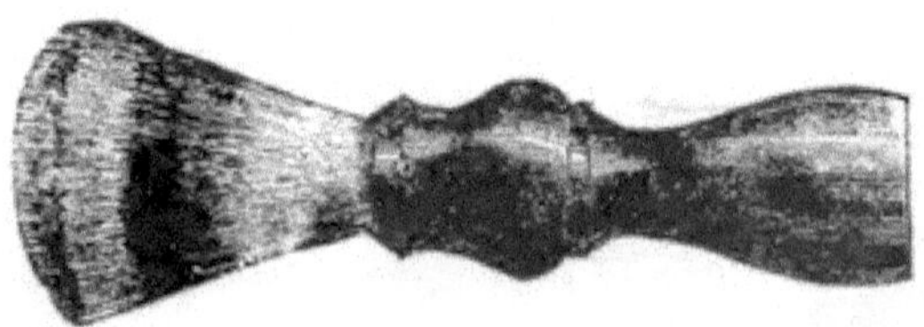

Gent's Badger Lather	$6 50
Barbers' " " (larger)	7 50

Badgers, all sizes and prices. Bristle, Bone or Horn Handle, all sizes and prices.

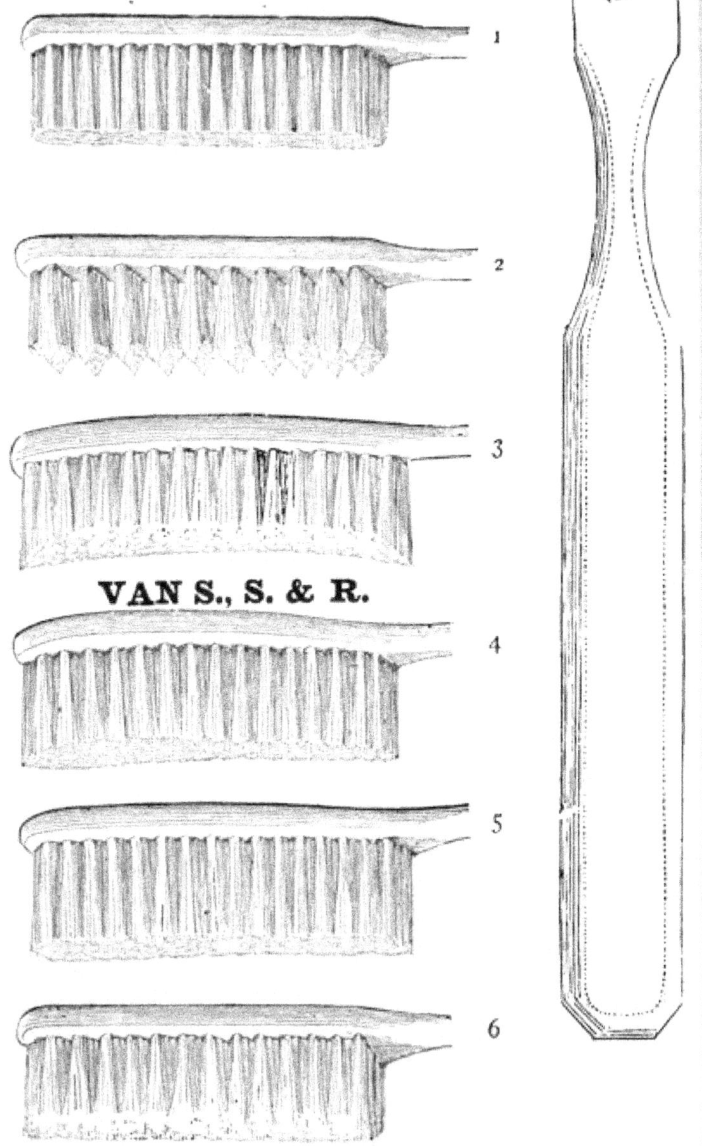

Styles of English Tooth Brushes.

Brushes, Tooth.

All 1 doz. in a box or assorted if so ordered.

	Per Doz.
English.	
4 Row, ordinary quality,	$1 50 to $1 75
3 " or 4 R., fine "	2 50 to 2 75
5 " fine quality	3 00 to 3 50
French.	
3 Row, ordinary quality,	75 to 1 00
4 " " "	1·25 to 2 00
4 " fine "	2 00 to 3 00
5 " " "	2 50 to 3 50
Pocket.	
Tulipw'd, Bone...., Buffalo.., Boxwood, } Closing on front or sideways	4 50
Double End..........	1 75 to 3 50
Badger Hair	2 00 to 4 50
Goat "	2 50 to 4 50
Fancy Patterns.......	3 75 to 6 00

Serrated and Palate in variety.
Children's Soft or Hard in variety.

English Tooth Brush.

French Tooth Brush.

Cachous, for the Breath.

	Per Doz.
Hooper's Metal Boxes..........	$1 50
" Paper "	1 25
" " " Ladies' ..	1 50
Jokes, Metal Boxes	75
" Paper Packets..........	50
Trix, " "	50

Capotes.

Special prices by gross.

	Per Doz.
Skin, Superfine................	$1 75
Rubber, White	75
" Pink	75
" in Roll	1 00
" Goodyear's Patent	1 50
" Caps White	50
" " Pink............	50

Chalk Balls. (See Lily Whites, page 234.)

Chamois Skins. (See Skins, page 202.)

Chest Protectors. (See Protectors, page .)

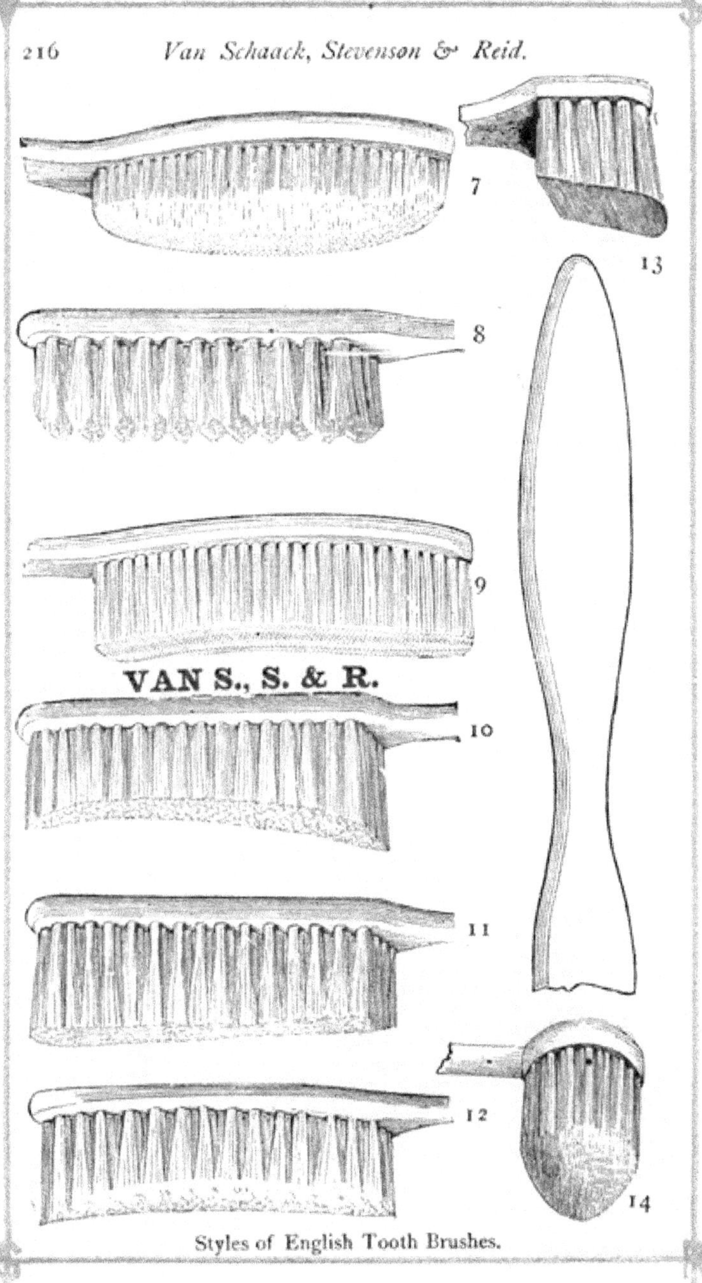

Styles of English Tooth Brushes.

Cards, Playing.

GOOD ALL TRADE MARK

	Per Doz.
No. 19, Steamboat	$ 1 90
" 23, Moguls	3 50
" 29, Great Moguls	4 50

	Per Doz.
No. 35, Saladee's Patent Round Corner Linen, for Whist or Poker	$ 6 00
No. 25, Decaturs; Plaid; Linen	6 00
Mount Vernons	6 00
Golden Gates	4 50
Euchre Packs	4 00
General Jacksons	3 50
Virginias	3 00
Moguls English, $11 00 and	8 75
Holly Backs, English	14 00
Fern " "	11 00
Shakespeare Backs, Eng.	11 00

Illuminated in variety.

Cases, Amputating
" Pocket,
" Vial, Physicians',
} See Surgical Instrument list, page —.

Cases, Vial.

	Each.
4-2 dram bottles, special for Vest	$1 00
16-1½ " " Sheepskin, with Strap	1 25
20-2 " " " "	1 50
16-2 " " Morocco and Gilt, with Strap	1 50
20-2 " " " " " " "	1 75
20-2 " " Red Calf Skin, with Strap	1 75
16-2 " " Wood Frame; Morocco covered; Gilt; with Lock	1 50
20-2 dram bottles; Wood Frame; Morocco covered; Gilt; with Lock	1 75
12-2 dram and 8-4 dram vials; Morocco covered; Lock	2 50

Peacock's.

18-2 dram vials, Peacock's Patent; covered Steel Springs hold in the bottles; (see Illustration)	2 50
10-3 dram vials; all one side; Red calf; Clasp	1 75

18–1½ dram vials; Morocco; (see Illustration); Steel Frame Portmonaie	$2 00
24–2 dram vials; numbered; Lock; Morocco; Gilt; (see Illustration, page)	2 50
6–½ oz. G. S. bottles, for Liquids; Morocco	2 50
20–4 dram vials; Morocco	3 00

Cold Cream.

Bazin's (see page 355.)		
Caswell, Hazard & Co.'s	Per Doz.,	$2 50
Patey's Rose, Imported	"	2 00

Cologne.

Burnett's, Half Pints, Cork Stop	Per Doz.,	$ 6 00
" Pints, "	"	11 00
" " Glass Stop	"	14 00
" Half Pints, Wicker Demijohn	"	8 00
" Pints, " "	"	14 00
" Quarts, " "	"	24 00
Farina, 4711, Long Bottles	"	6 00
" " Wicker, Half Pints	"	9 00
" " " Pints	"	15 00
" " " Quarts	"	27 00
Van Schaack, Stevenson & Reid's—		
Garden City	Per Gall.,	6 00
Floral	"	8 00
Frangipanni	"	10 00
Jockey Club	"	10 00

		4 oz.,	8 oz.,	16 oz.,	32 oz.,
Frangipanni / Jockey Club	Per Doz.,	$4 00;	$8 00;	$15 00;	$27 00

Combs, Raw Horn. Machine Cut.

First quality only.

Small Dressing, 1 inch teeth	Per Doz.,	$ 75
Medium " 1 "	"	1 00
Large " 1 "	"	1 25
Barbers' half inch teeth	"	1 50
Imitation Shell	"	1 50

Combs, Raw Horn. Hand Made.

Imported.

Irish White Horn	Per Doz.,	$6 00
English Gents' narrow teeth, all coarse	"	2 00
" " " " coarse and fine	"	2 00
French " " " all coarse	"	2 00
" " " " coarse and fine	"	2 00

The above make select styles for Barbers' use.

Combs, Blonde Horn. Nonsuch.

5½ inch Dressing	Per Doz.,	$1 25
6 " "	"	1 30
6½ " "	"	1 35
7 " "	"	1 40
7½ " "	"	1 50
8 " "	"	1 75
7½ " narrow teeth, Barbers'	"	1 75

Combs, India Buffalo Horn. French Manufacture.

Being nearly equal to Tortoise Shell, should always be recommended.

Quill. Back Straight.

7 inch, 7½ and 8 inch; Black	Per Doz.,	$6 00 to $7 50
7½ " Blonde	Per Doz.,	7 50
7 " Gothic; cut wedge shape between teeth to prevent pulling	"	9 00
Gents'; teeth all coarse	"	6 00
" 7½ x ½ inch; Black	"	4 50
" 7½ x ½ Blonde	"	4 50
6½ inch heavy straight backs	"	4 00
7 " " " "	"	5 00
7½ " " " "	"	6 00
8 " " " "	"	7 00

Combs, Tortoise Shell.

All varieties of Dressing and Fine Tooth.

Combs, Children's Long.

No. 3, Quill Back; plain	Per Doz.,	$1 10
" 7, " fancy	"	1 50
" 17, " heavy	"	1 25
" 25, " " fancy	"	2 25
" 32, " narrow	"	1 00
" 90, Flat Back; fancy	"	80
" 50, " plain	"	70

Combs, Children's Combination.

Teeth only at the ends.

No. 4, Flat Back; plain	Per Doz.,	$2 00
" 3, Quilled Back "	"	2 75
" 2, " fancy	"	3 00
" 10, Heavy Back "	"	3 00
" 11, Plain; for Ribbon	"	2 50
" 117½, Rosette; new	"	1 50

Combs, Hard Rubber. Dressing.

Straight Fancy Quill Backs.

6 inch	Per Doz.,	$2 25
7½ "	"	3 00
8½ "	"	3 50

Roached Fancy Quill Backs.

7 "	"	4 00

Very Heavy Arched Quill Backs.

7½ "	"	3 75
8 "	"	4 00

Metal (Tin) Backs.

6 "	"	75
7½ "	"	1 00

German Silver Backs.

6 "	"	1 50
7 "	"	1 75
8 "	"	2 00
No. 7060, Goodyear's 6 inch	"	1 50
" 7070, " 7 "	"	1 80
" 7080, " 8 "	"	2 10

Heavy Plain Square Backs.

7½ inch	"	1 50
8 "	"	1 75

Heavy Straight Quill Backs.

6 "	"	2 50
7 "	"	3 00
8 "	"	3 50

Cheap Plain Dressing.

6 "	"	60
7 "	"	75
8 "	"	80

Gents' Narrow Quill Backs.

(⅝ inch Teeth.)

7½ "	"	2 75
5 " in Case for Pocket	"	2 25

Massive Backs, Fine Large.

8 "	"	6 50
9 "	"	7 50

Combs, Fine Ivory.

(Common Width.)

Inches }	1⅝ to 2⅛	2⅛ to 2⅜	2⅜ to 2⅝	2⅝ to 2⅞	2⅞ to 3⅛	3⅛ to 3⅜	3⅜ to 3⅝	3⅝ to 3⅞	3⅞ to 4⅛
	No.8	No.9	No.10	No.11	No.12	No.13	No.14	No.15	No.16
Superfine ..	$1.10	$1.55	$2.20	$2.45	$2.70	$2.80	$2.95	$3.05	$3.20
S. S. " ..	1.45	2.05	2.60	2.90	3.20	3.35	3.60	3.83	4.00
S.S.S. " ..	2.00	2.75	3.45	3.85	4.25	4.85	5.25	5.50	5.60
NePlusUltra	3.25	4.05	4.85	5.70	6.10	6.80	7.30	7.70	8.10

(Medium Width and Thickness.)

	No.8	No.9	No.10	No.11	No.12	No.13	No.14	No.15	No.16
Superfine ..	$2.00	$3.00	$3.75	$4.25	$4.55	$5.00	$5.35	$5.55	$5.75
S. S. " ..	2.75	3.60	4.55	5.20	5.70	6.20	6.60	6.95	7.25
S.S.S. " ..	3.25	4.85	6.10	6.75	7.30	8.00	8.60	9.35	9.75

(Extra Width and Thickness.)

	No.8	No.9	No.10	No.11	No.12	No.13	No.14	No.15	No.16
Superfine ..	$3.25	$4.05	$5.00	$5.60	$6.10	$6.60	$7.10	$7.60	$8.10
S. S. " ..	4.25	4.85	6.10	6.75	7.30	8.00	8.55	9.20	9.75
S.S.S. " ..	4.85	6.50	8.10	8.95	9.75	10.55	11.35	12.20	13.00

Combs, Fine Hard Rubber.

Fine Combs.

Superfine.

Nos.	8	10	11	12
Per Doz.	$ 45	$ 55	$ 60	$ 70

Super Superfine.

Nos.	8	10	11	12
Per Doz.	50	60	65	75

Medium Superfine.

Nos.	10	12	14
Per Doz.	60	75	90

Medium Super Superfine.

Nos.	10	12	14
Per Doz.	70	85	1 00

Fancy Medium Superfine.

Nos.	10	12	14
Per Doz.	$ 70	$ 85	$1 00

Fancy Medium Super Superfine.

Nos.	10	12	14
Per Doz.	80	95	1 10

Fancy Superfine. Extra Wide.

Nos.	12	14	16
Per Doz.	1 10	1 30	1 50

Fancy Super Superfine. Extra Wide.

Nos.	12	14	16
Per Doz.	1 20	1 40	1 60

Combs, Infant, Assorted.

Per Dozen..$1 25 to 1 75

Combs, Lead, for coloring the Hair.

Combs, Magic.

Per Dozen..$7 50

Combs, Pocket.

	Per Doz.
Buffalo ; Black ; cased	$1 25 to $3 50
" B'onde ; "	1 25 to 3 50
" Black ; not cased	1 25 to 3 00
" Blonde ; " "	1 25 to 3 00
" Moustache ; coarse and fine	3 00
" " handled	3 00
Siamese Raw Horn	75
No. 3, Hard Rubber ; 3 in. long ; double	1 50
" 4, " 3½ " " "	1 75
" 5, " 4 " " "	2 00
" 7, " 3½ " " single	1 60
" 8, " half-moon shape	1 40
" 8, " " " in slides	1 00

Combs, Decorated Tube Back.

		Per Doz.
4	inch Redding	$ 50
6	" Dressing	75
7½	" "	1 00

Noyes' Patent Brass Back Combs.

Nos.			Per Doz.
000,	5	inch, with Teeth of ordinary width	$1 25
3160,	6	" " "	1 90
2770,	7	" " "	2 25
2780,	8	" " "	2 75
2965,	6½	" with 1 inch Teeth	2 50
2970,	7	" "	2 75
2975,	7½	" "	3 00
2980,	8	" "	3 25
4770,	7	" Improved Style, having Metal Teeth at both ends	2 50
070,	7	" Hotel Style, having Solid Tooth at one end and Loop at the other	2 50

Noyes' Patent German Silver Back Combs.

			Per Doz.
2050,	5	inch, with Teeth of ordinary width	$2 25
2675,	7½	" " "	3 25
2870,	7	" with 1 inch Teeth	3 25
2880,	8	" "	3 75

Noyes' Patent Barbers' Combs.

			Per Doz.
2480,	8	inch, German Silver Backs	$3 25
2575,	7½	" Brass Backs	2 25

Noyes' Patent Redding Combs.

			Per Doz.
16,	3	inch, German Silver Backs	$1 00
17,	3	" Brass Backs	75
1740,	4	" " "	1 00
6,		Rule Pocket, German Silver Backs	2 50

Combs, Fine Horn.

	Per Doz.
Clear Horn, S. S., Fine American	$1 00
" " " French	2 50
Raw Horn, S., " American	50
" " " French	3 50

Corkscrews.

(Handled.)

	Per Doz.
Wood Handle, with Brush on end	$1 50
Cocoa Handle; extra strong	1 75
1 dozen assorted on cards; ordinary	$1 00 to 2 00
1 " " " fine	3 00 to 4 50
Bronze; commercial	3 00
" universal	3 50
Bone, with Brush; ½ doz. on card	4 50

(Pocket.)

Hard Rubber; large; closing up	4 75
" small; "	4 00
German Silver; closing up	9 00
Extra fine Steel, with Key Ring; closing up	3 75
Common Steel; Harp	$1 50 to 1 75
Fine " "	2 50 to 3 50

The New Medicated Corn File.

This article will actually cure Corns, Excrescences, Calloused Places or Excrescent Bunions without pain. Never touches the sound flesh. Never causes the feet to bleed, and can be used on the sorest Corn.

Price	Per Doz., $1 75
Alexandra; imported	" 3 00

Corn Pencils.

DeGraves'	Per Doz., $1 75
Robinson's	" 1 75

Cosmetics.

	Per Doz.
Batchelor's new Cosmetic, applied by wetting the brush; with Brush and Mirror combined; for black or brown	$8 00
Bazin's small	80
" large	1 25
Coudray's Hongroise; in pot	4 50
" " in bottle	3 00
" 960	1 00
" 940	1 50
" 941	2 00
" 942	2 50
Italian, black or white; small	1 00
" " " large	2 00
Letchford's, 1 dozen on card; are not greasy	1 75
Lubin's, small	5 00
Rimmel's, large Rolls	6 00
" new Cosmetic	6 00

Dusters, Feather. (See list, page 103.)

Dusters, Toy.

Made of colored feathers........................per dozen, $4 50

Extracts, Handkerchief.

American, (price includes stamp.)

	Per Doz.
Bazin's, (see list, page 355.)	
Burnett's Florimel, 1 oz.	$ 7 50
Chinese Tea Rose	6 00
" Musk	6 00
Lundborg's Recherche, selected or assorted odors, ⅝ oz.	3 75
" Arcadian Pink, 1 oz.	7 00
" Ylang Ylang, 1 oz.	7 00
" Wood Violet, 1 oz.	7 00
" selected or assorted, 2 oz.	10 50
" Arcadian Pink, 2 oz.; one bottle, in fancy box	13 50
" Nilsson Bouquet, ⅝ oz.	3 50
" " " 1 oz.	6 50
Phalon's Improved,—	
" Dedication Bouquet, 1¼ oz.	7 00
" Flor de Mayo, 1¼ oz.	7 00
" Night Blooming Cereus, 1¼ oz.	7 00
" Paphian Bouquet, 1¼ oz.	7 00
" White Rose, 1¼ oz.	7 00
" Wood Violet, 1¼ oz.	7 00
" Assorted, in fancy chromo boxes, 1¼ oz.	7 50
" Pullman Palace Car Bouquet, ½ oz.	3 50
" " " " 1 oz.	6 50
Seely & Co.'s Victoria Regia, 1 oz.	6 50
Spencer's, (see page 227.)	
Tallman's Assorted, 1 oz.	1 50
" Musk, 1 oz.	1 75
" Prize Extracts, ½ oz.	3 00
" " " 1 oz.	6 00
" Egyptian Calla, 1 oz.	6 50
" Jockey Club, 1 oz.	1 75
Woodworth, (see page 225.)	

Imported, (Stamps charged extra.)

Atkinson's White Rose, 1 oz.	6 50
" Assorted, 1 oz.	6 50
" White Rose, 2 oz.	13 50
" Assorted, 2 oz.	13 50

WE OFFER A FULL LINE DIRECT FROM THE FACTORY.

C. B. WOODWORTH & SON,

Manufacturing Perfumers,

ROCHESTER, NEW YORK.

HANDKERCHIEF EXTRACTS.

Nos.	Holding.		How put up.	Per doz.
780	1 oz.	Nilsson Bouquet, a standard extract	½ doz. in box	$ 6.50
781	½ oz.	" " " "	½ "	3.25
791	1 oz.	White Rose, " "	½ "	6.50
792	½ oz.	" " "	½ "	3.25
716	1 oz.	Pitcher, extracts, assorted odors...	½ "	3.00
779		Corked, " with showy labels.	½ "	1.50

COLOGNES.

838	8 oz.	Handsome Pitchers of Cologne, assorted odors........	2 in box	6.00
980	10 oz.	Statuette of Beatrice, (new bust) assorted odors	3 "	4.00
956	5 oz.	Mirror Cologne, assorted odors, (new style)	½ doz. in box	2.50
955	2 oz.	Mirror Cologne, assorted odors, (new style)........	1 "	1.50
	2 oz.	Bell-shaped, paneled bottle, Cologne	1 "	1.50

HAIR OILS.

1025	10 oz.	Statuette of Beatrice, (new bust), filled with assorted oils	3 in box	4.00
1027		Large Pyramid, (new style, very good), filled with assorted oils...	4 "	4.00
1006	5 oz.	Mirror, assorted oils, (new style)...	½ doz. in box	2.50
1005	2 oz.	" " " ...	½ "	1.50
847	2 oz.	Bell-shaped, paneled bottle, Rose..	1 "	1.00
	4 oz.	Hand-shaped, hair oil, Nobby Youth, (grotesque figure)	½ "	2.00
		Scroll Hair Oil, (cheap)	1 "	1.00

POMADES.

4 oz. Tumbler Pomades, metal top...... 1 doz. in box 2.50

1 oz., 2 oz., 4 oz., 6 oz., 8 oz. Pomades, assorted, in great variety.

And beside the above, a large assortment of goods well suited to the wants of the trade.

Bayley's Essence Bouquet........Per Doz. $12 50
Crown Perfumery Co.'s Extracts, in 1 oz. bottles .... " 6 50
Coudray's Bouquet des Alpes...... " 18 00

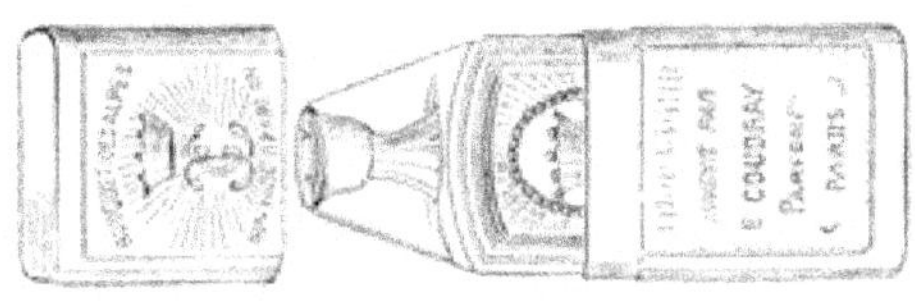

Per Doz.

Letchford's Floro Heraldic Extracts, illuminated label, assorted sizes ....$

Low Son & Haydon's, 1 oz., assorted........ 6 75

Lubin's, 1 oz. (see below list of odors,) in half dozen boxes......... 8 00

Assortment No. 1, each box contains

2 Jockey Club. 1 Patchouly.
1 Musk. 1 West End.
1 Violet.

Assortment No. 2, each box contains

1 Jockey Club. 2 Patchouly.
1 Ess. Bouquet. 1 Bouq. Caroline
1 New Mown Hay.

American Flowers.
Citronnella Rose.
Damask Rose.
Duchesse.
Fifth Avenue.
Fleur d'Orange.
Flowers of the West.
Forget-me-not.
Frangipanne.
Gardenia.
Geranium Rosat.
Hedyosmia.
Heliotrope.
Lily of the Valley.
Magnolia.
Marechale.
Michigan Avenue.
Mignonnette.
Mille fleurs.
Moss Rose.
Musc.
Mew Mown Hay.
Night Blooming Cereus.
Patchouly.
Pond Lily.
Rondeletia.
Rose.
Rosebud.
Rose Mousseuse.
Rose Musquee.
Rose, The.
Spring Flowers.
Sweet Briar.
Sweet Clover.
Sweet Pea.
Sweet Scented Shrub.
Tubereuse.
Upper Ten.
Verbena.
Violette.
Winter Blossom.
Ylang Ylang.

Bridal Bouquet.
Chicago.
du Jockey Club.
de West End.

Moulleron's American Flowers....................Per Doz., $ 9 00
" Assorted.......................... " 6 00
Piesse & Lubin's, Assorted .... " 9 50
" Frangipanni .... " 9 50
Rimmel's, 1 oz., 3 in each box.................... " 7 00

T. P. SPENCER & CO.,

MANUFACTURERS OF

Perfumery and Flavoring Extracts.

WESTERN DEPOT,

VAN SCHAACK, STEVENSON & REID,

92 & 94 LAKE STREET, CHICAGO.

HANDKERCHIEF EXTRACTS.

Nos.	Holding.		How put up.	Per doz.
125	2 oz.	Glass stopper, standard extracts, assorted, covered with satin, equal to the best imported............	½ doz. in box	$ 7.50
124	1 oz.	Same style and quality as 125.....	½ "	3.75
132	1½ oz.	Metal top flask, with sprinkler, assorted, the best bottle and finest extract in the market	½ "	6.50
817	1 oz.	Glass stop, chromo label, good quality extract, fancy wooden box...	1 doz. in fancy box. Very pretty for show case.	3.75
145	1 oz.	Tall glass stop, assorted extracts, fancy wooden box, labeled E. Coudray & Co., London	1 doz. in fancy box. Very pretty for show case.	3.75
814	2 oz.	Cork stop, Chinese label, assorted kinds	½ doz. in box	3.50
137	⅝ oz.	Cork stop, chromo label, wooden box, with drawer, assorted, tied in colors.....................	1 "	2.00
165	1 oz.	Glass stop, finest extract, silver plated stand, and twelve bottles, complete		Per stand. 7.00
166	1 oz.	Cork stop, assorted extracts packed in a black walnut bureau, with glass mirror, complete		Per bureau. 3.75

COLOGNES.

Nos.	Holding.		How put up.	Per doz.
216	8 oz.	German cologne, metal top, with sprinkler, warranted equal to the best imported, largest bottle in market	½ doz. in box	$ 7.00
907	1 oz.	Cork stop, sample size, German cologne	1 "	1.50

HAIR OILS.

Nos.	Holding.		How put up.	Per doz.
	10 oz.	Hair oil, bust of Henry Ward Beecher, patented	½ doz. in box	4.00
399	6 oz.	Oil of Magnolia blossoms, for the hair, photograph labels	½ "	3.50
303	12 oz.	Decanter, orange flowers hair oil	½ "	4.50
718	4 oz.	Flat oval, La Belle hair oil	½ "	2.00
723	2 oz.	" " "	1 "	1.00
360	4 oz.	West India Bay Rum oil, original quality	½ "	2.25
361	2 oz.	West India Bay Rum oil, original quality	1 "	1.25
425		Bear's Pomade, in jars, shaped like black and white bears	½ doz assorted colors in box.	3.00
300	4 oz.	Chinese silk hair oil, very attractive	½ doz. in box	2.00
310-11	10 oz.	Only Son hair oil, in the form of a "bust" of a crying baby, bottle patented June 9, 1874	½ "	4.00
313-14	10 oz.	Granger hair oil, bottle in the form of a bust of a Granger, patented	½ "	4.00
320	4 oz.	Hair oil, bottle in the form of a bust of a flower girl, patented	½ "	3.00
353	4 oz.	Bear's oil, bell-shaped, panel bottle	½ "	2.25
353½	4 oz.	Rose oil, " "	½ "	2.25
306	4 oz.	Cocoanut oil, flat oval	½ "	2.50
307	4 oz.	Hickory Nut oil, "	½ "	2.50
388	2 oz.	Cocoanut oil, "	1 "	1.50
392	2 oz.	Hickory Nut oil, "	1 "	1.50
350½	16 oz.	Oil and Cologne, assorted	2 bottles in box	6.00

Cunon.
Vanda.
White Rose.
Oriental Hyacinth.
Chinese.
Wood Violet.
Rimmel's Bouquet.
Nouvelle Marquise.
Grande Duchesse.
Frangipanne.
Jockey Club.

Ess. Bouquet.
Lily of the Vale.
Musk.
Patchouly.
Tea.
Coffee.
Jasmin.
New Mown Hay.
Heliotrope.
Rose Geranium.
Moss Rose, etc.

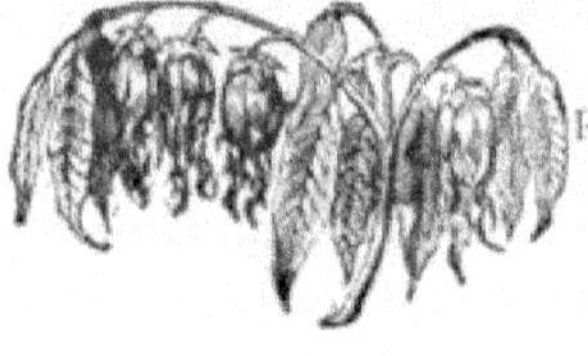

	Per Doz.
Rimmel's Ihlang-Ihlang, 1 oz.	$ 7 00
" Extracts, 2½ oz.	13 00
" " 5 oz.	24 00
" " 10 oz.	38 00

Extracts, Handkerchief. Bulk.

For list of odors see Lubin's list, page 226

	Per lb.
Chiris, all odors	$ 3 50
Lundborg's Ylang Ylang	4 50
" Musk	4 50
" Wood Violet	4 50
" all other odors	3 50

Fittings for Nursing Bottles.

Maw's 6d	Per Doz.,	$1 00
" 6d Pure Gum	"	1 50
" Alexandra	"	2 50
" Glass Tubes, separate	"	50
" All White Rubber	"	1 50
" Tubing for Nursing Bottles	Per Yard,	15

Flasks, Pocket.

	Per Doz.				
	¼ pt.	½ pt.	¾ pt.	1 pt.	½ pt. oval.
Wicker covered		$ 5 00	$ 6 00	$ 7 50	
Wattis' Patent, (¼ doz. in box.)					
Russet Sheep, with cup		11 00	13 00	16 50	$12 00
" " no cup		8 50	9 50	12 00	9 00
Morocco, with cup	$12 00	13 50	16 00	18 00	13 50

Julep Tubes.

Straight, per dozen, 25c. Bent, per dozen, 25c.

Glass Cologne Bottles.

Burnett's, cork stop.	4 oz.		Per Doz.	$	1 25
" "	8 "		"		1 50
" "	16 "		"		2 25
" "	32 "		"		3 00
" glass stop.	4 "		"		2 00
" "	8 "		"		2 50
" "	16 "		"		3 00
" "	32 "		"		3 75

Caswell's. Burnett's.

Caswell, Hazard & Co.'s, glass stop. 8 oz.	Per Doz.	$	2 50
" " 16 "	"		3 00
with glass label, Bay Rum, Cologne, *or* Spts. Camphor, 8 oz.	"		6 00
" " " " " 16 "	"		7 50

Glass Pens.

Briggs', (for marking linen)	Per Doz.	$	2 25
New style on cards	"		1 25

Glass Sachet Bottles.

8 oz. .... Per Doz. $ 8 00

Ground Glass Stoppers, with glass labels, assorted odors.

Glasses, Graduated.

(See p. 165.)

Glass Cologne Bottles.

DiamondCross.

Pitcher.

Octagon.

Cologne Bottles, cut glass........................Per Doz. $
A large assortment of Crystal and Bohemian Glassware; new and elegant designs....per doz., from $2 50 to $15 00
Pungents and Smelling Bottles, glass stop. " " —— to ——
" " " " " in leather cases.................... " " —— to ——
Pungents and Smelling Bottles, silver cap..........Per Doz. $
" " " " silver cap, gold plated "

All the above in crystal and decorated glassware, assorted colors.

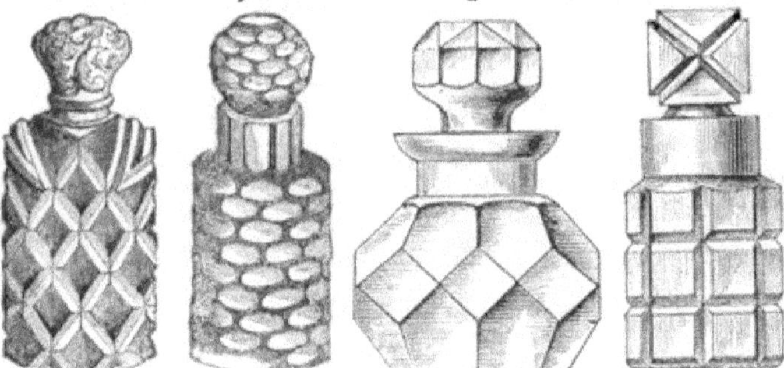

Cut Glass Union. Per Doz., $9 00.

No. 14.

No. 7.

No. 4.

These goods being of our own direct importation, we can make lowest possible prices.

No. 262.

Essex's Improved Nickel Silver Atomizer.

(See pages 203 and 204.)

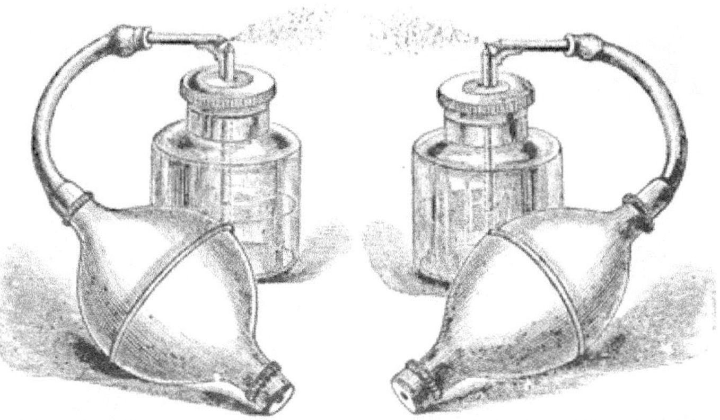

This Atomizer possesses many advantages over the ordinary instrument with Bulb on the top or on the side of the bottle, having a Nickel Plated Tube and Cap, and a vent and drip return passage combined. Wherever it has been introduced, it has superseded the sale of all others.

Bags, Felt, for Filtering, &c.

Large. Medium. Small.

Bulbs.

For Syringes, one hole. | For Syringes, two hole.

Hair Oils, &c.

	Per Doz.
Assorted, 2 oz., square panel	$ 1 50
Barlow's Restorative	7 50
Bazin's, (see page 355.)	
Dorin's Cocoa Dressing	3 75
Letchford's Cream of Limes	5 00
Morse's Luxurene, (choice)	4 00
Piesse & Lubin's Pestachio Nut Oil	9 50
Rimmel's Lime Juice and Glycerine	8 00
Phalon's, (see page 349.)	
Rowland's Macassar Oil	15 00
Schock's Hair Tonic, (see page)	7 50
Society Hygienique Oil, all leading odors	6 00
Spencer's, (see page 227)	
Tallman's Bay Rum Oil, 2 oz.	1 25
" Cocoa Oil, 2 oz.	1 25
" Bay Rum Oil, 4 oz.	2 25
" Cocoa Oil, 8 oz.	6 00
" Golden Oil	4 50
Woodworth's, (see page 225.)	

Hair Pins, Hard Rubber.

Nos. 1 and 2 .... Per Gross, $1 60

Hair Powders.

Blonde	Per doz., $2 00	Gold	Per doz., $2 00
Diamond	" 2 00	Silver	" 2 00

Face Powders.

	Per Doz.
Cascarilla	$ 1 00
Chardin's Talc de Venice	7 50
Coudray's No. 2499, assorted	1 50
" " 2507, Rice Flour	2 50
" " 2536, " oval box; fine	4 50
Dorin's Talc de Venice	6 00
" Blanc de Perle, No. 4	2 50
Elgin's Phantom Powder; small	2 00
" " " large	3 50
Harrison's Parian White	4 00
Hobb's Genuine Meen Fun	1 00
Hunt's Court Toilet Powder, in triangular boxes	3 75
Lewenberg's Pastiles de Florence	3 75
" " " Rose	3 75
Lubin's Rose, Nursery	2 50
" Violet, "	2 50
" Rice Flour	10 50
" Blanc de Perle	7 50
McArthur's Lily White	1 75
Miller's Antheo	3 50
Miniature Mother of Pearl	1 50
Opera Pearline	1 50
Piesse & Lubin's Pestachio Nut Powder	9 50
Potain's Blanc de Marguerite	3 00
" " " Rosiere	4 00
" " " Beaute, in glass, shape of a hand	4 00
Pozzoni's, white or flesh; small	3 75
" " " large	7 50
Rimmel's ½ lb. Violet	1 75
" ¼ lb. highly scented assorted odors	2 00
" Rose Leaf	4 00
Saunder's Pink Bloom of Ninon	3 50
" White	3 50
Shand's Alabaster Tablet	1 35
" Chalk Balls	75
" Rouge "	1 00
Schock's Creme de Lys, dry	3 50
" " liquid	7 50
Shattuck & Co.'s (Glenn's) Magnolia Tablet	1 00
" " " Pearl Powder, (face)	2 25

Tetlow, (see page 382.)

Toiletine, pink or white 3 50
Turner & Wayne's Snow Flake........................ 3 75

And in addition to above a large assortment of goods suitable for show case and fine trade, from which we shall be glad to send samples.

Tetlow's Lilywhites, Rouges and Toilet Preparations

LILYWHITES—in boxes.

No.		Per Gross.
1	Mountain Pearl	$30 00
2	Snowdrop	18 00
5	Cascarilla de Persia	12 00
6	Large Round	9 50
7	Medium "	7 50
8	Small "	6 50
9	Large Magnolia	10 00
10	Small "	8 00
100	Lisbon Powder	24 00
113	Myrtle Tablet	27 00
152	Cream Tablet de Ninon—round	18 00
153	Cream Tablet de Ninon—oval	18 00
162	South Sea Pearl—square	51 00
163	" " round	30 00

ALASKINE.

27	Florida Rose—oval	21 00
31	Opera Gems—round	16 00
32	Alaskine—oval	15 00
33	" "	12 00
34	" round	15 00
35	" "	12 00
36	" "	10 00

FARNESE TABLETS.

16	Farnese Tablet—oval	48 00
17	" " "	42 00

JAPONIQUE TABLETS.

147	Japonique Tablet	48 00
148	" "	36 00
149	" "	27 00

GEORGIAN CERYLOSA.

No.		Per Gross.
114	Georgian Cerylosa	$48 00
115	" "	36 00
116	" "	27 00

EXCELLOS.

183	Excellos—oval	39 00
184	" "	33 00
186	" "	15 00

LILYWHITES, WITH ROUGE,
in boxes.

75	Opera Bouquet	$27 00

LILYWHITES, WITH ROUGE.
EYEBROW PENCILS AND TOMPONS—in boxes.

21	Mountain Rose	$51 00
23	Cupid Secret	30 00
30	Lily Sceptre	51 00

ROUGES.

13	Rose of Cashmere—round	12 00
37	Opera Rouge, "	14 00
14	Castillian Rose Tint—oval	13 00
38	Imperial Rouge, "	15 00
76	Coral —diamond	18 00
158	Shell Pink Rouge	24 00

Key Rings.—A great variety.

Lip Salve.

Bazin's, (see list page 355)........................Per doz., $1 15
Glenn's .... " 1 00

Seidlitz Measures.

Wood, (polished) ....Each, $ 25

PATENT ADJUSTABLE

TOILET MIRRORS.

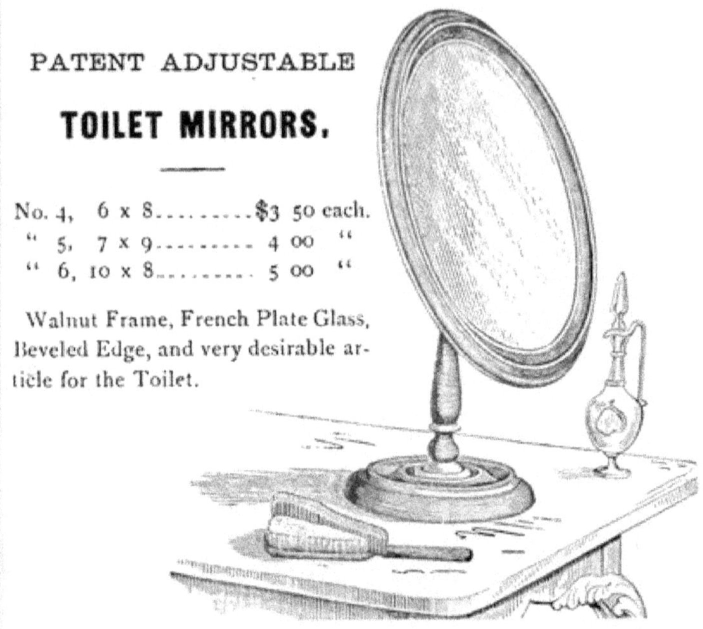

No. 4, 6 x 8..........$3 50 each.
" 5, 7 x 9.......... 4 00 "
" 6, 10 x 8.......... 5 00 "

Walnut Frame, French Plate Glass, Beveled Edge, and very desirable article for the Toilet.

Mirrors, Lionite. Florence.

No.		Size		Per Doz.
5,	Oval Patent Glass	3½	inches,	$ 3 50
6,	" " "	4	"	4 50
7,	" " "	4½	"	5 50
8,	" " "	5	"	6 50
9,	" " "	5½	"	7 50
10,	" " "	6	"	8 50
11,	" " "	6½	"	10 00
8,	Oval French Plate Glass, beveled edge	5	"	12 50
9,	" " " " "	5½	"	14 50
10,	" " " " "	6	"	18 00
11,	" " " " "	6½	"	22 00
7½,	Round Patent Glass, with Loop Handle to hang up by.	3½	"	5 50
8½,		4	"	6 50
9½,		4½	"	7 50
10½,		5	"	8 50
10½,	Round French Plate Glass	5	"	18 00
20,	Square, Standing			10 00
40,	Oval, "			12 00

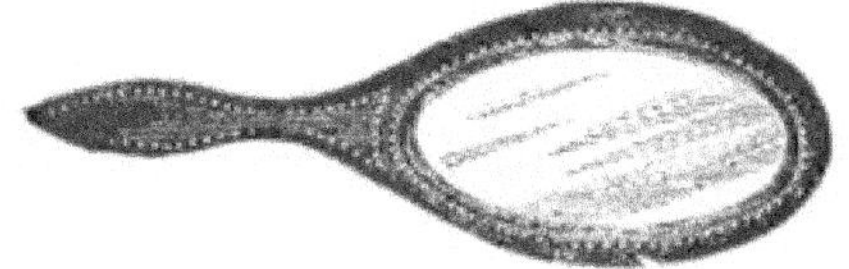

Mirrors, Leather Covered.

New Patent. Zinc Frame.

6 inch Oval, imitation Russia Leather	Per doz.,	$ 8 00
6½ " " beveled edge, (French Plate,) imitation Russia Leather	"	21 50

Mirrors, Pocket.

Common, Wood cased	Per doz.,	$ 60
Zinc, "	"	75
Small Oval, for Vest	"	2 50
Round Florence	"	1 75

Mittens and Glo es—Rubber.

No. 1. Mittens.	No. 2. Gloves.	No. 3. Gauntlet Gloves.
Per doz., $16 00	Gents' per doz., $14 00	Gents' per doz., $14 00
	Ladies' " 11 50	Ladies' " 13 00

Nipples, Rubber. (See page 172.)

Black—Pure Gum.

	Per Doz.
Ordinary Style	$ 35
" " extra large	75
Seamless, heavy	50
Maw's Small	35
" "D" extra pure	50

White.

	Per Doz.
Ordinary Style	$ 25
" " extra large	50
Seamless, heavy	35
Plug	75
Davidson's Patent	50
Maw's Small	25
" " Leechbite	25
" large Swanbill	75

Paper.

Fayard's, in Rolls	Per doz.	$ 2 75
Gayetty's, Water-closet	"	4 50
Jones' P.P.P. Packets, Water-closet	"	60
Mortimer's, Water-closet	"	2 25
Singer's, Gravel for Bird Cages, Round	"	2 00
" " " Square	"	2 00
Star Water Closet Paper	"	3 00

Pastiles.

Fumigating, Bulk, Black	Per lb.	$ 90
" " Red	"	1 00
" Coudray's, in boxes	"	1 50
" Hooper's "	"	1 50

Lewenberg's Pink, } (See Face Powders, page 233.)
" White, }

Pencils, Camels' Hair.

Hair, ½ inch to 2½ inches long.

	Ordinary. Gross.	Rose. Gross.		Ordinary. Gross.	Rose. Gross.
No. 1	$	$	No. 6	$	$
" 2			" 7		
" 3			" 8		
" 4			Asst. 1 to 8		
" 5				doz.	doz.
			" "		

Common Assorted, 1 gross boxes	Per box, $	
Extra, Goose Quills, Assorted	Per gross,	
" Swan " "	Per doz.	
Throat Pencils, Bent Handles	"	75
Corn, (for removing Corns, Warts, etc.)	"	1 75
Eye Brow	"	1 50
Freckle	"	1 75
Indelible, Clark's	"	2 25

Clark's Indelible Pencil.

Jeweler's Plate Brush.

Per Doz., $3 00

Pomades.

	Per Doz.
Bazin's (see page 355.)	
Bear's Pomade	$ 3 00
Coudray's ¼ Assorted, in Cans	5 50
" ½ " "	10 50
" de la Cour, finest	9 00
Fluted Jars Pomade	6 00
Graham's Philocome	1 75
Hurlbut's Cocoa Cream	2 00

	Per Doz.
Low's, Assorted, in Blue Box	$ 4 50
Savage's Ursina	6 75
Shattuck & Co.'s Transparent	5 50
Spencer's (see page 227.)	
Woodworth's (see page 225.)	
Glass stopp'd Fancy Bottles, 4 oz., 6 oz., and 8 oz. varieties in stock ....from $3 50 to	12 00

And many other choice patterns.

Plasters, Corn, Bunion, Court, Etc.

		Per Doz.
Bunion, Felt, thick		$1 75
" " thin		1 50
Corn, " "		1 25
" " thick		1 50
Soft Corn Felt		1 75
Court, London	all colors	18
" Tablets, 3 sheet		75
" Tallman's Arnica	all colors	50
" " Glycerine	"	50
" Dickman's Arnica	"	75
" " Tablets, 4 sheet		1 75
Rubber and Isinglass on Silk-Novelty Rolls	per roll,	65
Taffeta	"	50

Probangs.

Plain ....per doz., $1 25 | Tipped.........per doz., $1 50

Protectors, Chest and Lung. (See page 249.)

No.			Per Doz.
2,	Chamois Chest	Single,	$ 4 00
3,	" "	"	5 00
4,	" "	"	6 00
3,	" Lung	Double,	10 00
4,	" "	"	12 00
	Extra large Chamois Lung	"	18 00
2,	English Felt Chest	Single,	5 50
3,	" " "	"	6 50
4,	" " "	"	9 00
2,	" " Lung	Double,	15 00
3,	" " "	"	18 00
4,	" " "	"	21 00
3,	Felt Jacket, Sleeveless		24 00
4,	" " "		27 00

Puffs, Powder.

			Per Doz.
Small select		$1 25 and	$1 50
Medium "	$1 75,	2 50 "	3 00
Barbers' "	2 50,	3 00 "	4 50
Extra large, select		6 00 "	9 00

Pungents. (See Glassware List, page 231.)

Razors, Imported.

Width of Blade.	Genuine Joseph Rodgers & Son's.	Shape of Point.	Per Doz.
¾ inch,	The American Pride	Concave,	$ 7 00
⅝ "	Ivory Handled, India Steel	"	13 50
¾ "	" and the Shank cased in Ivory,	Round,	21 00
⅝ "	Celebrated Hollow Ground	Convex,	9 00
⅝ "	Plainer "	"	7 50
	Genuine Wade & Butcher's.		
⅝ "	Plain, not Hollow Ground	Concave,	5 00
⅝ "	Hollow Ground	"	7 50
¾ "	Wide Hollow Ground, very heavy	"	12 00
¾ "	Diamond Edge	"	9 00
⅝ "	Ivory Han'ld, Hollow Ground, Gilded Shank,	Convex,	12 00
	Genuine Wostenholm, (I. X. L.)		
½ "	Whisker, Hollow Ground	Round,	7 50
⅝ "	Light Hollow Ground	Convex,	7 50
⅝ "	Pat. Heavy Back, Gents', Celebrated	Concave,	12 00
⅝ "	I. X. L. Gilded Blade, Hollow Ground	"	9 00
⅞ "	Double Hollow Ground, Celebrated Pipe	Square,	12 00
⅝ "	Medium " "	"	9 00
⅝ "	Plain, not Hollow Ground, Pipe	"	4 50
¾ "	" " "	Concave,	6 00
1 "	Wide Hollow Ground, heavy	"	12 00

Salts, Smelling.

Ammonia Mono-Carb, (Preston Salts) .... per lb., $1 75

Bottles for same ;

Crystal, plain	per doz.,	1 75
" No. 262 (see page 231.)	"	4 50
" Similar but smaller	"	4 00
" Ruby, Cross Stopper, (see illust. near middle page 231.)	"	3 00
" White, " " (see illust. near middle page 231.)	"	3 00
" " " " large, (see illust. near middle page 231.)	"	6 00
" " and Amber	"	3 00
" Pine Apple, (see page 231, next to the Cross Stopper)	"	3 50
" Pine Apple Gilt	"	4 50
" No. 4 (see page 231)	"	3 00
" Assorted in Colors	"	3 00
" Extra Fine Cut Styles	per doz., $6 00 and	7 50
" No. 7, with Silver Cap; Gold Plated (see p. 231)	per doz.,	30 00
" Unions, (see page 231)	per doz., $7 50 to	9 00
" Finger Colored Pungents ; Bohemian ; Metal Cap	per doz.,	6 00

Bazin's, (see page 355.)

Littlefield's ; filled ; Screw Cap	"	2 00
" " Glass Stop	"	2 00

Razor Strops.

	Per Doz.
Emerson's Genuine	$ 6 00
" Imitation, (good)	3 00
Goldschmidt's Russian Leather ; small	9 00
" " " large	15 00
Hale's Magic ; square	3 00
Imported ; fine ; square	6 00
Hunt's Flexible Patent	4 50
" Combination, with Hone	6 00
Torrey's ; square	2 00

Emerson's Razor Strop.

Rouge.

Bazin's, (see list, page 355)

Coudray's Theatre, No. 2541	Per doz.	$2 50
" Vinegar, " 2575	"	3 00
Dorin's, Dry, No. 4	"	2 75
" Liquid, No. 42	"	1 75
Glenn's, Lemon, large	"	2 25
" " small	"	1 25
" Vinegar, large	"	2 25
" " small	"	1 25
Gourand's Vegetable	"	4 00
Lubin's Vinegar	"	13 50
Tallman's Lemon	"	1 25
" Vinegar	"	1 25

Sachets.

Atkinson's (for list of Odors, see Atkinson's Extracts)	Per doz.	$ 4 00
Bazin's (see list, page 355)		
Coudray's, in ¼ lb. Bottles	Per lb.	5 00
Lubin's, in Paper Packets	Per doz.	8 00

Oiled Silk.

In five yard Rolls	Per yard,	$1 25

Skins, Chamois, Plaster, etc. See page 202.

Soaps, Creams, etc., Shaving.

Bazin's, (see page 355.)		
Brown's Bars, (2 lb. in Bar)	per lb.,	$ 20
" (see page 240.)		
Buchans's No. 10 Carbolic	per doz.,	2 00
Glenn's Compound ; Mugs	"	3 50
Guerlain's ; Cake	"	4 50
" Cream ; Jars	"	5 00
Indexical, (10 bars in Box)	per box,	2 75
" Cakes ; each cake wrapped in foil	per doz.,	1 00
Lloyd's Euxesis	"	5 50

Soaps, Creams, etc., Shaving—(continued.)

Pear's Transparent Sticks; small	per doz.	$4 50
" " " large	"	6 00
Pioneer, (10 bars in Box)	per box,	2 50
" Cakes	per doz.,	50
Sarg's 30 per cent. Glycerine; Boxes	"	5 00
Vroom & Fowler's Military; Cake	"	1 25

Williams' Combination Mug (see Illustration)	per doz.,	6 00
" Barbers' Bar, (10 bars in Box)	per box,	3 00
" " Favorite; Cakes	per doz.,	75
" Clipper; Cakes	"	65
" Pocket Style	"	1 50
" Yankee (genuine)	"	1 00
" Verbena Cream Tablet	"	1 75
Winters' Rhypophagon; Cakes	"	4 50

Soaps, Toilet, etc.

Bayley's Spermaceti Tablet	per doz.,	$7 50
" Violet "	"	4 50
Bazin's, (see page 355.)		
Brown's Opodeldoc	per lb.,	20
" David S. & Co., (see page 240.)		

Buchan's Carbolic:

	Per Doz.
" " Laundry	$2 25
" " No. 1, large; perfumed	3 50
" " " 2, " plain	3 00
" " " 3, small; perfumed	2 25
" " " 4, " plain	2 00
" " " 5, Transparent Glycerine	4 50
" " " 6, Camphorated	3 50
" " " 7, Perfumed Bath	1 75
" " " 8, Medicinal (25 per cent.)	4 50
" " " 9, Dental	2 00
" " " 10, Shaving	2 00

Soaps, Toilet, etc.

	Per Doz.
Buchan's Carbolic:	
" " No. 11, Disinfecting; Cakes	$1 75
" " " 12, " Bars	5 50
" " " 13, Perfumed American	1 75
Calvert's Carbolic Nursery	2 25
" " Toilet	2 25
" Medicinal	4 50
Caswell, Hazard & Co.'s Juniper Tar	2 50
Constantine's Tar	1 75

Coudray's Glycerine, — Each cake wrapped, in boxes of three cakes each.	$5 00
" Jockey Club, — Each cake wrapped, in boxes of three cakes each.	5 00
" Lettuce, — Each cake wrapped, in boxes of three cakes each.	5 00
" White Rose, — Each cake wrapped, in boxes of three cakes each.	5 00
" Savon Voyage; Metal Box	7 00
Gelle Freres' Assorted	3 00
Gouraud's Italian Medicated	4 50
Indexical Soap Co.'s:	
No. 40, Alpine	1 25
" 86, Bee Hive Bath	1 25
" 3, Brown Windsor	1 50
" 19, Glycerine	75
" 2, " pure	1 35
" 24, " white	1 50
" 18, Honey	75
" 112, " hotel	40
" 11, " pure	1 35
" 235, " white	1 50
" 31, Oatmeal	1 50
" 60, Palm	70
" 5, Pumice	1 00
" 6, Sand	85
" 36, Silver	1 00
" 84, Tar	75
" 120, Turtle Oil	1 50
Low, Son & Haydon's:	
Old Brown Windsor, 3 Cake	1 25
Magnum Brown Windsor, No. 4, 3 Cake,	3 50
Musk " 3 Cake	1 75
Brown Windsor, ½ lb. Bars, 2 doz. in box	2 50
" 1 " 1 "	5 00
Elder Flower, No. 4, 3 Cake	2 25
" ½ lb. Bars, 2 doz. in box	2 50

16

Soaps, Toilet, etc—(continued.)

	Per Doz.
Low, Son & Haydon's:	
Elder Flower, 1 lb. Bars, 1 doz. in box	$5 00
Glycerine, No. 4, 3 Cake	2 25
" " 4, 3 doz. in wood	2 00
" ½ lb. Bars, 2 doz. in box	2 50
" 1 " 1 "	5 00
Honey, No. 4, 3 Cake	2 25
" " 4, 3 doz. in wood	2 00
" ½ lb. Bars, 2 doz. in box	2 50
" 1 " 1 "	5 00
Turtle Oil, No. 4, 3 Cake	2 25
" " 4, 3 doz. in wood	2 00
" ½ lb. Bars, 2 doz. in box	2 50
" 1 " 1 "	5 50
Lubin's large Assorted	12 00
" medium "	8 00
" small "	5 00
" large Rose	13 00
" medium "	9 00
" small "	5 50
Oakley, Jesse & Co.'s:	
No. 1 S, Transparent Bar	65
" 1, " "	70
" 5, " "	95
" 9, " "	2 00
" 2, } Roughly finished {	1 00
" 3, { Glycerine Transparent Balls. }	1 35
" 3, Polished Glycerine Transparent Balls, extra scented	1 75
" 65, Transparent Glycerine Cake	70
" 71, " Gauntlet "	1 00
" 90, " Glycerine "	1 25
" 210, Magnum Honey, 3 Cake	1 25
" 212, " Glycerine, 3 Cake	1 25
" 214, " Rose, 3 Cake	1 25
New Styles, Crown Honey	60
" " Glycerine	60
" " Brown Windsor	60
" " Variegated	60
Pioneer Soap Co.'s:	
No. 51, Bouquet, 3 Cake	1 50
" 52½, Cocoa, 3 "	1 25
" 48, Glycerine, 12 "	60
" 47, Honey, 12 "	60
Rieger's Transparent Prize Medal, Cake; contains 40 per cent. Glycerine	3 75
Rimmel's Assorted, 1 lb. bars, 12 bars in box ....per box,	6 00
Sarg's Patent Glycerine Soaps:	
" Liquid; 40 per cent. Glycerine	5 50
" Toilet; 33⅓ " " Boards	1 75
" Traveling; Metal Box; oval	3 00
Society Hygenique, Savon Dulcifie, sans odeurs, ½ lb. bars; fine	6 00
Wright's Alconated Glycerine, Tablet	4 00

Sponges. (See page 202.)

Suspensory Bandages. (For Rawson's, see page 252.)

	Per Doz.
Cotton, with Strings	$1 50
" " Elastic	1 50
Linen, " "	2 50
" hand-made, with Self-adjusting Strings	3 50
Lisle Thread	3 50
Silk, woven or crochet	5 00
Compression, Patent, for Varicocele	13 50
Suspensory, with Gon. Pouch attached	6 00
Gon. Pouch, pure Gum ; seamless, with Bands	6 00

Rawson's, see illustration and prices, page 252.)

The celebrated J. P. Suspensory, small, medium and large ; one only in a box, viz. :

Cotton, at $3 00, $3 50 and $4 00 per dozen.
Silk, at $6 00 and $7 50 per dozen.

Show Cases.

(See page 380.)

A large assortment, our own direct importation.

Shoulder Braces.

Ladies', Gents', Girls', and Boys'.

		Per Doz.
Common Sense	all sizes,	$10 00
Cutter's Expanding	"	12 00
Lateral	"	15 00
Leather Back (see Illustration, page 280.)	"	10 50
London	"	10 50
Patent (see Illustration, page 280.)	"	10 50
Pivot Action	"	8 00

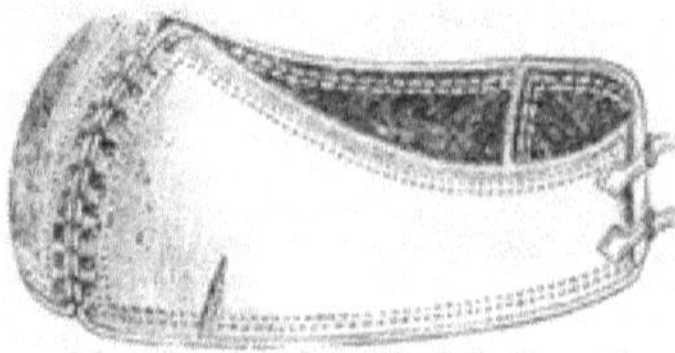

Mrs. Bett's Abdominal Supporter.

Supporters, Abdominal.

	Each.
Betts, Mrs., White Leather ; Laced	$2 50
Bow, Steel ; Leather cover'd	2 50
Elastic Truss Co.'s	9 00
" with Pessary attached	11 25

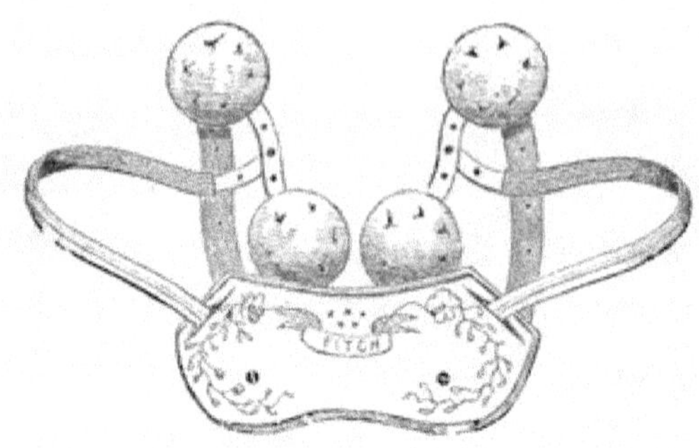

	Each.
Fitch's Polished Steel (see Illustration)........	$2 50
" Silver Plated ; Self-adjusting	3 50
London (see Illustration page 286)........	2 25
Silk Elastic Web	7 50

Supporters, Uterine.

		Retail Price.	Wholesale P.
Babcock's, see page 285	each,	$25 00	$12 50
Macintosh's........	"	12 00	6 00
" with Perineum..	"	12 00	6 50
Shannon's	"	15 00	8 00
Wadsworth's..	"	10 00	5 00
Wadsworth's Extras.			
Cup and Stem.	"	2 50	2 00
Belt	"	1 50	1 00
Socket	"	1 75	1 25

Macintosh's.

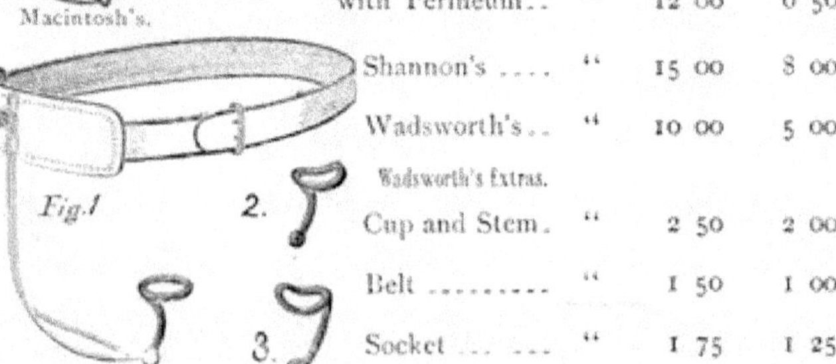

Shannon's.

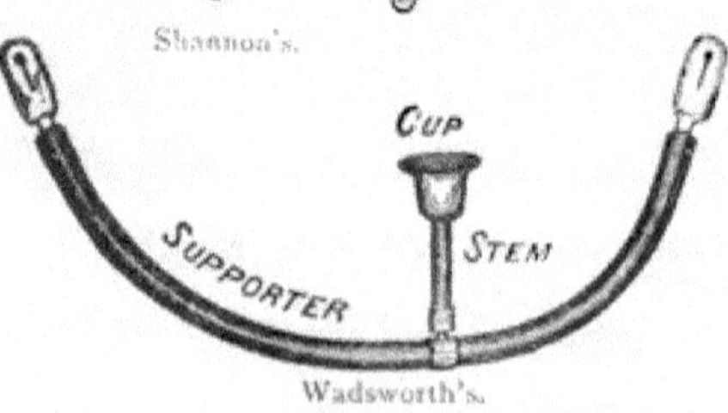

Wadsworth's.

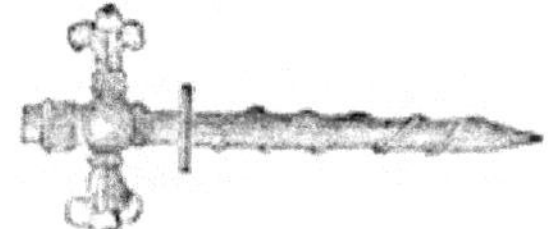

Tap for Champagne, Cider, and Aerated Waters.

Prince of Wales ....each, $2 50

Teething Pads.

Maltese Cross.....per doz., $ 50 | Lincoln ....per doz., $1 75

Teething Rings.

India Rubber, White	$ 50
" Black	75
Ivory	1 00

Thermometers. Fever or Body.

Bent, Plain		each, $	2 50
" Self-registering		"	3 00
Straight, "		"	3 00
Spiral, "	improved, to prevent the running together of the Quicksilver	"	5 00

Thermometers–Fancy.

8 inch Mahogany	per doz., $	5 00
10 " "	"	6 00
8 " Black Walnut, polished.	"	8 50
10 " Black Walnut, polished.	"	10 00
Black Walnut Mantle Stands	"	7 50
" Window	"	5 50
Parlor Styles	per doz., $4 50 to	12 00
8 inch, Kendall's Nickel Plated Patent Dairy, (see page 313.)	per doz.,	5 50
10 " Kendall's Nickel Plated Patent Dairy, (see page 313.)	"	6 00
Japanned Tin, see page 183.		
Chemical to 212° F., 400° F., 600° F.,	each, $1 50, $2 00	2 50
" " 100° C., 200° C., 360° C.,	" 1 50, 2 00	2 50
Brewers'	each, $2 00 to	5 00
Distillers'	" 2 00 to	5 00

The Great Health Preservers.

Protectors, Chest and Lung, (see page 238.)

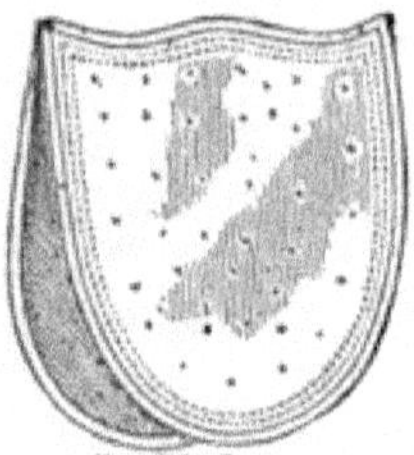

Double Lung.

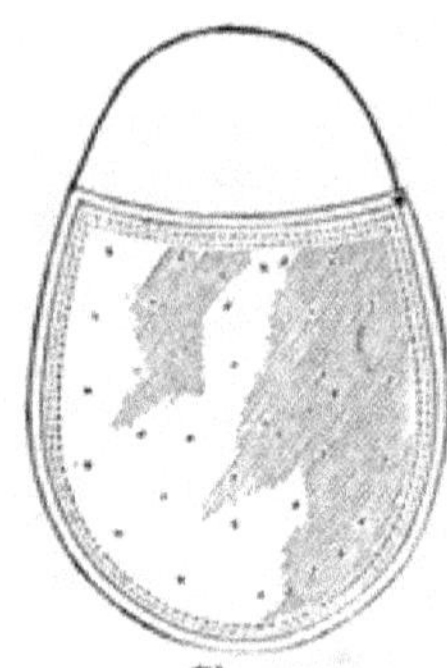

Chest.

Perforated Chamois Under Clothing.

See page 251

Hair Brush and Comb (combined)

Tooth Preparations.

	Per Doz.
Bazin's, (see page 355.)	
Bonn's Eau Dentifrice........	$7 50
Brown's Camph. Dentifrice...	1 75
Buchan's No. 9, Carbolic Soap	2 00
Burnett's, (see page .)	
Calder's Saponaceous Dentine..	1 90
Caswell's Dentine (Powder)...	3 00
" Formodenta (Paste).	5 00
Colgate's Odontique Tablet...	1 35
Delluc's Eau Angelique.......	8 00
Ennis Tablet................	1 75
French Tooth Powder (bottles)	1 50
Glenn's Rose Tooth Powder; in fancy glass boxes........	2 00
Glenn's Aromatic Rose Tooth Paste (pots)	3 00
Gosnell's Cherry Tooth Paste....	6 00
Jewsbury & Brown's Oriental Paste; large	9 00
Jewsbury & Brown's Oriental Paste; small	5 50
Lamplough&Campbell's (Camphorated)	3 00
Price's Tooth Lozenges	3 00
Rimmel's Coral Paste	6 00
Rimmel's White Rose Soap	2 25
Saunders' Smokers (Powder)	4 00
Tallman's Tooth Powder	1 50
Tallman's Rose Tooth Paste	3 00
Tallman's Charcoal Tooth Paste....	3 00
Thurston's Tooth Powder; large..	3 50
Thurston's Tooth Powder; small..	1 75
Thompson's Tooth Soap Tablet....	1 75
Vincent's Dendan, sprinkler top, (see Illustration)..........	3 75

Tooth Preparations—(continued).

	Per Doz.
Vosburg's Charcoal Tooth Paste	$1 75
" Rose " "	1 75
Water's Tooth Soap	1 25
Wright's, with Sprinkler Top	2 00

Tooth Picks.

Ivory, 3 blade, (3 doz. in box)	per doz.,	$ 25
Quill, large, No. 5	per 1000,	2 00
" medium, No. 4	"	1 65
" small, No. 3	"	1 50
Decorated, (5 bunches in box)	per box,	50
Wood, Double Pointed, (2500 in a box) see page 249,	per doz.,	2 25

Trusses, Adult.

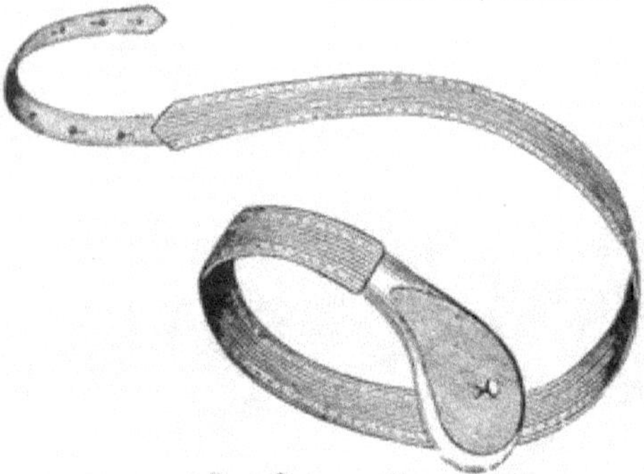

	Per Doz.
Best Common, Single	7 50
Best Common, Double	15 00
Chase Impr., Single	15 00
Chase Impr., Double	30 00
Chase Impr., Single, Nickel Pl'td Spring, and Cedar Pad	30 00

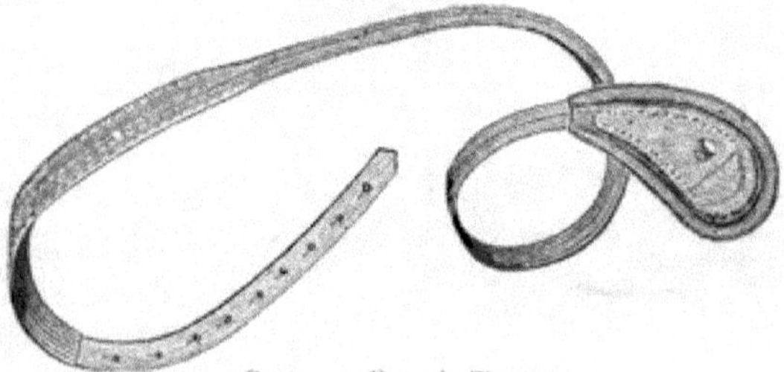

Best Common Truss.

Common French Truss.

Common French, Single	12 00
" " Double	24 00
" Sense, Single	24 00
" " Double	45 00
Hard Rubber, Single	36 00
" " Double	48 00
" " Extra Pads for Front or Back	6 00
Improved French Self-adjusting, single, light Spring, oval Pad	21 00
Rachet, Single	18 00
Radical Cure, Single	42 00

Rawson's Patent Elastic Self-Adjusting

U. S. ARMY SUSPENSORY BANDAGES.

The only perfect article ever offered to the public. NO MAN SHOULD BE WITHOUT ONE.

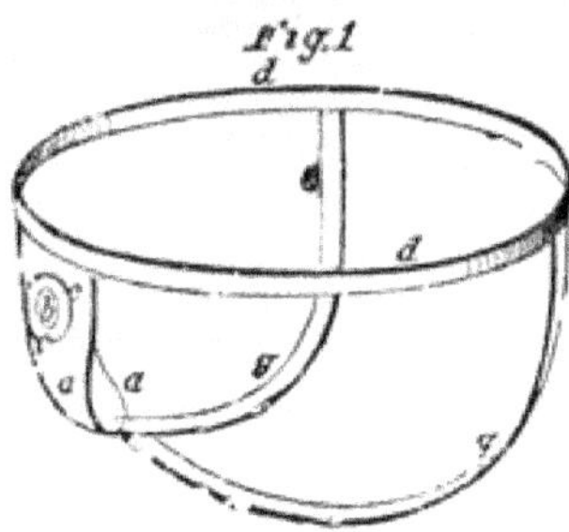

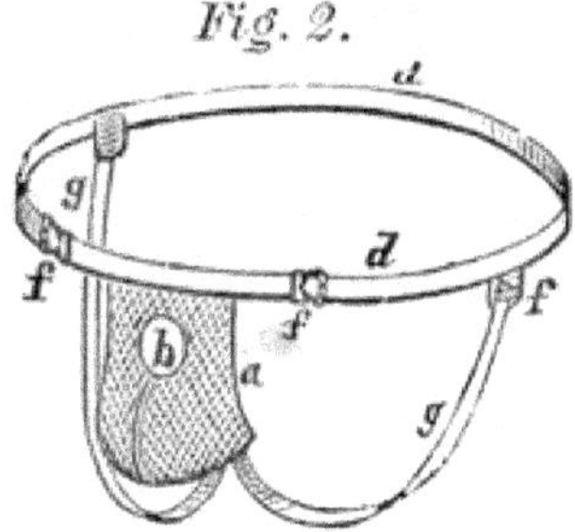

Fig. 3 | Fig. 4

PRICE LIST.

	Retail.	Per doz.
No. 1	$ 1 50	$10 00
" 1½	1 75	11 00
" 2	2 00	14 00
" 3	3 00	18 00
" 4	4 00	20 00

EXPLANATION OF THE CUTS.

Fig. 1—*a*, sack of silk, linen or cotton net, which will not interrupt the action of the respiratory organs of the skin; *b*, an opening, with an elastic ring or band, *c*, which keeps the material of the sack closed around the penis and exterior portion of the testicles; *d*, an elastic band, which encircles the body and passes around the hips downwards towards the penis, as in Fig 3, is made either with or without buckles, as in Fig. 2, or permanently, as in Fig. 1; *g*,*g*, elastic straps, passing over the buttocks and attached to sack, *a*, beneath, which keeps the sack, *a*, in position, so as not to pinch the testes; and in whatever position the person may put himself, the sack will remain in place—the straps, only, yielding, and accommodating themselves to the movements.

The Trade supplied at Manufacturers' prices by the Agents,

Van Schaack, Stevenson & Reid.

To the Druggists of the Northwest:

We feel assured the Trade fully appreciate our being the *Pioneers* in issuing a full descriptive and illustrated PRICED DRUG LIST. We hope they will find the present issue a still more complete book of reference.

"EXCELSIOR" IS OUR MOTTO.

We are constantly adding all the NEW DESIGNS, etc., necessary for a COMPLETE outfit for Drug Stores.

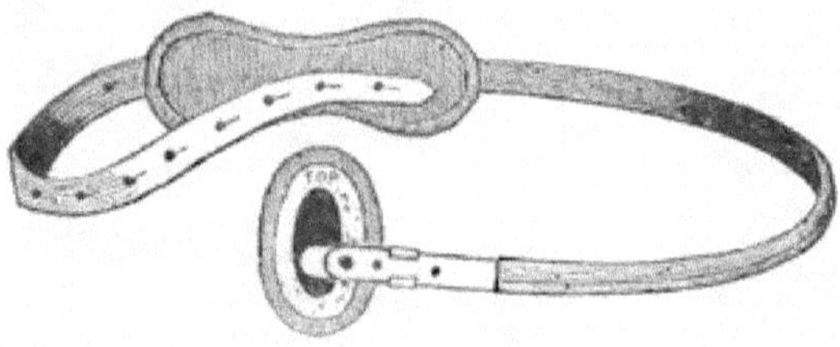

Self-Adjusting Truss.

	Per Doz.
Self-adjusting, Marsh Style, Single	$18 00
" " Double	36 00
Set Screw, Single	18 00
" Double	36 00
U. S. Army, Single	15 00

Trusses, Children's.

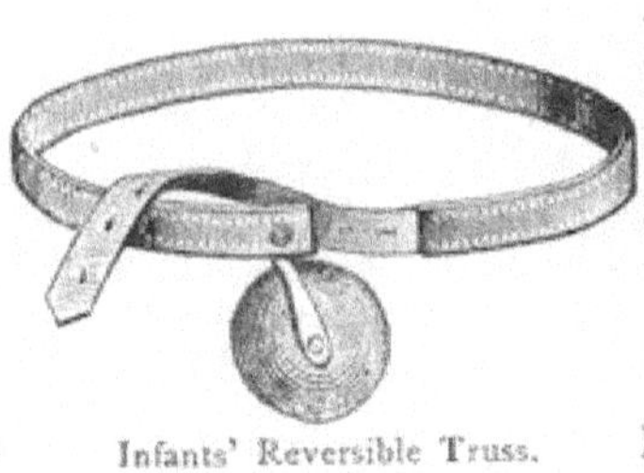

Infants' Reversible Truss.

Youths' Self-Adjusting Truss.

French Style, Single	$ 7 50
Hard Rubber, "	21 00
Improved Chase; Reversible; single (see Illust.)	7 50
Self-adjusting, Single	12 00
" Double	21 00

Trusses, Youths'.

French, Single	$12 00
Improved Chase, Reversible	15 00
Self-adjusting, Single	18 00
" Double	24 00

Trusses, Umbilical or Navel.

ADULTS' Bow, (Covered Steel)	each,	$ 2 50
" Elastic Truss Co.'s	"	3 75
" Hard Rubber	"	4 00
CHILDREN'S Bow	per doz.,	12 00
" Elastic	"	12 00
" French	"	12 00

Towels.

		Per Doz.
Turkish, Brown	from $4 00 to	$9 00
" Striped	" 7 50 to	15 00
" White	" 7 50 to	20 00

Tweezers.

Combination, English's.........................Per doz. $4 00

Common Steel, on Cards, per doz.	....	75
Fine, " "	....	1 50
Superfine, " "	....	3 00
Gifford's Patent, "	....	75

Vinegar Aromatic Toilet.

Bazin's....... $3 60	Bonn's........$7 50	Bully's.......$7 50
Coudray's 4 00	Rimmel's 7 50	S'y Hygienique 7 00

Wafers, Medicinal.

Boxes, Round, per doz.....$1 25	Sheets, Bulk, per 100 ....$1 50
" Square " 1 25	

Waters.

Barnes' Magnolia........................	...per doz.	$ 7 50
Bazin's Toilet, (see page 355)		
Bell Lavender	"	5 50
Chiris's Orange Flower, Half Pints................	"	3 00
" " Pints	"	4 50
" " Quarts	"	5 50
" Rose Flower, Half Pints................	"	3 00
" " Pints	"	4 50
" " Quarts	"	5 50
Coudray's Lavender Water, No.	"	6 00
Coudray's Verbena Water, Large, No.1745	"	16 50
" " Medium, " 1744	"	9 00
" " Small, " 1743	"	6 50
Lanman & Kemp's Florida Water................	"	6 75
Lubin's Lavender, Amber................	"	10 50
" " Distilled	"	9 00
Tallman's Florida................	"	6 75
California (J. M. Lundborg's)................	"	6 75

Verbena Water.

Burnett's Cologne Bottle, Glass Stop.

Lavender Water.

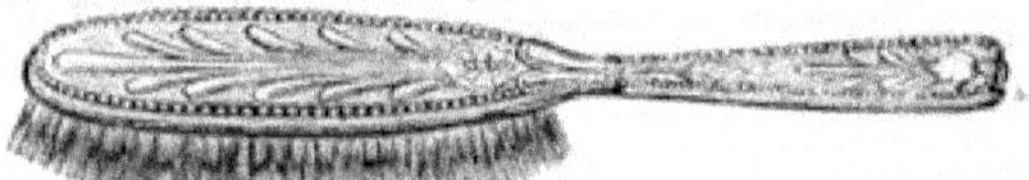

The Florence Hair Brush. [See page 218 for prices.]
Shape of Nos. 935, 825 and 795.

SPRINKLE TOP BOTTLES.

No. 1.

No. 2

No. 3.

No. 1.	1 ounce Lubin Bottle	per doz.,	$1	00
" 2.	2 ounce Square Panel Tooth Powder Bottle	"	1	25
" 3.	4 ounce Cologne Bottle	"	1	50
" 4.	8 ounce Cologne Bottle, same style as No. 3,	"	1	75

All the above are packed in boxes of one dozen each.

THE CELEBRATED

PRATT'S ATOMIZING COLOGNE,

(AND ATOMIZER.)

THIS new and elegant perfume is now being placed before the public with a feeling of certainty by its manufacturers, that it will stand at the head of all Colognes ever before in the market, and that all who use it will accord in saying that it is a reliable article. It is made by an entirely new process, and from the finest imported materials, rendering it everything that is required,

A FINE AND LASTING PERFUME,

and not expensive. This fact, in connection with the elaborate way in which it is put up, cannot but induce lovers of perfumery to at least try it.

In connection with the above inducement, the manufacturers have invented an ATOMIZER, so simple that they attach it to each bottle of their Cologne, making both complete in the 50 cent size. This Atomizer is acknowledged by all who have used or examined it, as FAR superior to any ever in use, being simple, compact and durable, and can be used wherever the old and expensive ones are, Everybody should have one of them, as they are a great saving in the use of Perfumes, and of great benefit in the sick room for spraying disinfectants. Remember, the small bottle put up in box with Atomizers, retails at 50 cents, and the Cologne separately, small bottle, 25 cents, large, $1.00. Early orders will *secure* a package of scented sample cards. Wholesale prices: Small cologne, $1.75 per doz.; *small cologne, with Atomizer with each bottle*, $3.75 *per doz.*; large cologne, $8.00 per doz.

For Sale by the Sole Western Agents,

Van Schaack, Stevenson & Reid,

92 and 94 Lake Street, cor. Dearborn, Chicago.

Ready-Made Shop Drawers.

The Trade supplied by Van Schaack, Stevenson & Reid.

Prices from Factory,

These Ready-Made Drawers of Pine bodies, Walnut or Ash fronts, mitered joints, are made by machinery. They are neat and strong. The Fronts have a Beveled Edge all around, and are finished in oil. They can be varnished by the purchaser, without other previous preparation.

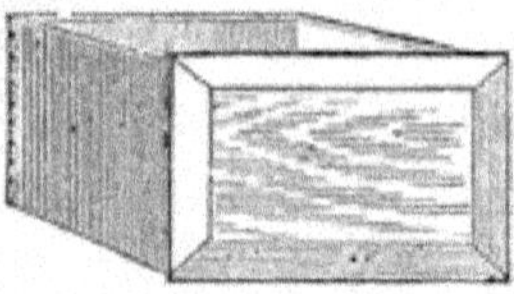

FIG. No. 1.

Pulls and Labels to fit them are adjusted at once

Size No. 1 weighs about 35 lbs. per doz.	Size No. 3 weighs about 45 lbs. per doz.
" " 1½ " " 45 " "	" " 4 " " 50 " "
" " 2 " " 55 " "	

The sizes, per inside measurement, are as follows:

No.	Wide	Long	Deep	Per dozen
No. 1	—9 inches wide,	10 inches long,	5½ inches deep, per dozen, - -	$5 00
" 1½	—10 " "	11 " "	6½ " " "	6 25
" 2	—11 " "	12 " "	7½ " " "	8 00
" 3	—19 " "	10 " "	3 " " "	5 00
" 4	—21 " "	11 " "	3 " " "	6 50

All orders must specify whether Ash or Walnut fronts are desired. Other kinds of wood can be furnished, and special rates given upon application.

We prefer to furnish Cases; with drawers in perfect fit, and easy working order. We offer them in sections, as per Figs. 2 and 3. Any number of sections can be placed end to end, giving as many drawers as are required. The joints meeting the ends of the sections, can be finished with a narrow moulding. The cases are made of pine, without paint or stain, subject to such finish as will harmonize with the general appearance of the pharmacy.

No base boards being attached to the cases, they can be made to fit close to the floor, or finished with base board if desired.

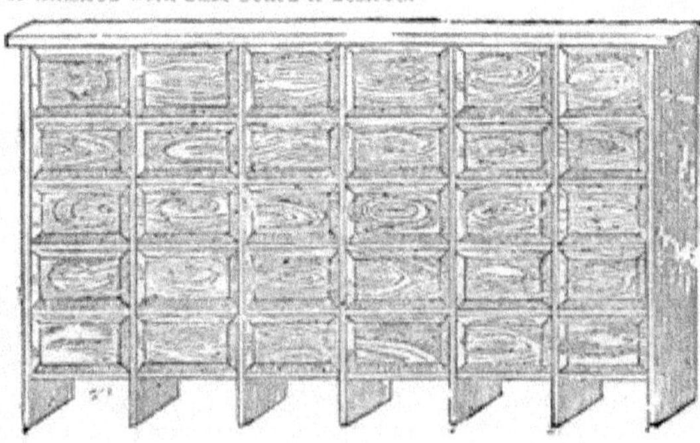

FIG. No. 2.

Case No. 1, (Fig. 2,) containing 30 No. 1 Drawers, per Section, - - - $25 00
Size of this Section is 5 ft. 8 in. long, 12 in. deep, 3 ft. 4 in. to 3 ft. 10 in. high.

Case No. 2, (Fig. 2,) containing 30 No. 1½ Drawers, per Section, - - $29 00
Size of this Section is 6 ft. 2 in. long, 13 in. deep, 3 ft. 8 in. to 4 ft. 2 in. high.

Case No. 3, (Fig. 2,) containing 24 No. 2 Drawers, per Section, - - - $28 00
Size of t iis Section is 6 ft. 8 in. long, 14 in. deep, 3 ft. 6 in. to 4 ft. high.

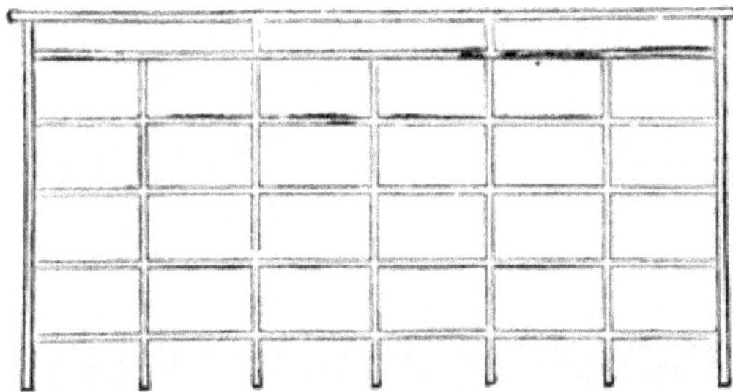

FIG. No. 3.

Case No. 4, (Fig. 3,) containing 24 No. 1 and 3 No. 3 Drawers, per Section, - $23 50

Size of this Section is 5 ft. 8 in. long, 12 in. deep, 3 ft. 1 in. to 3 ft. 7 in. high.

Case No. 5, (Fig. 3,) containing 24 No. 1½ and 3 No. 4 Drawers, per Section, $27 00

Size of this Section is 6 ft. 2 in. long, 13 in. deep, 3 ft. 5 in. to 3 ft. 11 in. high.

The top of each Case projects at the front and ends, subject to finish as required.

READY-MADE DISPENSING COUNTERS.

The Trade supplied by Van Schaack, Stevenson & Reid.

The Dispensing Counters are finished on the outside in panels of Black Walnut, oiled; the drawers and shelving of Pine or Poplar, stained; tops of White Marble. Drawers, shelving, compartments, and open spaces for pill machines, graduates, mortars, scales, vials, corks, labels, spatulas, pill tile, paste pot, and every requisite for a dispensary, are suitably arranged.

This is a beautiful piece of furniture.

FIG. No 4.

Fig. No. 4, Largest Size, Front View—Price, - - - - - $130 00

Size, 7 ft. long, 2 ft. 3 in. wide at the floor, top shelf 11 in. wide, and height over all, 5 ft. 6 in.

The front of this Counter has a Glass Perfumery Case with four doors, and shelves inside 3 inches wide. The back of the show case can be finished with mirrors, at an extra cost of $15.00.

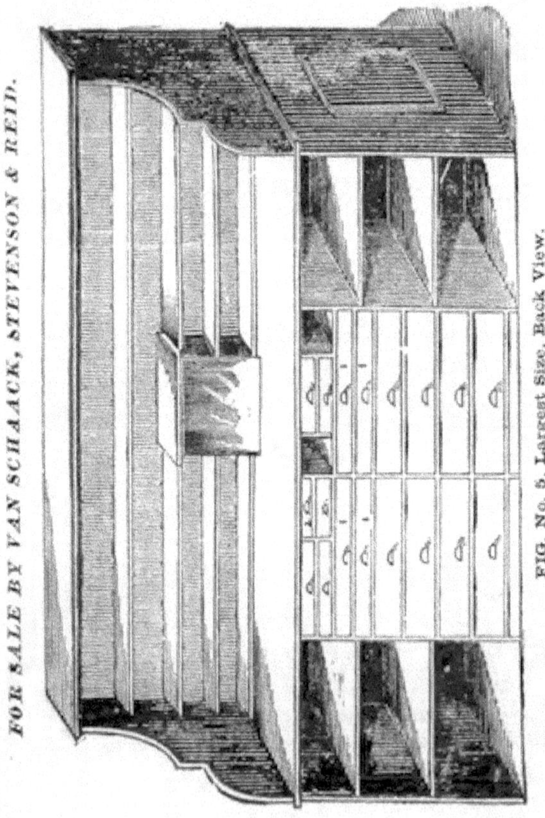

FIG. No. 5, Largest Size, Back View.

This Counter has a Scale Case with a glass door, shelving for dispensing bottles, with room for 10 pint, 20 half-pint, 28 four-ounce, and 34 two-ounce bottles. The shelving can be arranged to suit the purchaser. Poison Closet with glsas door, under lock and key, can be attached over the Scale Case at an extra cost of $5.00.

FIG. No. 6, Medium Size, Front View, - - - - - - - - - $95 00

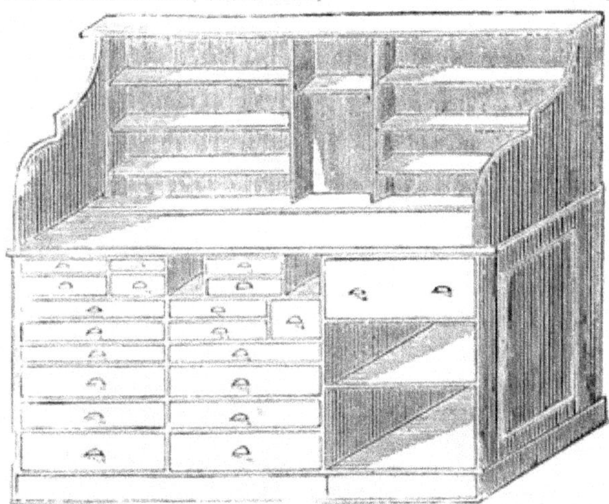

FIG. No. 7, Medium Size, Back View.

This Dispensing Counter (Figs. 6 and 7) is 6 ft. long, 2 ft. 3 in. at the floor, top shelf 7½ in. wide, height over all, 5 ft. 6 in.; and the shelving is arranged for 7 pint, 20 half-pint, 24 four-ounce, and 27 two-ounce bottles. No glass door to scale case. In place of the walnut panels in the front, we can substitute glass, in either the center one or all three. We can supply for the center panel a beautiful design—a Mortar, with the words "Prescription Department" ground in the glass—at an extra cost of $5.00. The other panels can be filled with gilded and decorated glass, in such designs as "Coats of Arms," or of the "Association of Pharmacies," or "Prescriptions Carefully Compounded," or other designs of any kind, at an extra cost of $7.00 for each panel.

FIG. No. 8.

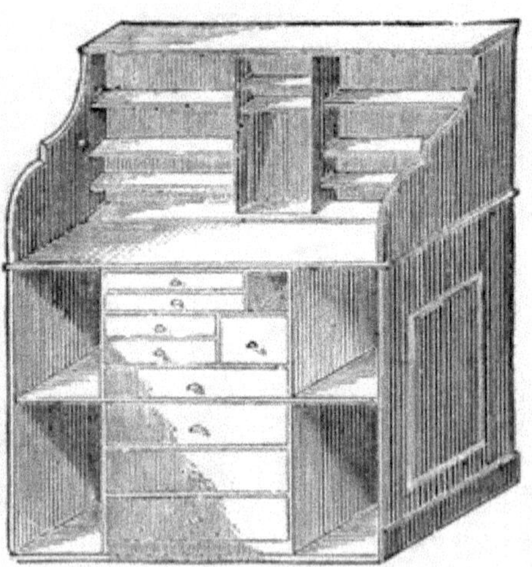

FIG. No. 9.

Fig. No. 8, Smallest Size, Front View, - - - - - - - - - $80 00

Fig. No. 9, Smallest Size, Back View.

This Dispensing Counter (Figs. 8 and 9) is 5 ft. long by 2 ft. 3 in. at the floor, top shelf 7½ in. wide, and same height as others, containing shelving to hold 5 pint, 16 half-pint, 20 four-ounce, and 23 two-ounce bottles. The glass panels can be substituted in this as in the larger sizes, at the same extra cost.

FIG. No. 10.

Fig. No. 10 is a fair representation of a Dispensing Counter with all Glass Panels. Either of these Dispensing Counters can have attached to the top beautiful Scroll Work, with or without Clock. (See Fig. 11.)

FIG. No. 11.

Scroll Work, with Eight-Day Clock in center, for either size Counter, extra,	$15 00
Scroll Work, without Clock,	5 00

If you desire the upper part of a Dispensing Counter, to place on the top of your shop counter, we will furnish them in any style or size desired, with the Show Case, Decorated Glass, or Walnut Panels, and other conveniences enumerated before, upon short notice. Special quotations given.

☞ We desire to call the attention of those about embarking in the Drug Trade to the elegance, convenience, rapidity of putting up, and cheapness of these Drawers, Counters, etc.

Van Schaack, Stevenson & Reid

HAVE ALWAYS A FULL SUPPLY OF

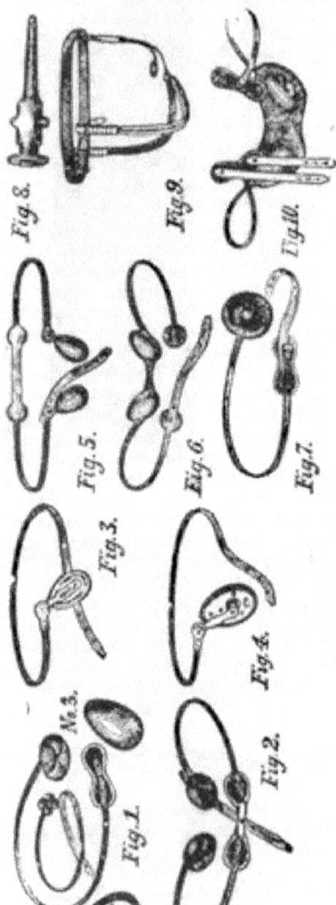

I. B. SEELEY'S

HARD RUBBER TRUSSES,

Supporters, Etc.

Comfortable, Light, Cleanly and Durable. Never Rusts, Breaks, nor Soils.

STEEL SPRING covered with HARD RUBBER.
(Used in Bathing.)

Fig. 1. ADULT AND INFANT'S SINGLE TRUSS, patent, grooved pad, relieving cord from pressure, used with or without strap. Nos. 3 and 8. Pads of this figure (convex plain, convex ball and socket set-screws) are applicable on Figs. 1 and 2 style Truss. Fig. 3. TRUSS applied from *side* of *Rupture*, strap attachment, pad reversible, with set-screw. Fig. 4. Same as Fig. 3 Truss, ball and socket, pad attachment, with set-screw. Fig. 5. DOUBLE TRUSS, of Figs. 3 and 4 style. Fig. 6. "Hood" Pattern Truss. Fig. 7. UMBILICAL TRUSS. Fig. 8. Hard Rubber PILE PIPE. Fig. 9. PROLAPSUS ANI SUPPORTER, with Hard Rubber Pad and Springs. Fig 10. Hard Rubber ABDOMINAL SUPPORTER.

Lightest, safest, neatest, most effectual and best Mechanical Supports known.

DWYER'S

PECTORAL SYRUP

For the cure of the various diseases of the Lungs and Bronchial Tubes, such as Coughs, Asthma, Hoarseness, Bronchitis, Whooping Cough, Influenza, Croup, Colds, and the incipient stages of Consumption.

Prepared only and sold by the Proprietors,

Van Schaack, Stevenson & Reid,

WHOLESALE DRUGGISTS, CHICAGO.

MEDICAL SADDLE BAGS.

Made of the Best Russet Bridle Leather.

Patent Leather Covers. Space under Covers for Instruments, etc. Pattern Mahogany Drawers in the ends of the lower part.

☞ Solid Leather drawers, One Dollar Extra.

No. 7, containing 20 1 oz., 4 1½ oz., g'd stoppered bottles, $14 00

No. 13, containing 16 2 oz. glass stoppered bottles........... 12 00

Made of the best RUSSET BRIDLE LEATHER, with Patent Leather Covers. Drawer of Polished Mahogany, Velvet Lined.

No. 1, containing 12 1½ oz., 12 ¾ oz.—24 bottles, $13 50

" 2, containing 10 1½ oz., 10 ¾ oz.—20 bottles, 12 50

" 3, containing 8 1½ oz., 8 ¾ oz.—16 bottles, 11 50

Flat Pattern, Two Flaps.

(See Cut.)

No. 14, containing 10 1½ oz., 10 ¾ oz.—20 bottles, - - 12 50

" 16, containing 12 1½ oz., 12 ¾ oz.—24 bottles, - - - 13 50

" 17, containing 12 1½ oz., 21 ¾ oz.—33 bottles, - - - 15 00

The Trade supplied by

VAN SCHAACK, STEVENSON & REID, CHICAGO.

MEDICAL SADDLE BAGS.

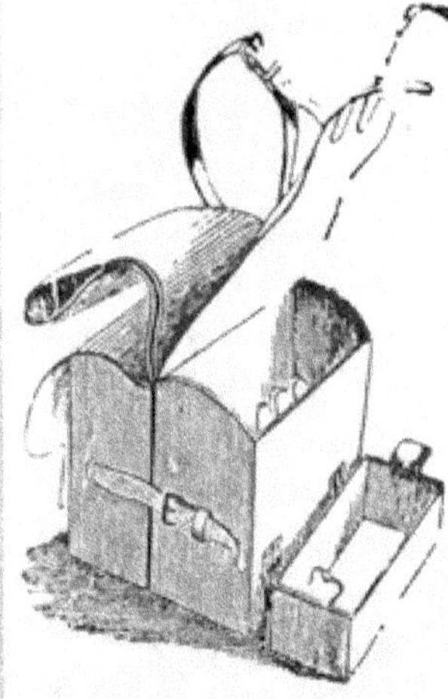

BOX PATTERN.

No. 4, containing 24 ground stoppered bottles.... $13 50

" 4, Extra, with Pockets.. 14 50

" 8, Extra, with Pockets.. 13 50

Plain Saddle Bags, containing 20 cork stoppered vials..... 9 50

Mahogany Medicine Chests.

WING PATTERN.

No. 3. Same as above cut, slide in back, one large offset Drawer at bottom for 3 Jars, and space for Instruments; small Drawers, inside, for Mortar, Measure, Scales, etc., containing 8 4 oz., 12 2 oz.—20 bottles.................. $21 00

Box Pattern, with Lid and Drawers under.

No. 7, containing 10 4 oz., 12 2 oz., 7 1 oz.—29 bottles...... $16 50
" 10, " 3 4 " 10 2 " —13 " 9 50

MEDICINE CHESTS FOR PHYSICIANS.

MADE OF THE

Best Russet Leather.

Containing the following Square Glass Stoppered Bottles.

In Mahogany Trays, Mortar, Graduated Measure,

Four Jars, Tray for Scales, and Space for INSTRUMENTS under Bottles.

		oz.	oz.	oz.	oz.	bottles.	price.	Inches. length.	width.	height.
No. 1,	containing	4 4,	16 2,	18 1,	6 ½,	44,	$22 50,	14,	9¾,	9¾
" 2,	"		14 2,		42 ½,	56,	25 00,	12¾,	7¾,	11
" 5,	(see cut,)	2 4,	14 2,	16 1,		32,	20 00,	11,	8¾,	9¾
" 6,	containing	2 4,	12 2,	13 1,		27,	18 00,	9¾,	8¾,	9¾

The following without Mortars, Measures or Jars.

					bottles.	price.	length.	width.	height.
" 7,	containing	4 4 oz.,	10 2 oz.,	6 1 oz.,	20,	13 50,	9½,	7,	8¾
" 8,	"	15 1½ oz. bottles,	-	-	-	12 00,	8¾,	5¾,	7¾

Tray in front of Bottles.

						bottles.	price.	length.	width.	height.
" 9,	containing	10 2 oz.,	4 1 oz.,	-	-	14,	11 50,	9,	6¾,	5¾
" 10,	"	5 4 oz.,	7 2 oz.,	6 1 oz.,		18,	14 00,	10,	7¾,	5¾
" 11,	"	4 4 oz.,	9 2 oz.,	12 1 oz.,	6 ½ oz.,	31,	15 50,	13,	9¾,	5¾

All the above with Locks.

			price.	length.	width.	height.
" 13,	Buckle and strap, 10 1½ oz. bottles,	- -	5 50,	7,	3⅝,	4¾
" 14,	containing 2 4 oz., 6 2 oz., space in front,	8,	7 50,	7,	5¼,	5

Complete Set of Latin Labels, Twenty-Five Cents.

APPLIANCES FOR JEROME KIDDER'S BATTERY.

	Each.
End Sponge Holder	$1 25
Side " " for use under a dress	1 75
Long Side Sponge Holder, 14 in.,	2 00
Sponge Holder, with Interrupter,	3 00
The same, smaller	2 00
Ear Electrode	1 50
Eye "	1 75
Metallic Hollow Ball Electrode	75
Metallic Brush for Anæsthesia	1 50
Throat Electrode	1 50
Tongue Electrode, Silver Plated	1 50
Rectal " " "	1 50

	Each.
Uterine Electrode	$1 50
" " Bell Shaped	2 00
Vaginal Electrode	1 75
Urethral "	1 50
Electrode for Paralysis of the Bladder	2 50
Needles, Steel & Platina, $1 00 to	1 50
Cautery Burners & Cutting Loops	4 00
Cutting Loop for Larynx	2 00
Burner " "	2 00
Adjustable Electrode Plate, 4¾ in.	1 25
Smaller do., 75c. Strap for do.	50
Hinged Copper Foot Plate	1 25

Physicians' Visiting Electro-Medical Apparatus—six currents, double stopper battery, with rubber stoppers, coils arranged as in No. 4; can be carried without spilling the fluid. Polished mahogany or walnut case, 8 inches long, 6¼ wide, and 6¾ deep. Price, including Handles and Sponge-holder, $25.

Kidder's Tip Battery.

Kidder Platina,	$2 00
" Zincs, pair,	1 00
" Clamp,	75
" Cup,	75
" Cord, yard,	30
" Appliances, full set in case, (See page 266.)	15 00

D. & K. Appliances, each,	$1 00
Kidder Battery, Acid,	18 00
D. & K. " Crank,	8 00
Galvanic " 8 Cells,	20 00
" " 16 "	35 00
" " 32 "	60 00

DAVIS & KIDDER'S Magneto-Electric Machine. (See page 157.)

DR. ABERNETHY says: "Electricity is a part of surgical practice that may be considered *unique*."

THIS MAGNETO-ELECTRO MACHINE produces galvanic electricity, by simply turning a crank. It requires no battery, no solutions, and no friction. It develops electricity by what is called the inductive influence of a permanent magnet, and it gives a current of the same character of the Electro-Magnetic Machines which require a battery to actuate them.

By turning the crank the armature is made to revolve rapidly on an axis of its extremities, passing close to the poles of the magnet, become charged with magnetism, which they lose as they leave the poles, and become charged with the opposite magnetism, as they approach the opposite poles at each half revolution. The electricity thus induced is powerful in proportion to the rapidity of the revolution, but is not directly available to give shocks, without a break piece, or spring, on the axis, which alternately breaks and completes the circuit of the coils, at which moments a very energetic reactive current is excited, which is amply sufficient for all medical purposes. The apparatus thus described, is precisely the same as used in *Davis & Kidder's Patent Magneto-Electro Machine.*

At each end of the machine are the conductors; at the farther corner of the box on the left is represented a sliding rod which controls the position of the straight armature, in relation to the poles of the magnet. By moving this on or off the poles, the strength of the shocks may be regulated.

SIZES OF GLOBULES.

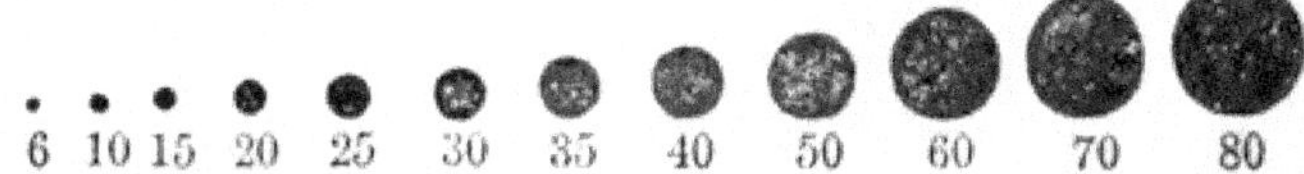

6 10 15 20 25 30 35 40 50 60 70 80

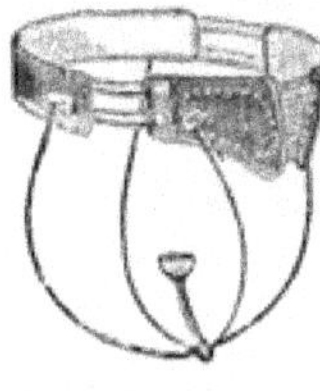

Dr. McIntosh's Natural Uterine Supporter.

Anatomical Models,

PATHOLOGICAL and OSTEOLOGICAL PREPARATIONS, &c.,

FOR SALE BY

VAN SCHAACK, STEVENSON & REID.

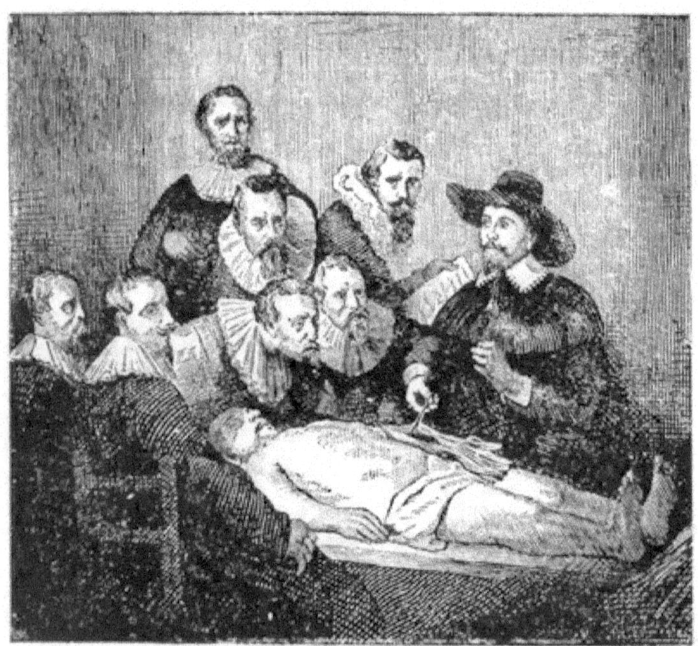

(Rembrandt's Dissecting Room.)

MODELS representing cases that have come under the observation of eminent Professors and Physicians, and have been formed in compliance with their immediate instruction.

No. 1.	Posterior wall of the bladder, freely exposed	$ 4.00
" 2.	Dilation of the heart	15.00
" 3.	Thyroid gland enlarged, or crop with deficiency of left artery	5.00
" 4.	Wolf's jaw	2.50
" 5.	Fracture of the collar-bone, with the entire chest	8.00
" 6.	Fracture of the ribs	1.00
" 7.	Glanders, (of the horse)	5.00
" 8.	Fissure of the uterus	12.00
" 9.	Cancer on the right extremity of the stomach, with diseased mesenteric glands	15.00
" 10.	Anchylosis of the elbow joint, with arm showing caries	3.00
" 11.	Inflammation, suppuration, gangrene of the small intestine	7.00
" 12.	Hernia umbilicalis, exposure of posterior wall of the bladder, deficiency of the ureters, short imperfect penis	15.00
" 13.	Small fatty tumor on the surface of the upper part of the jejunum,	1.50
" 14.	A part of the inflamed, jejunum	7.00

Anatomical Models—continued.

No. 15.	A stomach poisoned with sugar of lead	7.00
" 16.	A foot with three toes	7.00
" 17.	Recent fracture of the neck of the femur	3.00
" 18.	Fracture of the collar-bone a, b, each	1.00
" 19.	Fracture of the tibia	2.00
" 20.	Fracture of the lower extremity of the femur	5.00
" 21.	Fracture of the upper extremity of the femur, with anchylosis of head of the femur, with acetabulum	13.00
" 22.	Dislocation of the femur	9.00
" 23.	Caries of the left lower jaw	4.00
" 24.	Immoderately enlarged clitoris	14.00
" 25.	Extraordinarily thick (skull) cranium	12.00
" 26.	Tibia and double fracture of fibula	7.00
" 27.	Fracture of the humerus, with artificial joint	3.00
" 28.	Lower jaw, with tumor	4.00
" 29.	Hottentot's apron (elongation of the nymphoe)	6.00
" 30.	Caries of the right lower jaw	4.00
" 31.	Hermaphrodite, (2 parts,) each,	12.00
" 32.	Cranium, with caries	7.00
" 33.	Caries of the vertebræ	8.00
" 34.	Fœtus in fœtu, which was discharged with the feces by a girl ten years old	4.00
" 35.	Lungs inflamed and covered with tubercles	18.00
" 36.	Anchylosis of the knee joint	10.00
" 37.	Caries of the whole lower jaw	7.00
" 38.	Alveolar extension of the right upper jaw in consequence of the action of phosphorous vapors	2.00
" 39.	Resected upper jaw, with tumor	6.00
" 40.	Subject born with one testicle	16.00
" 41.	Calculi, (3 pieces,)	2.00
" 42.	Cow udder, with pox on the teats	12.00
" 43.	Representation of the positiun of the fœtus in the uterus	
" 44.	Deficiency of the anterior wall of the bladder with free opening of ureters	16.00
" 45.	Phosphorous necrosis of the entire lower jaw	10.00
" 46.	Heart, to be taken into four parts	8.00
" 47.	Oblique fracture of the upper part of the femur	5.00

Stands for Show Cases, etc., of Iron, Japanned, Green and Gold, 31 in. high.

Factory Price, each, $12.00.

These are very handsome, and make a firm support. The Braces, running from A and B to E, are arranged to have one end of each screwed to the bottom of the Case, which rests on the cross bars C and D—to which it is also attached. Can be used with Cases of different lengths. When packed for shipment the two braces are unscrewed at A and B, making a compact package.

HUMAN OSTEOLOGY.

The cause of the difference in price of these preparations will be found to exist in the quality of the same, whether the bones contain in their extremity more or less grease, or none at all.

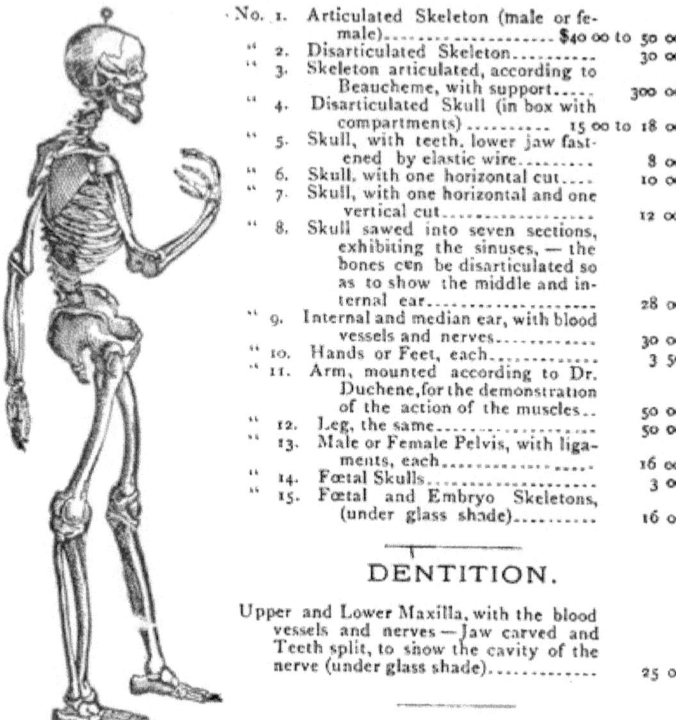

No.	Article	Price
No. 1.	Articulated Skeleton (male or female)	$40 00 to 50 00
" 2.	Disarticulated Skeleton	30 00
" 3.	Skeleton articulated, according to Beaucheme, with support	300 00
" 4.	Disarticulated Skull (in box with compartments)	15 00 to 18 00
" 5.	Skull, with teeth, lower jaw fastened by elastic wire	8 00
" 6.	Skull, with one horizontal cut	10 00
" 7.	Skull, with one horizontal and one vertical cut	12 00
" 8.	Skull sawed into seven sections, exhibiting the sinuses, — the bones cen be disarticulated so as to show the middle and internal ear	28 00
" 9.	Internal and median ear, with blood vessels and nerves	30 00
" 10.	Hands or Feet, each	3 50
" 11.	Arm, mounted according to Dr. Duchene, for the demonstration of the action of the muscles	50 00
" 12.	Leg, the same	50 00
" 13.	Male or Female Pelvis, with ligaments, each	16 00
" 14.	Fœtal Skulls	3 00
" 15.	Fœtal and Embryo Skeletons, (under glass shade)	16 00

ARTICULATED SKELETON.

No. 1, first qua'ity, bound by Elastic Wire, admits of suspension by a Ring fastened through the Skull. For Price, see this List.

DENTITION.

Upper and Lower Maxilla, with the blood vessels and nerves — Jaw carved and Teeth split, to show the cavity of the nerve (under glass shade)	25 00

OBSTETRICS.

Obstetrical Manikins from $35 00 to	350 00
Pelvis, in normal state (Papier Mache) with Fœtus Phantom	20 00

COLLECTION OF MODELS.

The same can be obtained separate, at the price opposite each, or in sets; material—Plaster Paris. They appear in their natural colors, and are now recommended by a number of prominent Colleges. A number of testimonials received.

(A). MAGNIFIED MODELS.

No.	Article	Price
1.	HUMAN HEART, front part to be taken off, showing the four Chambers of the Heart, together with their respective openings and valves	$8 00
2.	" EYE, the upper part of the Pupil (with a microscopic illustration of the Retina, to be taken off, so as to show the Cornea, Iris, the Vitreous Body and Chrystalline Lens	6 50

PRICE LIST OF SURGICAL APPLIANCES AND INSTRUMENTS

FOR SALE BY

VAN SCHAACK, STEVENSON & REID.

Amputating Instruments.

Knives	$3 00 to 5 00
Catlings	2 00 to 5 00
Capital Saw, Saterlee's	5 00
Butchers' Saw	7 00
Saws, Medium Size Movable Back	2 50 to 5 00
Finger Saws	2 00
Chain "	8 00 & 10 00
Lewer's "	3 00
Surgeon's Mallet Lead	3 00
Bone Forceps, small	2 50
" " medium	3 00
" " large, with spring	5 00
Bone Forceps, Saterlee's	2 50 to 3 00
" Gouging Forceps	2 50 to 3 00
" Forceps, Lion Jaw	3 50
Sequestrum Forceps	2 50 to 3 50
Parker's Retractors, per pair,	2 00
Artery Forceps, with Spring Fenestrated	2 00
Artery Forceps, with Slide Fenestrated	3 00
Artery Forceps, plain, with Spring	1 50
Serrafines, each	25 to 50
Tourniquets, Spiral	2 00
" Mott's	2 00
" Field	1 00
Bone Gouges	1 50
" Chisels	1 00 & 1 50
Trephines, Straight	4 00 & 5 00
" Gault's conical	4 50 to 5 00
Heys's Saw	1 75
Trephine Elevators	1 50

Amputating Cases.

Amputating Cases, Ebony Handle	$32 00 & 33 00
Amputating & Trephining Case, Ebony Handle	40 00
Amputating & Trephining Case, Ivory Handle	45 00

Chest Instruments.

Stethoscopes, Cedar, plain	$ 60
" T. & Co.	1 00
" with rubber ring	1 25
" Ebony, with rubber ring.	1 75
" Cammann's jointed,	6 00 to 7 00
" " with Allison's improvem't.	8 50
" with Knight's improvem't	9 50
Stethoscopes, Ebony, Ivory tip, with pleximeter and hammer	$3 50
" flexible rubber tips	1 00
" Ebony, sectional	1 [illegible]
Flint's Hammer	1 50
" Pleximeter	50
Hutchinson's Trocar	4 00
Spirometer, Brown's	10 00

General Operating Cases.

Buck's	$190 00
Post's	1[illegible] 00
Mott's	75 00
California, Ebony Handle	7[illegible] 00

Miscellaneous Cases.

Trephine Case	$16 00
Minor Operating Cases	18 00 to 25 00
Post Mortem Case	25 00
Bone Case, complete	66 00
Dissecting Cases	4 50 to 7 50
" Case, Finnel's.	12 00
Cupping Cases	8 00 & 10 00

Ear Instruments.

Ear and Eye Cases	$33 00 & 55 00
Ear Speculum, Simrock's	5 00
" " Kramer's plain	2 50
" " " G. S., Steel handles	3 50
" " Bivalve, G. S.	2 50
" " Wild's Silver, 3 in set	4 50
" " Wild's Plated, 3 in set	2 25
" " Toynbee's Silver 3 in set	5 00
" " Glass Mirror	75
" " H. R., set	1 50
" " Spier's	4 00
" Douch, Lucas's	1 50
" Spout	[illegible]
" Trumpets	3 00 to 6 00
" Auricles	5 00
Conversation Tubes, Cot.	3 00
" " Silk	[illegible]
Ear Forceps, Wild's	1 50
" " Toynbee's	1 50
" " Bumstead's	4 00
" Snare, Wild's	3 50
" Spoon, Gross's	1 00
" Caustic Carrier	2 00
Clark's Illuminator	[illegible]
Webber's Otoscope	[illegible]
Brinton's " H. R., 3 tips	1[illegible] 00
" " S. P., 1 tip	12 00
" " S. P., 3 tips	16 [illegible]
Toynbee's Otoscope, set	2 50

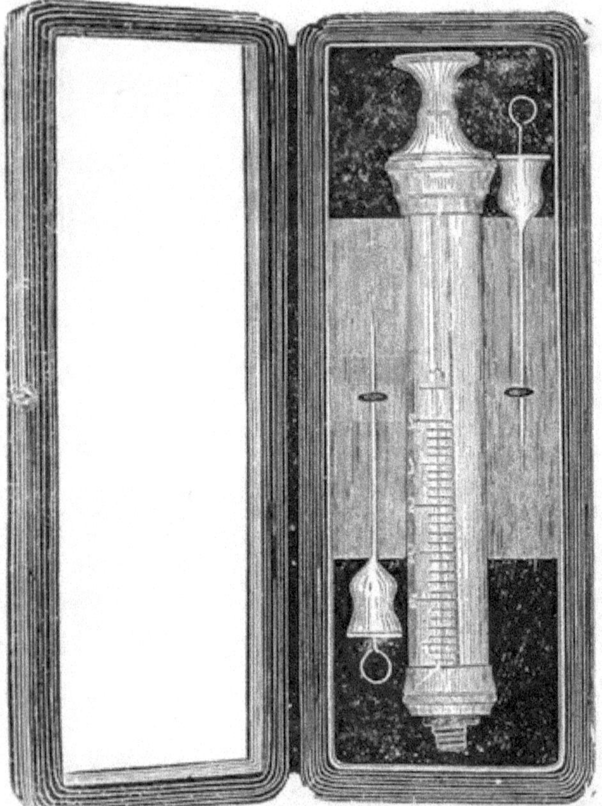

Hypodermic Syringe—Glass, Graduated.

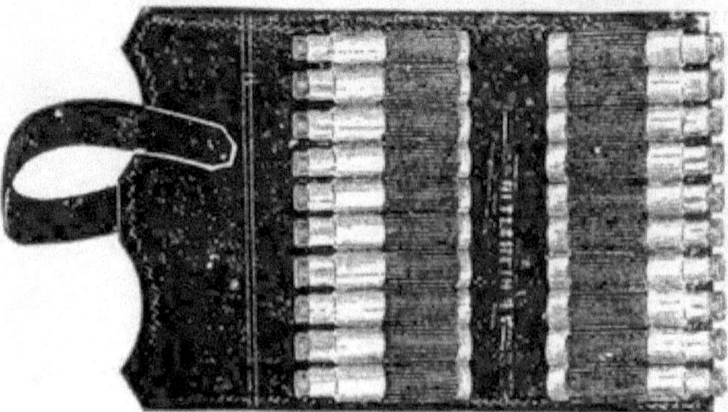

Physician's Case—Morocco, 20 two dram vials.

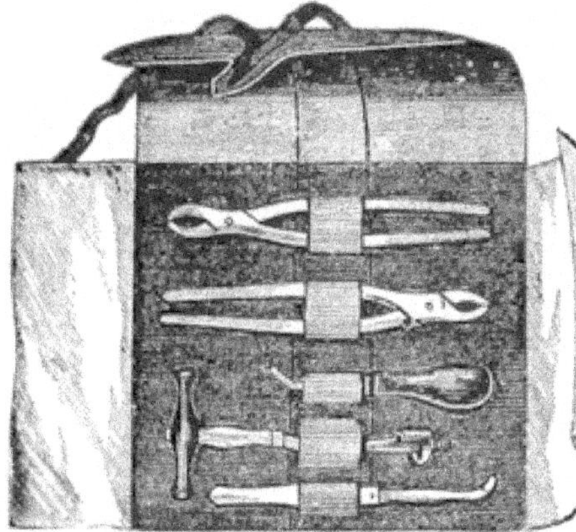

One Turnkey, three hook,	$2 50
One pair Tooth Forceps, curved	2 25
One pair Tooth Forceps, straight	2 25
One Elevator	75
One Gum Lancet, Ebony Handle	1 00
A neat folding leather case, portable	2 00

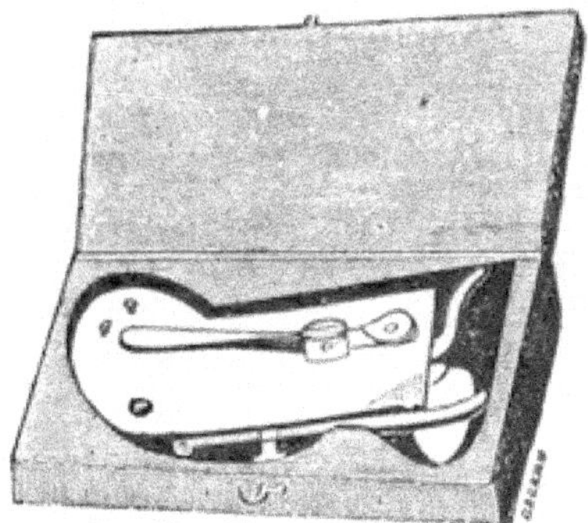
American Horse Spring Lancet.

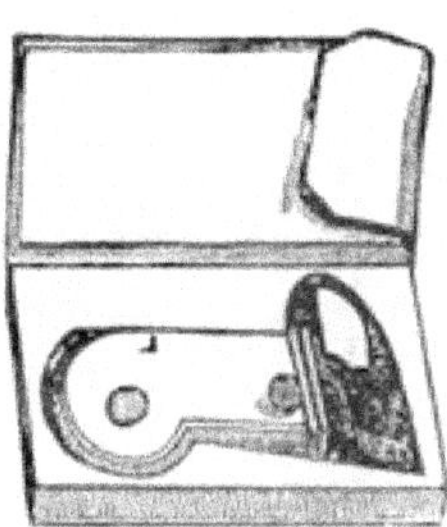
American Silver Spring Lancet, Button Trigger.

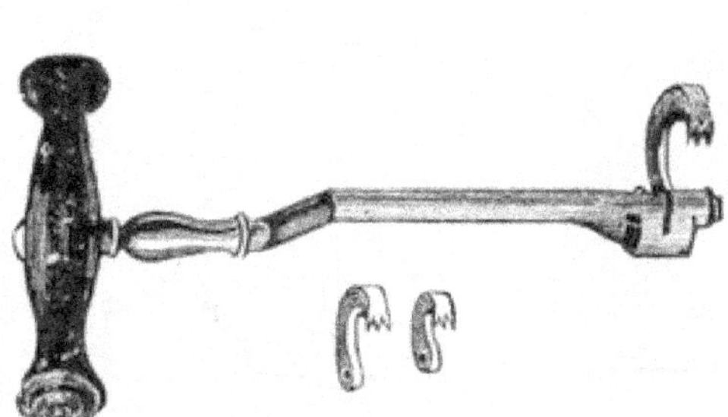
Tooth Key—Ebony Handle.

Tooth Forceps, Children's.

Eustachean Catheters, pure si.	$2 00
" " H.R.	1 00
Tympanums, Wire & Rub.	30
" all Rubber..	50
Trolche's Mirror, 2 inch..	3 00
" " 2 1-2 " ..	4 00
" " 3 " ..	6 00
Ear Speculum-Holder	1 00

Throat Instrum

Laryngoscope Cases...... 4 00 to	25 00
" Mirrors, wi head band..	12 00
Throat Mirrors..........	1 50
Trachea Tubes, Silver.... 4	5 00
" " " Doub. 6 0	8 00
" Hard Rubber....	4 00
" Knife	7 00
Tongue-Depressors, H. 2 00 to	2 50
" " " folding	2 00
" " Steel " 2 00	3 00
" " G.S. " 2 00	3 00
" " G.S. ivory h'dle, fol'g 3 50	4 00
" " Gilt, ivory h'dle, fol'g	5 00
" " Church's.	6 50
" " Green's. 1 50 to	2 50
" Forceps, T. & Co.'s	6 00
Tonsil Instrument, Fahnestock's, plain	6 00
" " Fahnestock's best in case..	12 00
" " T. & Co.'s..	15 00
Uvula Scissors, with claws	5 00
Laryngeal Brush, set.....	2 50
Powder Blower.......... 2 00 to	2 25
H. R. Spray Tube, Rev..	4 00
Plated " " " ..	3 00
Probangs, Sponge.........	25
" Richardson's, in case 3	4 00
" Granger's..........	3 00
" Bristle.............	3 00
Throat Forceps,.......... 2	4 00
" " Burge's..	5 00
Tobold's Lantern.........	25 00
" " with student's	30 00
Laryngeal " C. & S...	4 00
Throat Brush, Childs'....	1 50
Catarrh Syringe, H. R...	1 50
" " new style	2 00
Laryngeal Eccraseur ...	3 00
Tobold's Cauterizer......	5 00
Laryngeal Galvanizer....	3 50
Œsophagus Bougies, Eng.	2 50
Simrock's Rhinoscope, in ca	8 50

Eye Instrum

Eye Cases.............. 11 75 to	35 00
Strabismus Cases....... 8 25 to	10 00
Eye Scissors, straight....	1 00
" " curved or flat	1 50
" " knee bent...	1 25
" Styles, solid..........	50
" " hollow........	75
" Speculum, plain, small	75
" " " large	1 00

Speculum, Lawrence's	$3 00
" Critchet's.	2 00
" Greafes's..	2 50
Shades .... 50 to	4 00
Lid Elevators........ 1 50 to	2 00
Spatulas .... 75 to	1 25
l's Probe ....ea.	50
Forceps, straight.....	1 00
" curved......	2 00
Lash Forceps	1 00
tropium " with Knife, T. & Co...	8 00
" " with Knife, Morton's..	7 00
" " Desmarre's	3 50
wman's Probes, set.... 2 50 to	3 50
" " with director in case	5 00
" " with direc. and knife	6 50
ringe, Anel's, Silver....	12 00
" " Hard Rub.	6 00
hthalmoscope, Liebriec' 4 00 to	6 50
" Coscius's Lary	45 00
" Coscius's..... 5 00 to	7 50
" Desmarre's... 6 50 to	8 00
" Pocket	[illegible]
" Common ...... 5 00 to	[illegible] 00
" Reflector......	3 50

Obstetrical Instruments.

tetrical Cases, empty, short	3 50
" " empty, long	4 00
" " filled..18 00 to	45 00
" Forceps, Elliott's with screw	9 00
" " Miller's..	9 00
" " " with joint	[illegible]
" " Bedford's.	9 00
" " White's.	[illegible] 00
" " Hodge's $7 00 to	8 00
" " Thomas's	6 00
" " Simpson's short	7 00
" " Simpson's long	8 00
" " Davis's...	7 50
" " Meigs's..	7 50
" " Denmann's	7 50
" " Comstock's	8 00
" " Wallace's.	8 00
pson's Cranioclasp....	10 00
forator, Hodge's	6 50
" Thomas's	8 00
" Blot's	6 00
" Holmes's Lever	5 00
" " double crossing	5 00
" Nageli's.......	4 00
" Simpson's.....	3 00
" Bedford's	3 00
" Smelley's	2 50
acenta Forceps, plain...	2 50
" " d'ble cros'ng	3 50
" Hook and Lever, Carey's	1 00
" Forceps, Loomis's	7 00
raniotomy " Thomas's	6 00
" " Ramsbotham's	6 00
" " Meig's str. 3 00 &	4 00
" " " cur. 3 50 &	4 50
ectis .... 2 00 &	2 50

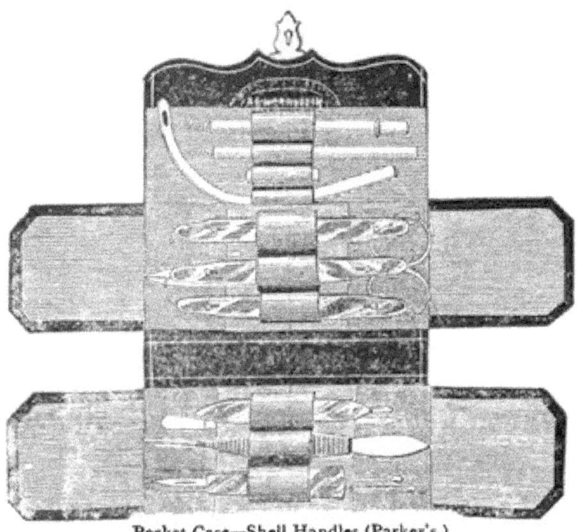

Pocket Case—Shell Handles (Parker's.)

Hypodermic Syringe—Fenestrated.

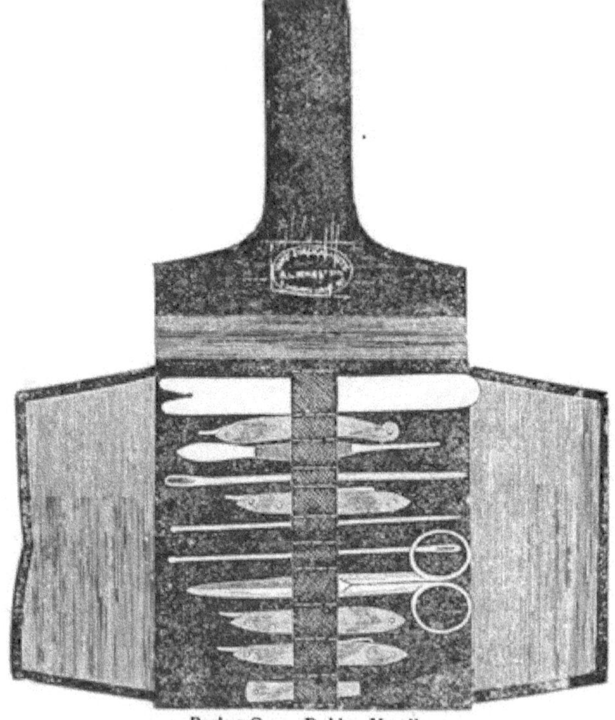

Pocket Case—Rubber Handle.

Tooth Forceps—Incisors.

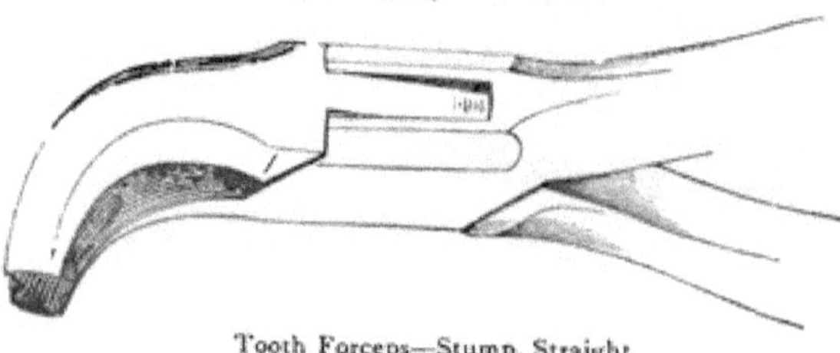

Tooth Forceps—Stump, Straight.

Blunt Hooks and Crotchet comb'd $1 50
Blunt Hooks and Crotchet combined, guarded, Budd's 5 00
Crotchets, Ebony handle.. 1 50 & 2 50

Rectum Instruments.

Speculum, Weise's........ $12 00
" Van Buren's... 5 00
" Bovin's, 2 blades and hinge...... 6 00
" Bovin's, 3 blades and hinge...... 9 00
" O'Reiley's bi-valve, steel.... 5 50
" bivalves, Ger. sil. 5 00
" Matthews'..... 14 00
" glass mirror.... 1.00
Pile Clamp, Smith's....... 5 50
" " " ivory m'td 6. 50
" Needles, Bush's....set 5 00
" Supporter, Boulton's. 1 50
" " Reed's ... 1 50
Rectum Bougies, English, straight 1 00
" " English, conical 1 50
" Ointment Syringes 2 50 to 5 00
Trocar 3 00
" Suppository Syringe, H. R.... 75
" Suppository Syringe, glass....... 25
" Dilator, steel..... 6 00

Urethral Instruments.

Stricture Cutter, Peter's.. 6 00
" " White's. 15 00
" " Charrier's 15 00
" " Maisoneup's 15 00
" " Civial's.. 12 00
" " Thompson's 18 00
" " Gouley's. 14 00
" " Trelot's.. 15 00
" Dilator, Priestley's 15 00
" " Stearn's.. 6 00
Urethral Sounds, steel, Nos. 1 to 12, ea.. 1 00
" " steel plat'd, Nos. 1 to 12, each 1 50
" " nick. plated, Nos. 1 to 18, VanBuren's conical, each 1 75
Bougies, Eng., plain.... " 25
" extra ... " (13 to 18) 50
" olive tips " 75
" conical.. " 60
" filiform. " 1 00
" French, olive tip " 75
" plain... " 15
" conical. " 1 00
" Wax........... " 50
" Aboule .... " 1 00
" Metal, flex. Nos. 1 to 12......... " 50
Catheters, Eng. .... " 25
" " extra... " (13 to 18) 50
" French....... " 15
" Silver .... " 2 00
" " plated. " 1 00
" Metal, flex... " 75
" Gouley's 4 50
" Squires's vertebrated $6 to $8

Gouley's Guides.......... $ 60
Lalleman's porte caustic silver, 5 00
Bumstead's Insuflator.... 5 00

Lithotomy Instruments.

Lithotomy case............ 35 00 to 45 00
" staves.......... 2 00 to 2 25
" bistouries 2 00
" scoup & gorget 3 50
" forceps......... 2 50 & 3 50
" bistouri cache . 22 00
" Thompson's lithotrite....... 35 00
" stone crushing forceps........ 8 00

Uterine & Vaginal Instrum'ts.

Uterine Case, Gardner's.. $50 00
" " Elliott's .. 25 00
" Instrument, Gardner's manifold..... 10 00
" Canula, with Trocar 5 00
" " " Claws. 5 00
" Sounds, Simpson's plain 1 75
" " " graduated 2 00
" " " grad. joint. 2 50
" " Sims's..... 1 50
" " Elliott's elevat'g 8 00
" Probes, Lent's, 1 bulb 1 25
" " " 2 " 2 00
" " Emmet's ... 1 25
" " Sims's...... 1 25
" " Budd's 60
" Aplicator, Emmet's 1 75
" Knife, Emmet's, 2 blades, with caustic holder.............. 10 00
" Caustic Carrier, Byford's, silver blades. 2 50
" Caustic Carrier, Byford's, platina blades 3 50
" Caustic Carrier, Gardner's Gilt..... 3 00
" Caustic Carrier, Gardner's Silver... 2 50
" Caustic Carrier, Sanger's 2 00
" Dressing Forceps, Elliott's 2 00
" Dressing Forceps, with lock.......... 2 25
" Polypus Forceps... 4 00 & 6 00
" Caustic " ... 2 50 & 3 00
" Galvanizer, Hammond's 7 00
" Dilators, Barnes, for air...............set 7 50
" Dilators, Barnes', for air or water.....set 9 00
" Dilators, Atlee's... 3 00
" " Nott's.... 3 00
" Syringe, " 3 50
" " H. R..... 1 50
" Catheter, double current, silver........ 5 00
" Catheter, double current, silver, Nott's. 4 00
" Tenaculum, " .. 3 00
" Sponge Tent Carrier 1 25
" Hysterotome, Simpson's 8 00
" Sound, Gidden's... 2 50

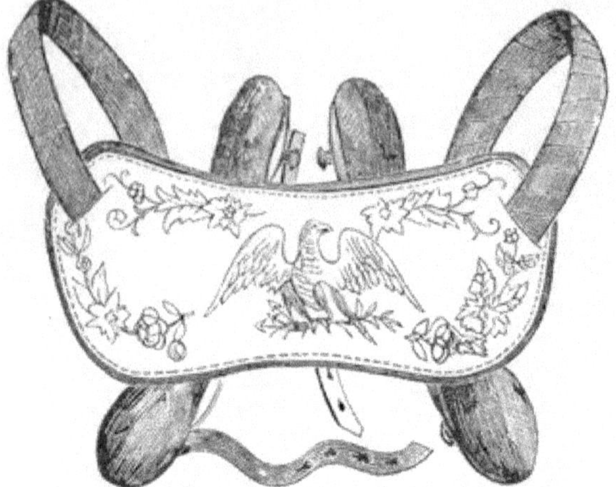

Supporter—Fitch's Kid or Velvet Pad.

Tonsil Instrument, Fahnestock's

Caustic Pocket Holder—Silver Handle.

The New Style Drug Mill. Strong and effective, can be readily taken apart and cleaned.

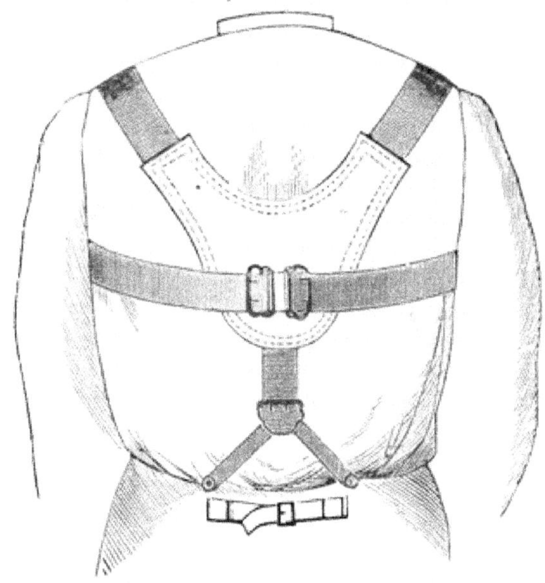

Shoulder Brace--Leather Back, Female or Male, per dozen, $10 50.

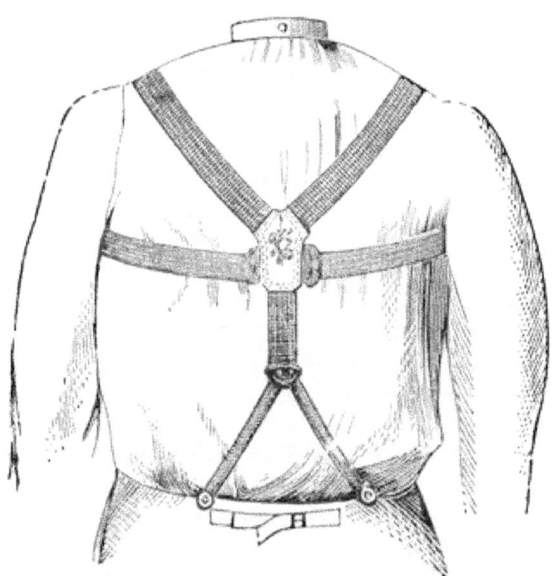

Shoulder Brace--Patent, Female or Male, per dozen, $10 50.

Uterine Eccraseur, Warren's	$6 00
" Speculum	8 00
" Scarificator, Pinkham's	7 00
" Canula, Gooches', silver, straight	6 00
" Canula, Gooches', silver, curved	6 00
" Scarificator, Buttles'	2 00
" Dry Cupping Inst., Thomas's	2 00
" Porte Tampon, Thomas's	2 50
" Ointment Syringe, Lent's	8 00
" Ointment Syringe, Messinger's	8 00
" Ointment Syringe, Stohlman's	6 00
" Ointment Syringe, Barker's	2 50
" Ointment Syringe, Dixon's	2 50
" Tents, Sea Tangle, solid ... each	25
" Tents, Sea Tangle, hollow	35
" Tents, Sponge, with Carbolic acid	25
Pessaries, Gum Ring } see p. 154.	
" Hard Rub. concave }	
" Hodge's open end. }	
" " closed " }	
" Meig's watch spring }	
" Sims's metal ring	50
" Hoffman's soft rub.	2 00
" " hard "	5 00
" Frazier's " "	5 00
" Cutter's " " jointed	3 00
" Scattergood's	3 50
" Galvanic stem, plain	1 00
" " " Thomas's	1 50
" Inflating, long stem American	
" Inflating, long stem French, stop cock	1 50
" Inflating, long stem French, with stop cock and inflator	3 00
" O'Leary's ... 6 00 to	8 00
" Hornby's ... 6 00 to	8 00
" C. & S., stem & cup	6 00
" Babcock's silver	12 50
" Simpson's stem	7 00
Speculums, 4 valve T. & Co.'s	12 00
" 4 " " with plug	9 00
" 4 val., long h'dle with plug	12 00
" 4 val., long h'dle fold'g, with pl'g	12 00
" Bi-valve, G. S., plain	7 00
" Duck bill, Cusco's folding, hand	6 00
" Duck bill, Storer's	6 00
" " Hewitt's	6 50
" " Taylor's	8 00
Speculums, Thomas's latest	$22 00
" " Duck bill	6 00
" " Telescopic	8 00
" Emmet's	20 00
" Sims's ... 4 00 to	5 00
" Nott's ... 13 00 &	15 00
" Meadow's 3 val.	11 00
" " 4 "	13 00
" Smith's	9 00
" Hard Rubber	1 00
" Glass Mirror	75
" " " round end	1 00
" Opaque Glass.	75
" Plain Glass	50
Sims's Knife, rotating	5 00
" Knives, R. & L. per pr.	3 00
" Knife, straight	1 50
" Tenaculum	1 25
" Wire Adjuster	1 50
" Blunt Hook ... 1 25 to	3 00
" Sponge Holder	1 00
" Curette ... 1 00 to	1 75
" Seizing Forceps	4 00
" Twisting "	4 00
" Needle " ... 2 50 to	5 00
" Scissors, straight	4 00
" " cur. or flat	4 50
" " knee bent.	4 50
" Caustic Forceps	4 00
" Elevator	8 00
" Depressor	1 50
" Catheters, silver	2 00
" " metal	75
" " hard rub.	75
" Porte Tampon	5 00
" Vaginal Dilators, glass	1 00
" Vaginal Dilators, hard rub.	1 25
" Needles ... doz.	2 00
Silver Wire ... per coil	60
Platina Cups, Lent's	3 50

Pocket Case Instruments.

Shell Handle, 1 blade, plain	$ 1 00
" " 2 " "	1 50
" " 1 " catch back	2 00
" " 2 " "	3 00
" " 4 " "	7 00
" " Finger Saw.	2 25
Forceps, Artery ... 75 to	3 00
" Gunn's	3 00
" Hamilton's	3 00
Dressing Forceps, Scissor handles	1 50
Plain Forceps	1 00
Thumb Lancets, shell	75
Abcess " "	1 00
Exploring Needles, shell	1 00
Gum Lancets	1 00
Seaton Needle Shell	1 00
Scissors, straight ... 75 to	1 00
" curved or flat	1 50
" knee bent	1 50
Silver Probes	50
" " jointed	1 50
Steel Directors	75
Surgeon's Needles ... doz.	1 00
" Silk ... "	1 00
Spatula and Elevator	60

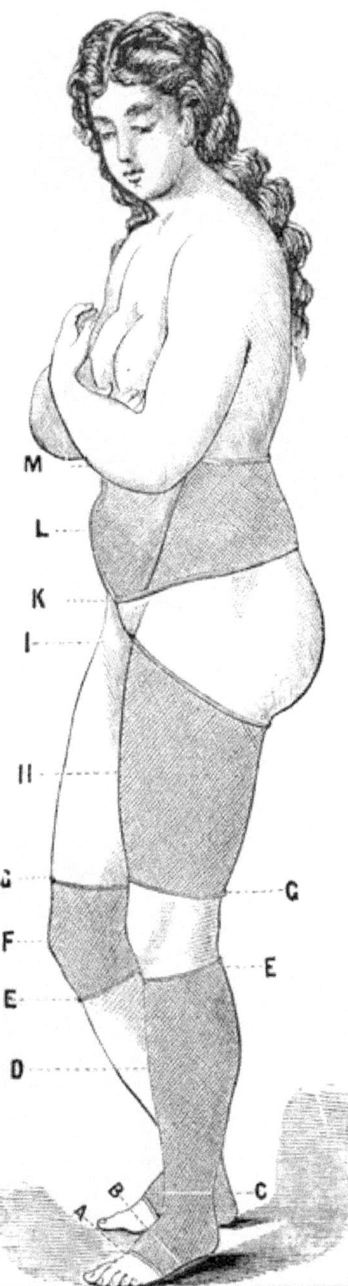

DIRECTIONS

— FOR —

Abdominal Belts, Thigh Stockings, etc.

Explanation of the Letters.

H—Centre of Thigh.
I—Uppermost part of Thigh.
K—Two inches below tips of Hips
L—Navel.
M—Two inches above the Navel.

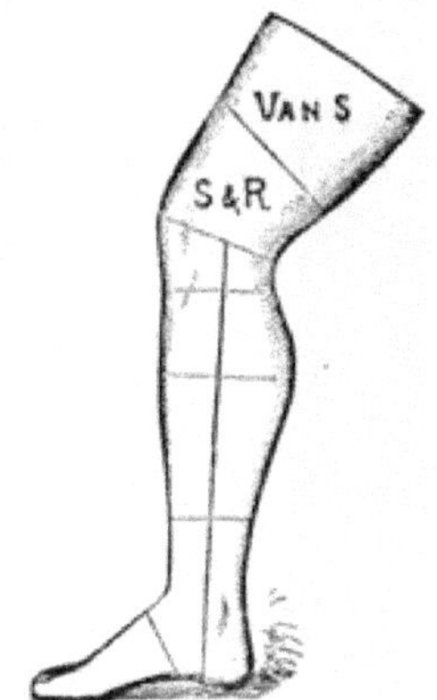

Directions for Measuring for Elastic Stockings.

Indicate the measure around the limb in inches, opposite the lines of the cut.

For Stocking to cover the entire limb, measure at I, H, G, F, E, D, C, B, A, also give length of limb from F to floor at B, and from F to I.

For Stocking to reach to H, measure at H, G, F, E, D, C, B, A, also from F to floor at B.

For Stocking to reach to E, measure at E, D, C, B, A.

For Anklet, measure at D, C, B, A.

For Knee Cap, measure at G, F, E.

Hypodermic Syringe, Pocket.

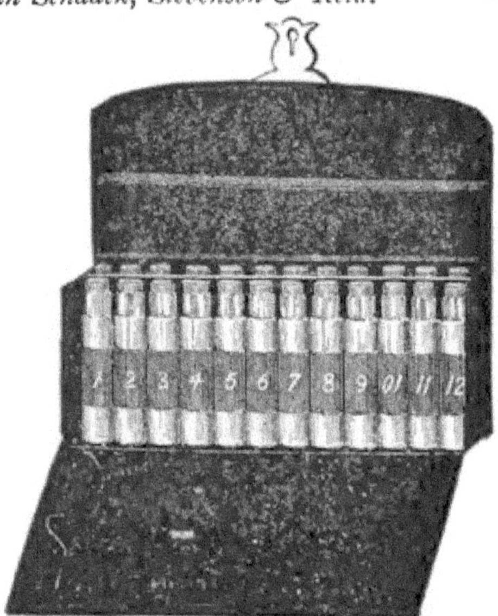

Physician's Vial Case, Morocco, 24 vials, 1½ dram.

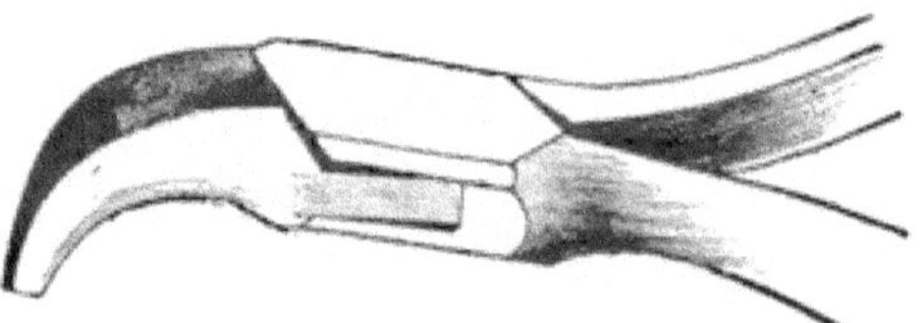

Tooth Forcep, Stump curved.

Stethoscope, Flexible.

Cupping Apparatus, Scarificator.

Caustic Holders, short, silver $2 50
" " H. R. & " 1 00
Catheters, Female, silver.. 1 00
" male & female " .. 3 50
" " " plated 1 50
" " " "
with Caustic holder 3 00
" silver, " " 5 00
Empty Pocket Cases, Morocco, 2 fold............ 1 25 to 2 50
Empty Pocket Cases, Morocco, 3 fold............ 1 50 to 3 00
Empty Pocket Cases, Russia, 2 fold............... 3 00
Empty Pocket Cases, Russia, 3 fold............... 4 00

Pocket Instrument Cases.

Plain, Rubber Handle....$7 50 to 10 00
" Shell " .... 9 00 to 18 00
Tiemann & Co.'s Patent Case. 33 00
Hamilton's, with Silver Catheter. 32 00
" without Silv'r Catheter. 27 00
Andrew's Case, Hard Rubber. 30 00
B. & S. Case, Silver Catheter, 27 00
" Plated Catheter 25 00
Parker's Case, Plated Catheter. 18 00
" Silver Catheter, 25 00
Van Buren's Case, Silver Catheter............... 22 00
Gunn's Case.............. 20 00
Smith's " 14 00

Miscellaneous Instruments.

Politzer's Bag............$2 00 to 2 50
" " with Buttle's Inhaler 5 00
Ovarian Clamp, Brown's.. 15 00
" Trocar, Wells's.. 12 00
Hare-Lip Pins, Silver Canula, 50
" " common, doz. 10
Plastic Pins............... " 15
Penis Speculums 1 00
Microscopic Knife......... 8 00
" Syringe...... 6 00
French Skeletons, wired.. } p. 271.
" " wired, best, }
Obstetrical Manikins..... 45 00
Polypus Forceps.......... 2 50
Penis Conjester.......... 7 00
Ecraseurs, Chain, medium, 18 00
" " small... 10 00
" " sml in case, 12 00
" Wire.......... 6 50
Cable Wire................ 50
Catheter Gauge.........., 2 50
" double current, silver, male..... 6 50
"D.C " female.. 5 00
Chilsom's Chlorof. Inhaler 3 00
Embalming Syringe, brass, 20 00
Stomach Pumps.......... 18 00
" Tubes, Eng. .... 1 50 to 2 00
Bumstead's Lamp........ 5 00
Phymosis Forceps, Fisher's 3 50
Respirators............... 2 50 to 5 00
Spermatorrhœa Rings.... 1 00
Bone Dr'ls, Hamilton's, set. 7 50
Hypodermic Syringes, glass, 2 50
" " fenestrated, 3 50

Hypodermic Syringes, Hard Rub. $2 50
Urine Test Case, small... 5 00 to 6 50
" " " complete, 10 00
" " " " with reagents, 22 00
Urinometers.... .... 1 00
Trocars, Rectum, silver canula, 3 00
" Hydrocele, " 2 00
" Abdom. " 2 00
" Exploring, " 2 00
Test Tubes...........doz. p. 162.
Litmus Paper....... sheet, p. 31.
Vaccinators............... 2 50 to 4 00
Spring Lancets, common.. 1 00
" " T. & Co., G.S. 2 50
Silver Wire, coil.......... 75
Bullet Forceps, best....... 3 00
" Prode, Nelaton's... 60
Needle Forceps, Sand's.... 5 00
" " Sims' .... 2 50 to 5 00
Mott's Needles.......set, 3 00
Volcellem Forceps........2 00 to 3 00
Universal Syringe, large.. 12 00
" " small . 8 00
Hydrocele Syringe, H. R., 4 50
Atomizers, Steam, complete, 6 00
" Hand Ball, Clark's 4 00
" " Shurtliff's, with face shield, 4 50
" Hand Ball, Clark's, with freezing tube, 6 00
Nasal Douche, Bottle Style, 1 25
" " Hale's 1 25
" " Nichol's ... 1 25
" " Potter's..... 1 50
" Canula, double silver 2 00
" Sound, Belloque's.. 3 50
Day's Splints.........set, 75 00
Welch's Splints .... " 40 00
Ahl's Splints......... " 30 00
Gutta Percha, for Splints, lb, 4 00
Pocket Vial Cases, see p. 283.
Saddle-Bags...... " " 264.
Medicine Chest.. " " 266.
Kidder Battery, acid......
D. & K. " crank ....
Galvanic " 8 cells ....
" " 16 " ....
" " 32 " ....
Kidder Platina...........
" Zincs.........pair
" Clamp ....
" Cup..............
" Cord.........yard,
} see page 267.
" Appliances, full set
" " ....each, see p. 266.
D. & K. " "
1 00
Tooth Forceps, all patterns, 2 25
Fever Thermometers, plain,
" self-registering. } p. 248.
Surface Thermometers.... 3 00
Cupping Cases...........8 00 to 10 00
" Glasses, see p. 155.
Scarificators............. 4 00 to 6 00
Urinals, Earthen, 1 00
Garrett's Disk, Circle.... 2 50
" " Long Leaf, 2 50
" " Button.... 2 50

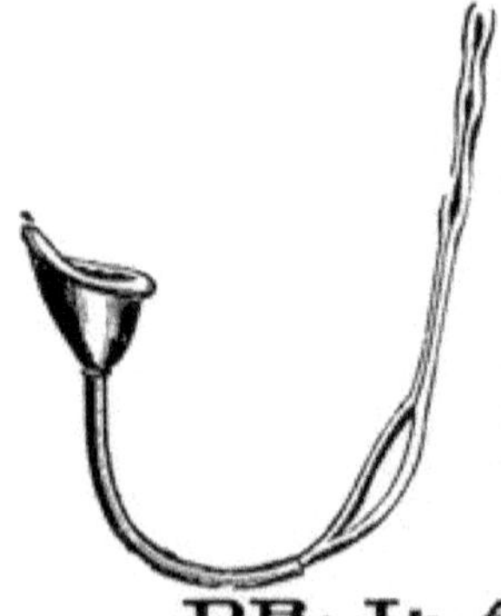

A side view of the Female Pelvis, showing the application of DR. BABCOCK'S UTERINE SUPPORTER, holding the Prolapsed Uterus up in its place without interfering with any other organ, or producing any irritation or inconvenience in wearing.

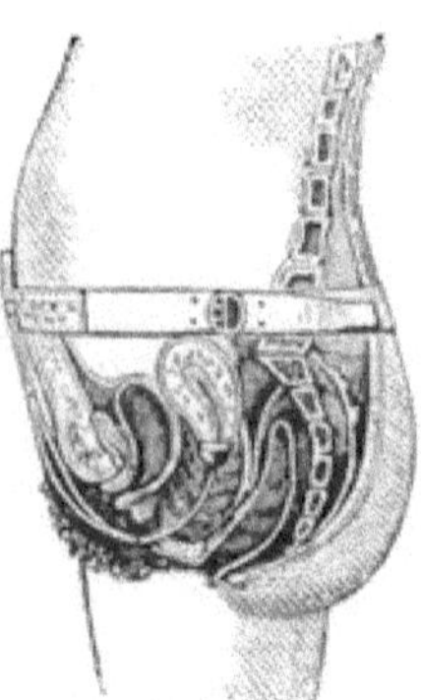

DR. L. A. BABCOCK,

INVENTOR, MANUFACTURER AND SOLE PROPRIETOR.

Dr. L. A. Babcock's pure silver Uterine Supporter is the best instrument now in use for any displacement of the womb, because it is perfectly simple in its structure and made to fit the parts exactly. It has no straps or strings to hold it in its place, and does not have to be taken off every time there is a movement of the bowels, or micturition. It has no rubber to bend by the natural heat of the body, or to chafe and irritate the parts, producing leucorrhœal discharge and weakness.

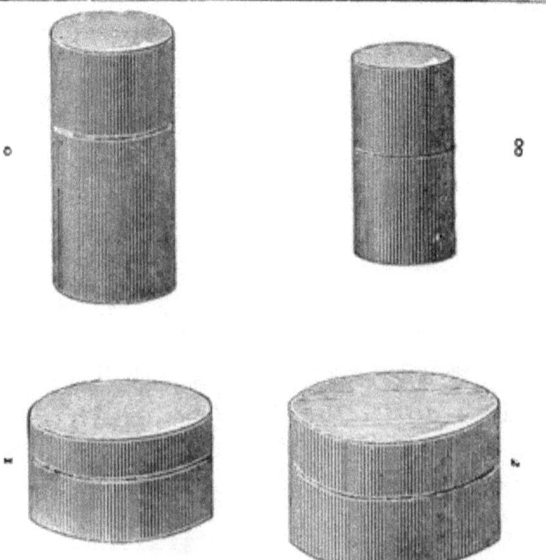

PRESCRIPTION BOXES--For Pellets or Triturations. Round; with shoulder; lined with pure white paper; covers fitting tightly; easily carried in the pocket; very much cheaper than vials and corks for prescribing pellets and powders; equal number white and red colors in each package. Exact size given in the cut.

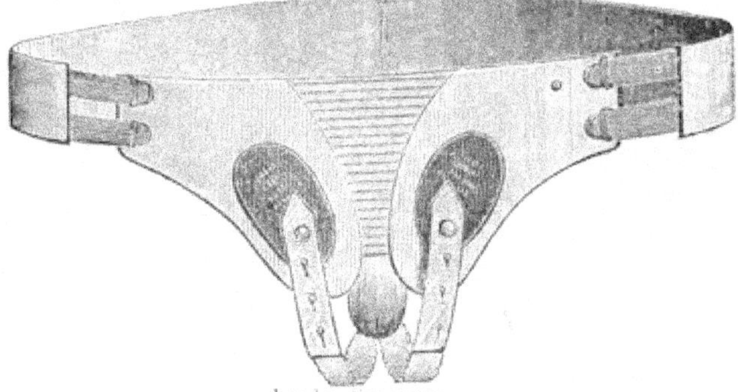

London Supporter.

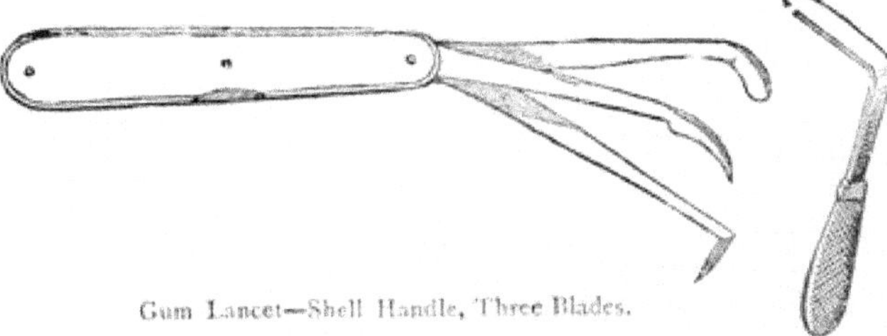

Gum Lancet—Shell Handle, Three Blades.

Tongue Depresser.

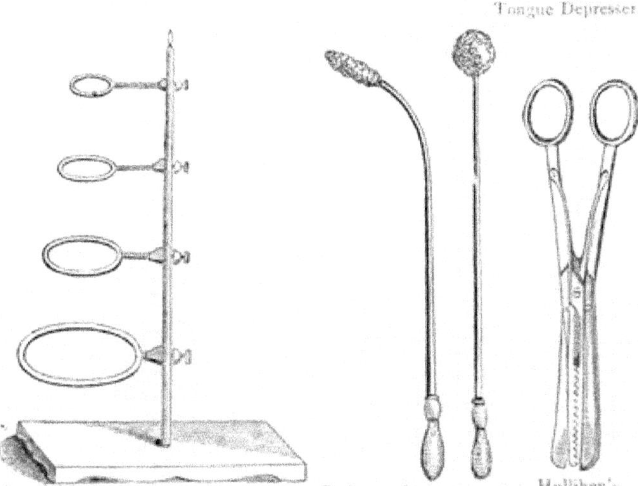

Retort Stands—Enameled Base, easily cleaned, etc., Four Rings. Price, each, $—.

Probangs, (see page)

Hullihen's Uvula Scissors

ADDENDA TO INSTRUMENT LIST.

Amputating and Operating Instruments.

	Each.
Parker's Amputating Saw.	$5 00
Wood's Circular Knife	4 00
Langenbeck's d'ble edged Catlings	2 50 to 5 00
Liston's Bull-Dog Forceps spring catch	1 50
Double Hook, Feruled Handle	2 50
Brainard's Bone Drills	3 50
Liston's Bone Forceps, straight edge	2 50 to 3 00
Liston's Bone Forceps, curved, on the flat, and angular	3 50 to 4 00
Trepanning Brush, to clean Trephines and Saws	25c. to 50
Tiemann & Co.'s double action Circular Saw for operations on the jaws and skull	12 00
Bone Shears, various sizes and patterns	4 00 to 7 00
Bone Gouges, for the hand or mallet, various sizes	1 50 to 2 00
Curved Hand Gouge	1 50
Bone Chisel, various sizes.	50c. to 1 50
Post's Toe-nail Forceps, for extracting inverted toe-nails	2 50
Raspatory with Ebony handle	2 00
Langenbeck's Hook for extracting pieces of bone	1 50
Mott's Set of Aneurism Needles	3 00
Plain Aneurism Needle	75
Hollow Needle for Wire Sutures	4 00
U. S. A. Bullet Forceps	2 00
Bullet Scoop	2 00
Tonsil Hook	2 00
Double-edged Staphyloraphy Knife	2 50
Staphyloraphy Knife, single edge, curved to the right	2 50
Staphyloraphy Knife, single edge, curved to the left	2 50
Cooper's Hernia Knife, Ebony Handle	1 00 to 1 25
Straight Finger Knife, or French Bistourie, Ebony Handle	1 00 to 1 25
Sharp and blunt-pointed Tenotome, Eb'ny handle	1 00 to 1 25
Concave and convex-edged Tenotome, Eb'ny Hndle	1 00 to 1 25

Eye Instruments.

	Each.
Laurence's Eye-lid Tourniquet	3 00
Jaeger's Plate Lid-holder	1 00 to 2 00
Cilia Forceps	1 00 to 1 25
Stoke's Eye-lid Compressor	3 00
Liebreich's Eye Speculum	2 50
Speir's Lachrymal Catheter, silver	1 00
Liebold's Subpalpebral Syringe	1 50
Galezowsky's Canalicula Dilator	2 50
Gensoul's Canula, for cauterizing the Nasal Duct	3 00
Desmarre's Scarificator	1 50
Desmarre's Cautery Iron, for Obliterating the Sac	3 00
Desmarre's Knife, for enlarging the Corneal Section	1 50
Walton's self-holding Iris Forceps	2 50
Beaumont's Concealed Canalicula Knife	8 00
Agnew's Canalicula Knife	1 75
Scalpels, for operating on the Lids and Face	1 50
Bistouries, for opening Lachrymal Tumors, Lachrymal Sac and Nasal Duct	1 50 to 1 75
Style, silver, 50 cts. to 75 cts.; gold	1 50 to 5 00
Lachrymal Canula, silver and gold	50c. to 5 00
Seatangle Tents, for dilating the Lachrymal Duct	25
Strabismus Tenaculum, with an eye near the point, for carrying ligatures	1 25
Double Hook, for fixing the Eye	1 50
Strabismus Bistourie, sharp and probe-pointed	1 75
Pupilometer	5 00
Optometer, plain	3 50
Trial (Test) Glasses, cylindrical, spherical and prismatic, in sets	25 00 to 100 00
Beer's straight Cataract Needle	1 25
Narrow, straight Cataract Needle	1 25
Walton's Grooved Needle, for soft Cataract	2 00
Instrument for Exhausting soft Cataract	2 25 & 4 50

	Each.
Up. De Graff's Instrument for Exhaustion of a soft Lens	$4 50
Livingston's Elastic Probe	2 50
Double Hook, sharp, for insertion into the sclerotica	1 50
Dix's Spud	1 25
Couching Needle	1 25
Angular Needle	1 25
Gouge, grooved	1 50
Richardson's Spray, silver tube for douching	6 50
Extra fine Eng. Douche	5 00
Clark's Eye Douche	2 50
Broad, straight Paracentesis Needle	1 25
Beer's Cataract Knife	1 50
Hall's Iris Scissors (sheep-shear-like)	5 00

Nasal Instruments.

Robert & Collins' Nasal Speculum, German-silver	$3 50
Bivalve German-silver Nasal Speculum	2 50
Thudichum's Nasal Speculum, plain	1 50
Nasal Polypus Forceps	2 25
Plain Polypus Dressing Forceps	1 50
Gross' Polypus Forceps	2 25
Simrock's Angular Polypus Forceps	3 00
Nasal Polypus Canula, silver	1 50 to 3 00

	Each.
Simrock's Rhinoscope	$8 50
Plain Rhinoscopic Mirror	1 25
Posterior Nares Syringe, hard-rubber	1 25

Veterinary Instruments.

Scalpels	$1 50
Pricking Knives	1 50
Bistouries	1 50
Bistouries, concealed	8 00
Dissecting Forceps	75
Scissors	1 00, 1 25, 1 50
Hoof Knives	1 50
Directors	75
Trocar and Canula	3 50
Seaton Needle, in 1 piece	75
" " 2 pieces	1 75
" " 3 pieces	2 50
Abscess Lancets	1 00
Tooth Rasps, guarded	3 50
Complete Pocket Case	22 00
Trachea Tubes	5 00
Spring Lancets	2 50 to 3 00
Fleams, 2-bladed	75
" 3-bladed	1 00
Stomach Pump Reeds	18 00
" " Lever	25 00
Castrating Clamps	10 00
" Ecraseur	25 00
Balling Irons	1 25
Cribbing Saw	1 25
Horse Catheters	3 50
Raveling Needles	25
Probes	75

Boxes in every variety—see pages 114 and 151.

CENTAUR LINIMENT

WHITE WRAPPER,

Will cure Rheumatism, Neuralgia, Bruises, Stiff Joints, Burns, without a scar, Swellings, Gout, Poisonous Bites and Stings, Salt Rheum, Caked Breasts, Sore Nipples, &c., and all flesh, bone and muscle ailments.

CENTAUR LINIMENT--YELLOW WRAPPER,

Will cure Spavin, Ring-bone, Wind-Galls, Scratches, Sweeny, Sprains, ordinary cases of Big Head, Poll Evil and Spring Halt, Harness-Galls, Screw Worms in Sheep, and all skin, flesh, bone or muscle ailments upon animals.

From the fact that some of the ingredients of this Liniment are equally adapted to the animal and human frame, it has been called after the Centaurs of old, who were half horse and half man.

The marvelous effects of Centaur Liniment have produced the natural consequences attending the introduction of a certain, sure and safe remedy. Wherever it has been tried, it becomes a matter of neighborhood conversation, and orders are pouring in for it perfectly surprising to us. No article ever acquired such a rapid endorsement and extensive use. Sells for 50 cents and one dollar per bottle.

CHILDREN CRY—for PITCHER'S CASTORIA. It regulates the stomach, cures wind colic, and causes natural sleep. It is a substitute for Castor Oil. Pleasant to take as honey. Retails at 35 cts. a bottle.

J. B. ROSE & CO., Prop's, New York.

The Trade supplied by the Agents,

VAN SCHAACK, STEVENSON & REID,

Old Salamander Wholesale Drug Warehouse, 92 and 94 Lake St.

CUTLER'S POCKET INHALER

And Carbolate of Iodine Inhalant.

A MOST WONDERFUL REMEDY

For all Nasal, Throat and Lung Diseases. Its curative properties have proved superior to anything hitherto used or known, affording relief in some cases in a few minutes.

IT WILL CURE CATARRH.

It has cured Catarrh, where all other remedies have failed. It is in fact the only real Catarrh remedy. Bronchitis yields to it, and Consumption, if taken in any kind of season. The character of the disease, upon using the Inhaler, is rapidly changed for the better.

In Asthma it has no equal. Every case is either cured or benefited.

Coughs, Sore Throat, Hoarseness, and Loss of Voice, are relieved at once. It is used by Nilsson, Kellogg, and Eustaphieve, and by hundreds of Clergymen and Public Speakers. It will correct the most offensive breath, rendering it as sweet as an infant's.

IT IS A PREVENTIVE OF DISEASE,

And with this Inhaler, one may bid defiance to the whole catalogue of infectious diseases, plagues and miasms. The Inhaler can be carried about the person as handily as a pencil case, and used with the same facility as a cigarette or a smelling bottle. All physicians approve it. Sells readily. Send for circular. FOR SALE BY ALL DRUGGISTS.

	Retail Price.	*Per Doz.*
One Pocket Inhaler, and one bottle of Carbolate of Iodine Inhalant, (sufficient for two months' use) neatly put up with full directions...............	$ 2 00	$ 12 00
Carbolate of Iodine Inhalant......................	50	3 00

W. H. SMITH & CO., Proprietors,

402 MICHIGAN ST., COR. EAGLE, BUFFALO.

The Trade supplied by the Agents,

Van Schaack, Stevenson & Reid,

OLD SALAMANDER DRUG WAREHOUSE, CHICAGO.

FOR

DRUGGISTS' GLASS WARE

AND

Druggists' Sundries

OF EVERY DESCRIPTION.

Having a large Stock o all Goods necessary for

Druggists', Perfumers and Chemists' Use,

I am enabled to fill all orders at short notice. Would call particular attention to my

Shop Furniture,

with the

IMPROVED GLASS LABEL

of different new designs,

SHOW BOTTLES

AND

PAINTED SHOW JARS.

Having control of

DR. HALE'S

Nasal Douche

my facilities for manufacturing the same, enable me to offer them at a great reduction in price.

Special quotations given when desired, and price lists furnished on application.

HENRY ALLEN,

258 Pearl Street, New York.

☞ Western Trade supplied with above by

VAN SCHAACK, STEVENSON & REID,

WHOLESALE DRUGGISTS, CHICAGO.

E. M. TUBBS & CO.,

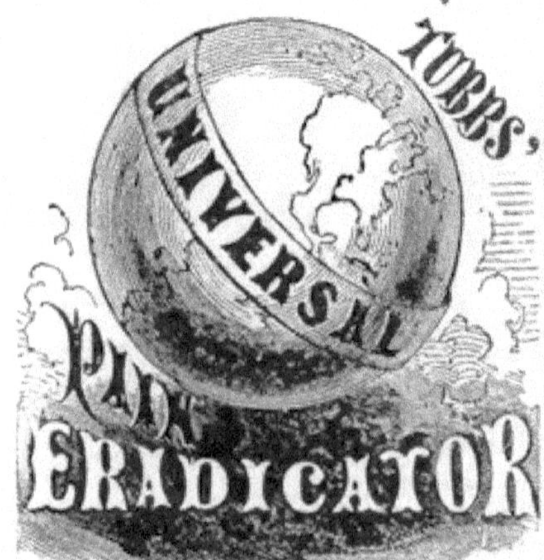

Is a Genuine Hair Restorative, changing Gray, Light, Red or Faded Hair to the *Dark*, *Lustrous*, *Silken Tresses* which so adorn youth or age. It will positively eradicate Humors and Dandruff from the scalp, and where there is life in the glands, will cause a new growth of Hair to put forth on bald spots. Thousands are testifying to the above.

This is what it purports to be—a genuine Pain Killer. It is unrivaled for the relief of Colds, Coughs, Rheumatism, Lumbago, Neuralgia, Sciatica, Cramps, Sprains, Dyspepsia, Diarrhœa, Colic, Sore Throat, Toothache, &c.

E. M. TUBBS & CO., Proprietors,

MANCHESTER, N. H.

The Wholesale Trade supplied by

VAN SCHAACK, STEVENSON & REID,

Wholesale Druggists, Chicago.

HALE'S HONEY OF HOARHOUND AND TAR.

FOR THE CURE OF

Coughs, Colds

Influenza, Hoarseness,

DIFFICULT BREATHING

And all affections of the

THROAT,

Bronchial Tubes and Lungs

Leading to Consumption.

Retail Prices, 50 cents and One Dollar.

Per Doz., $4.00 and $8.00.

Pike's Toothache Drops

The "One Minute Cure." 25 cents per bottle. Per dozen, $1.75.

HILL'S HAIR DYE.

Black or Brown, 50 cents.
The largest size, the best style, the best quality, and the best and largest sale of any **50** cent Hair Dye in the United States. Per dozen, $3.75.

KNOWLES'

POWDER GUN.

Chinnock's Patent, Aug. 10, 1869.

Retail, ...	$ .25
Per dozen, ...	1.75

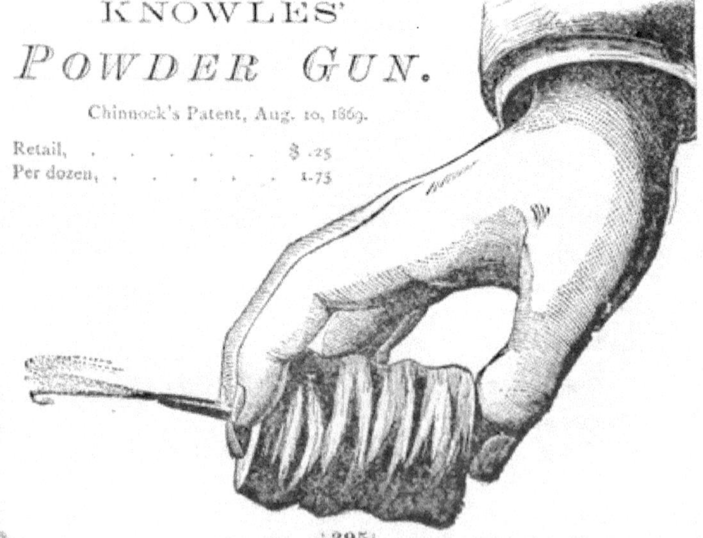

GOOD OLD DR. MOFFAT did not live in vain! He gave to the world

MOFFAT'S LIFE PILLS,

A purely Vegetable Pill.

MOFFAT'S PHŒNIX BITTERS

Are good old fashioned articles.

The Dyspeptic Dragon Driven Out.

WASHINGTON, April 27, 1872.

DEAR SIR—I took the MOFFAT'S PHŒNIX BITTERS according to instructions. Result, my awful *Dyspepsia* is gone. I feel like another man. I tell the glad tidings everywhere. You may do the same. Send me a case of BITTERS for a friend.

Faithfully yours, ELIOT P. MOOERS.

PAGE'S CLIMAX SALVE.

Useful and needed in every house.

Parties ordering PAGE'S CLIMAX SALVE by mail must enclose 40 cents. Postage is 15 cents, and Salve 25 cents. Price of MOFFAT'S LIFE PILLS, 25 cents, and Postage, 5 cents.

EXTRACTS FROM LETTERS.

"Moffat's Pills entirely purified my blood."
JAMES WILSON YOUNG, Troy.

"Page's Climax Salve cured me of Salt Rheum."
ALFRED L. TORREY, Geneva.

"Moffat's Pills never fail to cure Costiveness."
REV. EDWARD P. PARKS, Newark.

"Moffat's Pills are the great cure for Biliousness."
MRS. W. H. YOUNGS, Saratoga.

"Page's Climax Salve healed my Scrofulous Sores."
PETER K. SCHEIFFELIN, Bedford.

"I owe my life to Moffat's Pills. Case of General Debility."
WILSON PATTERSON, Brooklyn.

Volumes of such testimony might be added, if space permitted. The best recommend is a trial, which costs but very little, and will never be regretted.

BEWARE OF COUNTERFEITS.

All of Moffat's Life Pills and Phœnix Bitters, hereafter manufactured, will be wrapped with this paper, "*Moffat's Good Samaritan*," over which is the outside steel plate label that bears the portrait of Dr. Wm. B. Moffat, and the *fac simile* signature of John Moffat, of Wm. B. Moffat, and of J. T. Howland, 1866.

J. P. MILLS, Proprietor, 85 Liberty St., New York.

The Trade supplied by

VAN SCHAACK, STEVENSON & REID,

Wholesale Druggists, CHICAGO.

—MANUFACTURED BY THE—

San Antonio Meat Extract Factory

SAN ANTONIO, TEXAS,

Received the Medal of Merit at the Vienna Exposition, 1873. Also the GRAND SILVER MEDAL at the American Institute Fair, 1874.

Proven by analysis to be the best preparation in the market.

SCHOOL OF MINES, COLUMBIA COLLEGE, NEW YORK, Feb. 4, 1874.
I would further say that the results of the analyses are in favor of the SAN ANTONIO MEAT EXTRACT FACTORY'S product in almost every particular. Very respectfully yours, C. F. CHANDLER, Ph. D., Professor of Analytical and Applied Chemistry.

I have made several analyses of Meat Extracts for the purpose of comparing them with the San Antonio Meat Extract. In all the particulars that Liebig has pointed out as forming a good Meat Extract, the San Antonio excels all others.
SAM. R. PERCY, M.D., Prof. Materia Medica,
NEW YORK, November, 1874. 47 West 38th St., N. Y. M. C.

The SAN ANTONIO MEAT EXTRACT answers perfectly all demands made on good Meat Extract in regard to taste and flavor.
WIESBADEN, GERMANY. DR. R. FRESENIUS.

The analyzed Extract is to be considered at least equal to the Fray-Bentos Extract. Besides this, it dissolves clear in water, is free from unpleasant taste, contains no fat or glueish substance, and is therefore not more subject to "souring" than the Fray-Bentos Extract. A. OBERDŒRFFER, Analytical Chemist.
HAMBURG, GERMANY.

The analyzed Extract equals in its component parts Liebig's Company Meat Extract (Fray-Bentos Extract). As advantages of the broths prepared from the San Antonio Extracts may be mentioned the milder taste and the more agreeable roast-like flavor, and also that the analyzed Extract is much cheaper.
VIENNA, AUSTRIA. DR. J. H. POHL, Imperial Professor.

Testimonial of Prof. Dr. WILLARD PARKER:
I have examined the SAN ANTONIO MEAT EXTRACT, an analysis of which has been made by Prof. C. F. CHANDLER, of Columbia College, and for which Mr. R. DANNHEIM is General Agent. I regard it useful as a tonic and as an article of diet in Health and Sickness when properly prepared. I permitted a Jar of the Extract to remain exposed to the air at a temperature of 70° F. No decomposition took place, nor change in the agreeableness of its flavors.
WILLARD PARKER, M.D., Prof. Surgery, etc.,
NEW YORK, November 16th, 1874. 41 East 12th Street, New York.

R. DANNHEIM, General Agent, 100 Chambers St., New York.

The Trade supplied at Manufacturers' Prices by

Van Schaack, Stevenson & Reid,
WHOLESALE DRUGGISTS, CHICAGO.

CATHARTIC PILLS

For all the purposes of a Laxative Medicine.

Perhaps no one medicine is so universally required by everybody as a cathartic, nor was ever any before so universally adopted into use, in every country and among all classes, as this mild but efficient purgative PILL. The obvious reason is, that it is a more reliable and far more effectual remedy than any other. Those who have tried it, know that it cured them; those who have not, know that it cures their neighbors and friends, and all know that what it does once it does always—that it never fails through any fault or neglect of its composition. We have thousands upon thousands of certificates of their remarkable cures of the following complaints, but such cures are known in every neighborhood, and we need not publish them. Adapted to all ages and conditions in all climates; containing neither calomel nor any deleterious drug, they may be taken with safety by anybody. Their sugar coating preserves them ever fresh and makes them pleasant to take, while being purely vegetable no harm can arise from their use in any quantity.

They operate by their powerful influence on the internal viscera to purify the blood and stimulate it into healthy action—remove the obstructions of the stomach, bowels, liver, and other organs of the body, restoring their irregular action to health, and by correcting, wherever they exist, such derangements as are the first origin of disease.

Minute directions are given in the wrapper on the box, for the following complaints, which these PILLS rapidly cure:—

For **Dyspepsia** or **Indigestion, Listlessness, Languor** and **Loss of Appetite**, they should be taken moderately to stimulate the stomach and restore its healthy tone and action. For **Liver Complaint** and its various symptoms, **Bilious Headache, Sick Headache, Jaundice** or **Green Sickness, Bilious Colic** and **Bilious Fevers**, they should be judiciously taken for each case, to correct the diseased action or remove the obstructions which cause it. For **Dysentery** or **Diarrhœa**, but one mild dose is generally required. For **Rheumatism, Gout, Gravel, Palpitation of the Heart, Pain in the Side, Back** and **Loins**, they should be continuously taken, as required, to change the diseased action of the system. With such change those complaints disappear. For **Dropsy** and **Dropsical Swellings** they should be taken in large and frequent doses to produce the effect of a drastic purge. For **Suppression** a large dose should be taken, as it produces the desired effect by sympathy. As a DINNER PILL, take one or two PILLS to promote digestion and relieve the stomach. An occasional dose stimulates the stomach and bowels into healthy action, restores the appetite, and invigorates the system. Hence it is often advantageous where no serious derangement exists. One who feels tolerably well, often finds that a dose of these PILLS makes him feel decidedly better, from their cleansing and renovating effect on the digestive apparatus. As a DINNER PILL, this is both agreeable and useful. No Pill can be made more pleasant to take, and certainly none has been made more effectual to the purpose for which a dinner pill is employed.

Prepared by Dr. J. C. AYER & CO., Practical and Analytical Chemists, Lowell, Mass.

The Trade supplied by

VAN SCHAACK, STEVENSON & REID,
Old Salamander Wholesale Drug House, Chicago.

Notice to the Retailer.

The FRAZER AXLE GREASE is now recognized as the Standard Axle Grease of the United States; is sold in every State and County in the Union, and is to-day without a rival. So universally is this fact recognized that numerous imitations have been made, all claiming to be as good as the FRAZER, thus virtually admitting its superiority; some imitators even using the name to palm off a spurious article—yet as every package bears our Trade Mark, dealers and consumers will be able to distinguish the genuine from the imitation, and thus protect themselves against the intended fraud.

THE FRAZER GREASE is the highest price of any in the market; and yet to the consumer it is by far the cheapest, the wonder and admiration of all who use it; will cure Corns and Old Sores on men and horses; with it the Indians smear their bodies to keep off the Mosquitoes; while the black man of the plantation greases his mules to protect them from the gnats and flies. THE FRAZER GREASE will not melt when exposed to the intense heat of a summer sun, hence, unlike the Petroleum and imitation greases, remains in the boxes in perfect condition until sold.

The Trade supplied by

Van Schaack, Stevenson & Reid,

WHOLESALE DRUGGISTS, CHICAGO.

BURRINGTON'S

PATENTED

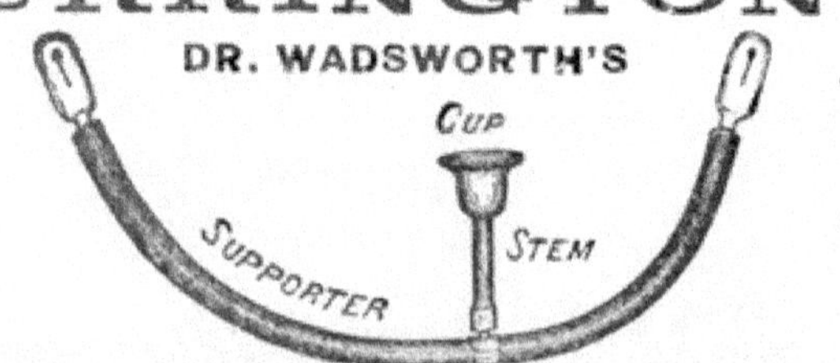

JAN. 24, 1860.

UTERINE ELEVATOR,

OR STEM PESSARY, IMPROVED.

The Instrument is made of a preparation of Caoutchouc, or India Rubber (not the common), that has no lead in it, and does not irritate at all the soft parts with which it comes in contact. There are THREE sizes of the Instrument—in nineteen cases out of twenty the middle size is the one wanted.

The subjoined is the deliberate voice of the Medical Profession of Providence, R. I., in favor of this Instrument:

The undersigned, having examined the UTERINE ELEVATOR, invented by JOHN A. WADSWORTH, M.D., of this city, freely certify that we believe it eminently adapted to the purpose intended, and superior to any other means within our knowledge for the reduction and cure of Prolapsus Uteri (Falling of the Womb).

Hervey Armington, M.D. S. A. Arnold, M.D. Walter E. Anthony, M.D.
George P. Baker, M.D. Ira Barrows, M.D. W. O. Brown, M.D.

CUNNINGHAM & IHMSEN,

Manufacturers of

GLASS AND GLASSWARE, &C.

FRUIT JAR PRICE LIST.

"HERO."

A First Class and Reliable Jar.

Quart, per gross, $
Half Gall. "

"QUEEN."

All Glass.

The Leading Jar in the market.

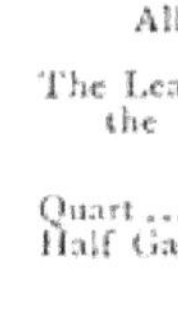

Per Gross.
Quart ....$
Half Gallon..

Stone's Patent

Or Tin Top.

This Jar is so well known, and so generally used, as to need no further mention.

Per Gross.
Quart ....$
Half Gallon..

The Trade supplied by

VAN SCHAACK, STEVENSON & REID,

Wholesale Druggists, Chicago.

ESTABLISHED 1825.

GOOD, ROOF & CO.,

No. 34 BROADWAY, NEW YORK,

—IMPORTERS OF—

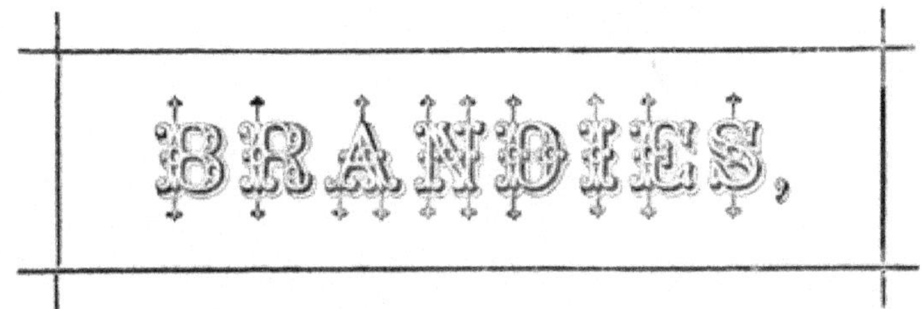

WINES, GINS, ETC.,

WITH SPECIAL REFERENCE TO THE

Requirements of the Drug Trade.

AGENTS FOR

PINET, CASTILLON & CO. COGNAC.

VAN DUEKEN & WIELAND, ROTTERDAM.

NUGUE RICHARD, BEZIER.

A. LOWERRA, BORDEAUX.

Sold by

Van Schaack, Stevenson & Reid,

—AT THE—

"Old Salamander Wholesale Drug Warehouse."

☞ No Druggist should fail to ORDER these Medicines, as they can handle no Medicines in their business that sell better. It is impossible to sell a bottle without securing a trade, which will prove a permanent one.

PRICE LIST

— OF EITHER —

Dr. A. Boschee's German Syrup

FOR COUGHS, COLDS AND CONSUMPTION,

— OR —

Green's August Flower

For Costiveness, Liver Complaint and Dyspepsia.

One Dozen of Either or Both.................... Per Dozen, $5.50

Twelve sample bottles of BOSCHEE'S GERMAN SYRUP or AUGUST FLOWER always sent free of charge with each and every dozen, when specially requested, for you to GIVE AWAY FREE to any adult customer wishing to try the medicine.

Circulars, Posters, etc.

Van Schaack, Stevenson & Reid,

Wholesale Agents for Chicago, Ill.

These Medicines were only introduced to our Trade about eighteen months ago, and we are having a lively demand for them.

VAN S., S. & R.

THE LADIES DON'T CRY

But they rejoice at the discovery of

Dr. B. C. PERRY.

Perry's Comedone

Cures Pimples, Fleshworms, &c., on the face.

FOR PIMPLES ON THE FACE

Blackheads and Fleshworms,

— USE —

Perry's Improved Comedone

And Pimple Remedy.

The Great Skin Medicine.

Prepared only by

Dr. B. C. Perry, Dermatologist,

No. 49 Bond Street,

NEW YORK.

Perry's Moth & Freckle Lotion

For Moth Patches, Freckles and Tan, use PERRY'S MOTH AND FRECKLE LOTION. It is *Reliable and Harmless.* Depot, 49 Bond Street.

Address orders to the Western Agents,

Van Schaack, Stevenson & Reid,

92 AND 94 LAKE STREET, COR. DEARBORN.

Opposite "Tremont House."

" "Commercial Hotel."

THE TRAVELER

On the cars, passing from Philadelphia to New York, is attracted by huge placards, placed on the elegant grounds of DR. SCHENCK, reading

"CONSUMPTION CAN BE CURED."

THE VISITOR

To Philadelphia fails to see the "sights" of the "City of Brotherly Love," without a visit to the beautiful building of

DR. J. H. SCHENCK,

An old and honored Physician of Philadelphia, who has yielded to the solicitation of the public, and put up his valuable Remedies in proprietary form.

Schenck's

PULMONIC SYRUP.

Schenck's

SEA WEED TONIC.

Schenck's

MANDRAKE PILLS

All have become household words—endorsed by the Medical Faculty and all who have used them.

The Trade supplied by

VAN SCHAACK, STEVENSON & REID,

Wholesale Druggists, Chicago.

WESTERN DEPOT

FOR THE CELEBRATED

"SURE POP"

ON RATS, ETC.

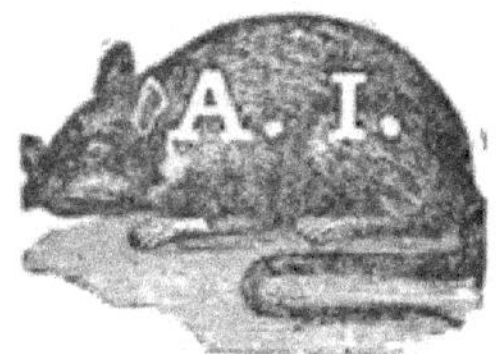

PROF. ISAACSEN

Has discovered the very best

PHOSPHORIC PASTE

For the IMMEDIATE destruction of

Rats, Mice, Cockroaches, Etc.

The Boxes are Double Size, Hermetically Sealed.

Read one from many Certificates.

WILLARD'S HOTEL, WASHINGTON, D. C., October 19, 1863.

MR. ISAACSEN—*Dear Sir:* It gives me great pleasure to testify to the gratifying results obtained at this Hotel, through using your Phosphoric Paste; it is now two years since I first heard of your remedy, and determined to give it a fair trial. That trial proved so successful, that not a trace of Rats or Roaches have since been discovered, although previous to that time we were completely overrun.

I remain, yours respectfully,

JOHN WOLFE, *Engineer.*

Adolph Isaacsen

The Trade supplied by the Western Agents,

VAN SCHAACK, STEVENSON & REID,

Old Salamander Wholesale Drug Warehouse,

92 and 94 Lake Street, Chicago.

A GREAT SUCCESS!

Leamon's Aniline Dyes

Magenta.
Scarlet.
Purple.
Brown.
Violet.
Blue.

Only Twelve Kinds

Solferino.
Crimson.
Yellow.
Green.
Black.
Mauve.

Made or Required.

Unequaled for SIMPLICITY, BEAUTY and ECONOMY.

Cut of the Black Walnut and Glass Case we give with the first order for one gross.

The Trade will be furnished with an ample supply of a pamphlet entitled "Aniline Dyes; what they are, and how to use them," and beautiful sample cards showing twenty-four different colors from the twelve dyes. To any one purchasing one gross we will give a handsome black walnut counter case, like engraving. This, in connection with a judicious distribution of the books, never fails to create an active demand.

PRICE, $1.75 PER DOZEN.

THERE ARE ONLY TWELVE COLORS, but over one hundred beautiful shades and colors can be made from these; the dealer can, therefore, have a full assortment, and be able to answer all calls, while carrying a small stock. The Dyes are warranted to give satisfaction in every instance, when used according to directions. We should be pleased to receive your orders, and assure you that you will find these most desirable goods, WORTHY OF YOUR CONSIDERATION.

VAN SCHAACK, STEVENSON & REID,

Sole Wholesale Agents for the Northwest.

Rolfe & Donaldson's

PATENT VAPORIZING INHALER

A Valuable Remedy for

CATARRH.

I'm off for one of them now. You should use Rolfe & Donaldson's Inhaler.

DIRECTIONS.

1st. Fill the bulb in the tube reaching to the bottom of the bottle one-half full of hydrochloric (muriatic) acid, great care being taken not to spill any of it on the outside of the tube, or in the inside of the bottle.

2d. Fill the bottle about one-third full of water, and in this put the medicine prescribed for the disease, and also into the same drop about five or six drops of aqua-ammonia, then it is ready for use. Great care must be taken not to blow through the mouth-tubes, or spill any of the acid into the alkali (the mixture in the bottle), as it would entirely stop the chemical action, and would not produce any vapor.

It is intended mostly for Lung, Throat and Catarrhal diseases, and is sure, in nearly every case, to cure.

Anything can be used in the Inhaler that yourself or any physician would prescribe, it being the best way of having the medicine reach the diseased parts.

A full supply always kept on hand at the

"Old Salamander Wholesale Drug Warehouse"

—OF—

VAN SCHAACK, STEVENSON & REID,

92 and 94 Lake Street, cor. Dearborn, Chicago.

USE CHAMBERLAIN'S

IMMEDIATE RELIEF!

NO CURE, NO PAY!

☞ One bottle of the *IMMEDIATE RELIEF* can be had of any one of my Agents, and the privilege given the purchaser of using one-half its contents on trial, and if it does not prove satisfactory, the money will be refunded.

This Medicine is used with great success for

DIARRHŒA,

DYSENTERY, BLOODY FLUX, CHOLERA

COLIC AND CHOLERA MORBUS.

ALSO FOR

FEVER AND AGUE

COLDS, SORE THROAT,

CATARRH, SICK HEADACHE and NEURALGIA.

It is an excellent Application for

Rheumatism, Burns, Bruises, Cuts, Sprains, Bee Stings

AND

TOOTHACHE.

THOUSANDS OF RECOMMENDS might be given here, but upon these terms I do not think them necessary, for I am confident that *your own trial of it* will do more towards convincing you of its worth, than whole pages of printed recommendations.

THIS MEDICINE IS PURELY VEGETABLE,

And free from all poisonous substances. It is wonderful in its effects, yet harmless as water to the most delicate person. The RELIEF has been used for a number of years, and has repeatedly cured the worst cases for which it is recommended, and physicians have pronounced it *one of the most valuable medicines known.*

Where a *fair trial* was given, it has been used with great success, and it has seldom, if ever, been known to fail when the directions were strictly followed.

☞ **Beware of Counterfeits.**—This medicine is now put up in large and small panel bottles, with its name and my address blown in them. **Price, 50 and 35 cts.** ☞ None genuine without my signature on the wrapper.

MY CATHARTIC PILLS Are peculiarly adapted to all BILIOUS AFFECTIONS, and, as a safe and reliable Purgative and Alterative, have no superior. They should be taken before the IMMEDIATE RELIEF in all Bilious Complaints.

A. N. CHAMBERLAIN, Sole Proprietor,
ELKHART, INDIANA.

☞ Also, Manufacturer of HORSE POWDERS, GREEN MOUNTAIN SALVE, OINTMENT, FLAVORING EXTRACTS, and ESSENCES.

The demands of the Trade promptly filled at very lowest price, by

Van Schaack, Stevenson & Reid,
WHOLESALE DRUGGISTS, CHICAGO.

THE DRUG TRADE ALL DELIGHT IN SELLING

THE CELEBRATED

HORSE MEDICINE,

BECAUSE IT IS THE BEST.

Safe! Sure! Mysterious!

WORKS LIKE MAGIC!

Promptly relieving every kind of Pain and Inflammation, and all External Diseases of Man or Beast.

Testimonials of the strongest kind can be furnished.

RETAIL PRICE, 50c. AND $1.00.

YOU WILL NOT REGRET A TRIAL ORDER.

GIFFORD & TOMLINSON, Prop's, Philadelphia.

Trade supplied by the Agents,

Van Schaack, Stevenson & Reid,

92 and 94 Lake Street, cor. Dearborn, CHICAGO.

HEGEMAN'S

GENUINE COD-LIVER OIL.

Warranted pure *Newfoundland Oil.* It has stood the test of twenty years' experience, and can be relied on in every particular.

THE MOST PERFECT IRON TONIC.

Hegeman's Ferrated Elixir of Bark, or Elixir of Calisaya Bark with Pyrophosphate of Iron.

The FERRATED ELIXIR OF BARK is a pleasant cordial, possessing the valuable properties of CALISAYA BARK deprived of its tannin and coloring matter, and contains eight grains of the PYROPHOSPHATE OF IRON in each fluid ounce; and in all cases where a mild and efficacious Iron Tonic is desired, will be found a most valuable preparation. As a preventive of fever and ague, and as a tonic for patients recovering from fever or other sickness, it cannot be surpassed.

HEGEMAN'S CORDIAL ELIXIR OF CALISAYA BARK,

(PERUVIAN BARK.)

For persons living in Fever and Ague Districts, it will be found invaluable as a preventive—half a wine-glass taken night and morning rendering the system much less subject to the unhealthy influence of the atmosphere.

Prepared by **HEGEMAN & CO.,**
NEW YORK.

The Trade supplied by
VAN SCHAACK, STEVENSON & REID,
Wholesale Druggists, CHICAGO.

Hear a Voice—but not from the Tomb!

DAWSON STATION, Pa., January 13, 1875.

MESSRS. R. E. SELLERS & Co.—Gents: I need *more* of your *splendid* Rheumatic Medicine, "Johnson's Rheumatic Compound." The medicine is superior to any Rheumatic Prescription I have ever known in practice. Send me at once my usual supply. J. E. DOUTHET, M. D.

MESSRS. R. E. SELLERS & CO.

(*PITTSBURGH, PA.*)

Are also the manufacturers of

Sellers' Cough Syrup,

Sellers' Liver Pills,

Sellers' Vermifuge,

Sellers' Worm Confections,

Lindsey's Improved Blood Searcher,

Bœrhave's Holland Bitters,

WHICH HAVE STOOD THE TEST OF TIME!

The Trade supplied by the Agents,

VAN SCHAACK, STEVENSON & REID,

Wholesale Druggists,

92 and 94 Lake Street, cor. Dearborn, Chicago.

Himrod's Cure

— FOR —

Asthma,
Catarrh,
Whooping Cough,
Croup,
Rose Cold or Hay Fever,
Incipient Consumption,
and Common Colds.

Adopted and Recommended by Physicians.

Used as directed it never fails in effecting a

Thorough and Permanent Cure.

☞ Public Speakers and Singers will experience great benefit by the use of this article a few minutes before their appearance.

Testimonials to the wonderful efficiency of this Remedy are received daily. From among them we call attention to the following from the distinguished editress of the "Victoria Magazine," London.

50 Norfolk Square,
HYDE PARK, LONDON, April 8th, 1874.

MR. HIMROD:

I cannot speak in terms sufficiently strong to convey my thankfulness for your wonderful remedy. I have suffered with Asthma since I was ten years old. I have tried all kinds of *so-called* remedies, but utterly without success until I became acquainted with yours. If taken early enough it enables me to stop an attack at once, and under all circumstances affords me such marvelous relief that I would not be without it on any consideration whatever. I know many Asthmatics who would be thankful for such a cure, and in their interest I entreat you to have an English agent. I have often wished to tell you the boon this cure has proved to me. Whereas I nearly died of Asthma on my sea voyage from Liverpool to New York, the use of your inhaling powder gave me comparatively an easy voyage home. Those who know the misery of Asthma will understand my saying, that had America yielded me no other pleasure, my visit there would have been rich in blessing, thanks to your invention.

Yours cordially, EMILY FAITHFULL.

For sale by all druggists.

Price, One Dollar per Box.

HIMROD MANUFACTURING CO., Proprietors, New York.

The Trade supplied by

Van Schaack, Stevenson & Reid,

92 and 94 Lake Street, cor. Dearborn, Chicago.

USE THE BEST MINERAL WATER!

Waukesha Water

MINERAL ROCK SPRING.

Cures Dropsy, Dyspepsia, Diabetes, Constipation, Gravel, Jaundice, Bright's Disease, Female Weakness, and all diseases of the Liver, Kidneys and Urinary Organs, both Male and Female.

Price, bbls., $12.00, half bbls., $7.00, including package; demijohns, jugs, and cans, 50 cents per gallon, package extra. MONEY MUST ACCOMPANY THE ORDER.

For orders for the Water, or pamphlets giving description of the Water, and the diseases that it will cure, address VAN SCHAACK, STEVENSON & REID, Wholesale Agents, Chicago, Ill., or C. C. OLIN & CO., Waukesha, Wis.

The following testimonials speak for themselves.

LETTER from Judge Hiram Barber: HORICON, WIS., July 7, 1874.

MR. C. C. OLIN, Waukesha, Wis.—Dear Sir:—Inclosed please find $1.25 for five gallons of Mineral Rock Spring Water—can sent to-day by express.

I am much pleased with the results from the use of your Spring water. I have used for the last two years other waters, but have never received from the use of others as decided results as from the use of that you sent me. It may be owing to other causes operating at the same time, which have aided in the favorable results from the use of yours; still I am not aware of any, as my habits have been the same, making use of no drugs or other medical treatment. H. BARBER.

CHICAGO, Oct. 5, 1874.

C. C. OLIN & CO.—Gents:—I commenced the use of Waukesha Mineral Rock Spring Water some two or three months since with little expectation that it would benefit me, but since I received the first can I have been steadily gaining in health. My habits are sedentary; I am closely confined to my office which deprives me of outdoor exercise, and for the past two years I have suffered from indigestion and dyspepsia, and have twice had inflammation of the kidneys. I have taken a great deal of medicine, but have received no permanent benefit; but since using your Waukesha Mineral Rock Spring Water I have been much better, and have great confidence in its curative powers, and shall be more particular in its use in the future, believing that I shall receive a permanent cure. I am satisfied that no medicine will reach the kidney difficulty, and it is certainly a great blessing to mankind that such pure and wholesome water should be discovered for the healing of the nation, if they will but drink. Too much value cannot be put upon such water, and there are tens of thousands in our land that would be benefited and their lives lengthened out for years, if they only knew of the healing qualities of this invaluable water. Respectfully yours, E. A. LESSEY.

WAUKESHA, WIS., May 23, 1874.

C. C. OLIN & CO.—Gents:—I came to Waukesha about two months since, very much out of health. I had no appetite; could not sleep; bowels very much constipated—so much so that no medicine would affect me in the least. All the food that I took into my stomach would not digest for hours, and I began to think that I must die.

But as I live near your spring I thought I would try the water. I must say it has had a wonderful effect upon my system. In these two months Mineral Rock Spring Water has regulated my bowels, given me a good appetite, the flow of urine has become natural, digestives entirely cured, and I am now so well that I can do a good hard day's work without fatigue, and I consider myself a cured man. I would recommend to all that have been afflicted as I have, to try the Waukesha Mineral Rock Spring Water and recommend it to their neighbors.

Respectfully yours, GEORGE GRULL.

The Trade supplied by the Agents,

VAN SCHAACK, STEVENSON & REID,

Old Salamander Wholesale Drug Warehouse, 92 and 94 Lake St.

Established in 1801. **Established in 1801.**

BARRY'S

Celebrated Toilet Preparations.

Patronized by the principal families of Asia, Africa, Europe and America.

These excellent articles are admitted to be the standard preparations for all purposes connected with the toilet.

Barry's Tricopherous

Imparts vigor, gloss and luxuriance to the hair, and removes scurf and dandruff; for cleansing and restoring the hair, and for all ailments of the head, it is a sovereign remedy.

Prices—Retail, 50 cts.; Trade, per doz., $; Per gross, $

Barry's Hair Dye.

Put up in two forms: first, Barry's Safe Hair Dye, in a single bottle, complete in itself, and which colors the hair any shade from auburn brown to black; second, Barry's Black Hair Dye, in a box, accompanied with a preparation, by the use of which, a jet black color is instantly brought.

Prices—Retail, $1.00; Trade, per doz., $; Per gross, $

Barry's Pearl Cream.

The most exquisite and delicious preparation for beautifying the complexion and preserving the skin. A single application removes wrinkles, freckles, sunburn, and tan, and turns the sallowest and darkest skin as pure as alabaster.

Prices—Retail, 50 cts.; Trade, per doz., $; Per gross, $

Barry's Marfilina.

For beautifying and preserving the teeth, perfuming the breath, and hardening the gums. It is the best article for removing tartar, canker, scurvy, and all other diseases or substances injurious to the teeth.

Prices—Retail, 50 cts.; Trade, per doz., $; Per gross, $

BARCLAY & COMPANY,

SOLE MANUFACTURERS,

No. 26 Liberty Street, New York.

Sole Wholesale Agents for the Northwest,

VAN SCHAACK, STEVENSON & REID,

92 and 94 Lake Street, cor. Dearborn, Chicago.

This standard remedy for all diseases of the hair and scalp is now so well known the world over, that it seems scarcely necessary to do more than announce that it is in the market; yet, lest there may be some who but partially understand its wonderful restorative powers, we will specify a few of the ills it alleviates.

GRAY HAIR, PRODUCED FROM WHATEVER CAUSE, IS ALWAYS RESTORED TO ITS ORIGINAL COLOR BY A FAITHFUL USE OF THE RENEWER ACCORDING TO DIRECTIONS. It *never fails;* and although it restores color to the hair, it may be freely applied with the hands without fear of the slightest stain to the skin.

It removes all Eruptions and Dandruff; and in cases of ITCHING OF THE SCALP it relieves immediately, giving a cooling, soothing sensation of great comfort.

By its tonic properties it restores the capillary glands to their normal vigor, preventing baldness, and making the hair grow thick and strong.

As a dressing, nothing has been found so effectual, or desirable.

It is also the most economical *Hair Dressing* ever used, as it requires fewer applications, and gives the hair a splendid, glossy appearance. A. A. Hayes, M.D., State Assayer of Massachusetts, says: "The constituents are pure, and carefully selected for excellent quality; and I consider it the *best preparation* for its intended purposes."

Sold by all Druggists and Dealers in Medicine.

BUCKINGHAM'S

DYE FOR THE WHISKERS

This elegant preparation may be relied upon to change the color of the beard from gray or any other undesirable shade to brown or black, at discretion. It is easily applied, being in *one preparation*, and quickly and effectually produces a permanent color which will neither rub nor wash off.

Manufactured by R. P. HALL & CO., Nashua, N. H.

Sold by all Druggists and Dealers in Medicine.

The Western Trade supplied by the Agents,

VAN SCHAACK, STEVENSON & REID,

92 and 94 Lake Street, Chicago.

HAGAN'S
MAGNOLIA BALM

A few applications make a

Pure Blooming Complexion.

It is Purely Vegetable, and its operation is seen and felt at once. It does away with the Flushed Appearance caused by Heat, Fatigue and Excitement. Heals and removes all Blotches and Pimples, dispelling dark and unsightly spots. Drives away Tan, Freckles, and Sunburn, and by its gentle but powerful influence, mantles the faded cheek with *Youthful Bloom and Beauty.* Sold by all druggists and fancy stores. Lyon Man'fg Co., depot, 53 Park Place, N. Y.

Was first known in America **30** years ago. It has the oldest and best record of any Liniment in the world, and has caused more cures of Rheumatism, Gout, Stiff Joints, Swellings, Burns, etc., in men, women and children, than all of the other Liniments put together. It has been in use over THIRTY YEARS, and has saved the lives of Thousands upon Thousands of Valuable Horses, and other Blood Stock, and is regarded by shrewd and observant Veterinary Surgeons, Horsemen, and Cattle Breeders, as indispensable to the proper treatment of a large class of disorders and disabilities to which beasts of burden are liable.

Orders promptly filled through

Messrs. VAN SCHAACK, STEVENSON & REID.

HOSTETTER'S STOMACH BITTERS

A SOVEREIGN HEALTH PRESERVATIVE,

—AND AN—

Invaluable Remedy for Sickness and Debility

THE PREVENTIVE resources of Medicine are, unfortunately, too seldom invoked. It is generally when the symptoms of a disorder are fully developed that a remedy is resorted to, and many a malady that might have been avoided by timely medication, is allowed to gain such headway through neglect, that it becomes quite or nearly impossible to check it. The antecedents of sickness are physical weakness and organic irregularity. Remove these, and the far more serious evils to which they give rise are surely prevented. To do this thoroughly and with a degree of promptitude most essential to a debilitated system hourly menaced by disease, a combined Tonic and Alterative is required, which will simultaneously and speedily reinforce the flagging energies of the body, and restore uniform and harmonious action to its internal mechanism. This double result is ensured when **Hostetter's Stomach Bitters** is the agent employed to effect it. That incomparable **Vitalizer and Corrective** quickly repairs the waste that has been going on in the system in consequence of imperfect digestion and assimilation, and puts a stop to the drain of strength by re-establishing good order among the organs which supply nourishment to the body, as well as those which carry off its refuse matter.

For all disorders of the liver, stomach and bowels; for indigestion, dyspepsia, costiveness, diarrhœa, intermittent, remittent, and other malarial fevers, the **Bitters** are an absolute specific, affording speedy and unfailing relief, and if persisted in, effecting a permanent cure.

That they are the best and most popular protective and remedy which can be used in regions where *miasma* impregnates the air and water, is borne out by the fact that they are in immense and constantly increasing demand by the inhabitants of every locality on this continent afflicted by malaria, and are also extensively used in those portions of the West Indies and South America where its most malignant types prevail.

When they are employed to build up a system shattered by sickness, or to endow with vigorous health a constitution to which nature has refused that blessing, the recuperative and strength creating power of the **Bitters** is signally shown.

The Trade supplied by the Agents,

VAN SCHAACK, STEVENSON & REID,

Old Salamander Wholesale Drug Warehouse, 92 and 94 Lake St.

ASTHMA AND CATARRH.

Having struggled twenty years between life and death with ASTHMA, I experimented by compounding roots and herbs, and inhaling the medicine. I fortunately discovered a wonderful remedy and sure cure for Asthma and Catarrh. Warranted to relieve severest paroxysm instantly, so the patient can lie down to rest and sleep comfortably. Druggists are supplied with sample packages for FREE distribution. Sold by druggists. Package by mail, $1.25.

D. LANGELL, Proprietor, Apple Creek, Ohio.

Read two from thousands:

SALINEVILLE, OHIO, Jan. 20, 1869.

D. LANGELL—Dear Sir: I have been afflicted with Phthisis or Asthma some twenty years, in its severest form. My suffering cannot be described, and I was never able to find relief, either from physicians or any other source. I saw your advertisement describing your own case, which compared so well with mine that I concluded to send for a package of your Remedy. I inhaled it according to directions; relief followed immediately, and I have not lost one hour's sleep, or one meal of victuals, or one day's work since I commenced using it. I recommend it to every person I know of afflicted with Asthma.

G. W. BOREING.

WOOSTER, OHIO, Nov. 4, 1871.

D. LANGELL—Sir: Having been severely afflicted with Catarrh in the head, I was persuaded to try your inhaling Remedy, which cured me with such astonishing success that I give it public testimony with the fullest confidence, as being a successful and speedy care for Catarrh.

C. V. HARD,
Cashier of the Wooster National Bank, Ohio.

VAN SCHAACK, STEVENSON & REID, Wholesale Agents,
92 and 94 Lake Street, cor. Dearborn, Chicago.

JAS. BUCHAN & CO.'S

TRADE PRICE LIST OF

CARBOLIC SOAPS,

AND SAPONACEOUS COMPOUNDS.

Buchan's Carbolic Laundry Soap.
1 oz. of Acid to the lb.

In 6 lb. paper boxes........................per lb. $ .16

Buchan's Carbolic Toilet Soaps.
(Five per cent. of Acid.)

		Doz.	Gross.
	Buchan's Carbolic Bath Soap.—(Five per cent. of Acid.)		
No. 7.	Boxes of 6 cakes, (paper)	$1.50	$16.50
	Buchan's Carbolic Medicinal Soap.—(25 per cent. of Acid.)		
No. 8.	Boxes of 1 doz. cakes, (paper)	4.00	45.00
	Buchan's Carbolic Dental Soap—(Ten per cent. of Acid.)		
No. 9.	Boxes of 1 doz. cakes, (paper)	2.00	21.00
	Buchan's Carbolic Shaving Soap.—(Ten per cent. of Acid.)		
No. 10.	Boxes of 1 doz. cakes, (paper)	2.00	21.00
	Buchan's Carbolic Balm Ointment.		
	1 oz. glass, boxes of 1 doz. (paper)	1.50	17.00
	Buchan's Carbolic Disinfecting Soap.		
No. 11.	Boxes of 1 doz. cakes, (paper)	1.65	18.00
No. 13.	Boxes of 6 lbs. (paper) 1 lb. bars	5.00	54.00
	Buchan's Carbolic Plant Protector.		
	Cans of 1 lb. (hermetically sealed)	6.00	

Trade 3 Mark.

The Trade suplied by

Van Schaack, Stevenson & Reid, Wholesale Druggists,
92 and 94 Lake Street, cor. Dearborn, Chicago.

WOODRUFF'S

FLY PAPER

IS GUARANTEED BY THE PROPRIETOR

To be superior to any

FLY KILLER, FLY PAPER, OR FLY POISON,

For the instant destruction of Flies. Our sales the past year were very large and satisfactory. By the patent process of the Inventor,

PROFESSOR LOREN B. LORD,

He succeeds in turning out a *uniform* superior quality—every sheet containing an equal amount of "destructiveness," and the paper being of superior make, is capable of absorbing a large amount of the ingredients, and hence its effectiveness. Ask for

"Woodruff's Fly Paper."

The Trade supplied by the Western Wholesale Agents,

Van Schaack, Stevenson & Reid,

92 AND 94 LAKE ST., CORNER DEARBORN,

CHICAGO.

CRUMB'S

POCKET INHALER

FOR THE EFFECTUAL TREATMENT OF

ALL CATARRHAL AFFECTIONS

By the use of Medicated Air or Vapor, prescribed and recommended by the first Physicians in the country.

Charged with the Inhalant, the Inhaler is carried in the pocket as easily as a penknife or smelling bottle, and is so constructed that, by simply breathing in common air through it, a highly medicated vapor is produced, which by the natural process of respiration, finds its way in a very agreeable and effective manner into every part of the

Throat, Lungs and Nasal Cavities,

and furnishes the only successful and really scientific method of treating the diseased mucous membranes of these organs. It may be used by Throat or Nostrils at any moment, regardless of the position of the body and without interfering with business or occupation.

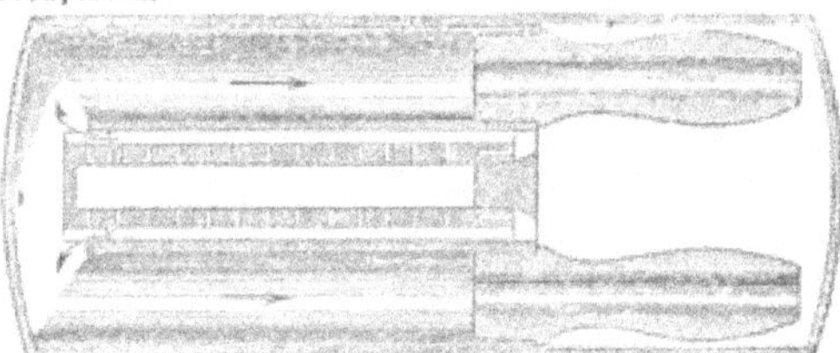

CRUMB'S

Carbolated Chloride of Iodine,

sufficient for 3 or 4 months use, accompanies each Inhaler, and by dropping in a few drops two or three times a week, the Inhaler will be kept sufficiently charged, even though used a dozen times daily.

Crumb's Peruvian Alterative, The Great Blood Purifier.

Crumb's Ague Killer.

Crumb's Compound Pills, Of Butternut and May Apple.

Humphrey's Homœopathic Worm Tablets.

Copland's Sweet Castor Oil.

A full supply constantly on hand at the Western Depot of

VAN SCHAACK, STEVENSON & REID,

Wholesale Druggists, Chicago.

THE CHAMPION

— OF —

American Perfumers!

X. BAZIN,

No. 917 Cherry Street, Philadelphia.

☞ Price List and Catalogue subject to **10** per cent. discount.

No.	SHAVING CREAMS.	Doz.
1	Unrivaled Premium Shaving Cream, Rose, small size	$ 2 75
2	" " " " large size	3 50
3	" " " Ambrosial, small size	2 50
4	" " " " large size	3 25
5	" " " Almond, small size	2 50
6	" " " " large size	3 25

No.	SAPONACEOUS COMPOUND.	Doz.
11	In Mugs	3 50

No.	SHAVING SOAPS.	Doz.
17	Military Shaving Tablet, square, 1st quality	1 50
18	" " " 2nd "	1 00
20	Walnut Oil " "	1 50
21	Barber Soap (10 cakes in box) per box	5 00

No.	FANCY TOILET SOAPS.	Doz.
22	Omnibus, Brown, small size	1 75
23	" " large size	2 75
24	" Palm Oil, small size	1 64
25	" " large size	2 50

PONCINE SOAP.

Especially adapted for cleansing, whitening and softening the hands, and for use in the bath.

No.		Doz.
52	One dozen in a box	2 00

No.	DETERSIVE PUMICE SOAP.	Doz.
53	One dozen in a box	1 75

No.	SUPERFINE TOILET SOAPS.	Doz.
66	Camphor and Glycerine Soap	2 50
77	Bazin's Superior Frangipanni Soap	3 50
78	Tooth Soap	1 35
79	Erasive Soap	67
83	Jockey Club Soap	3 50
86	Upper Ten Soap	3 50
87	New Mown Hay Soap	3 50
88	Moss Rose Soap	3 50
89	Geranium Soap	3 50
90	Tonquin Musc Soap	5 00
91	Superfine Violette de Parme Soap	3 50
97	Floral Soaps, assorted, six perfumes in a box	2 50
108	Olive Soap	1 37

No.	FRUIT SOAPS.	Doz.
122	Lemon Soap	1 63
123	Orange Soap	1 63
124	Pear Soap	1 37
125	Apple Soap	1 75

No.	TOILET WATERS.	Doz.
146	Hedyosmia Water, 4 oz. glass stoppers	$ 5 00
147	" " 4 oz. cork stoppers	4 50
148	" " 8 oz. glass stoppers	8 00
149	" " 8 oz. cork stoppers	7 00

FLORIDA WATER.

162	Florida Water, half pint bottles	3 13
163	" " pint bottles	5 50

AROMATIC VINEGAR.

166	Bazin's Vinaigre Aromatique, Cosmetique and Anti-Mephitique de Bully	4 00

LAVENDER WATER.

176	Bell Lavender Water, (Provost)	6 00

CELEBRATED EXTRACT for the HANDKERCHIEF.

335	Bazin's Extract of Musk, round bottle, 1 oz. glass stoppers	7 25
337	" " half oz. round glass stoppers	3 00
343	Pearl, 2 oz	4 00
356	Ylang Ylang, 1 oz. G. S. bottles	6 25

BAZIN'S HEDYOSMIA.

395	Hedyosmia, in 2 oz. oval bottles	6 50
396	" " " glass stoppers	7 00

TURKISH ESSENCE.

397	Crescent bottles, glass stoppers, small	4 50
398	" " " " large	5 75

PREPARATIONS FOR THE HAIR.

POMADE PHILOCOME.

420	Bazin's round Philocome, small, metal tops	2 25
421	" " " medium, "	3 00
422	" " " large, "	4 00
423	" oval red " small, "	2 25
424	" " " medium, "	3 00
425	" " " large, "	4 00

POMADES.

439	Camphor Ice	1 75
440	Diamond, with lip, wide mouth, glass stopper	10 00
441	Band Diamond, with lip, " "	10 00
442	Tulip, " " "	10 00
443	" no lip, " "	9 00
444	Diamond, " " "	9 00
445	Band Diamond, " " "	9 00
446	Pine Apple, small, with cover	5 50
447	" medium, "	9 50
448	" large, "	12 50
460	Medium Tulip, with lip	8 50
461	" " no lip	7 50

Aretusine and Ursina Canadian Bear's Grease.

514	Large round bottles, glass stoppers, (in cases)	8 00

POND LILY AND QUEEN OF FLOWERS HAIR OILS.

535	Pond Lily Hair Oil	3 00
536	Queen of Flowers Oil	3 00

MAY FLOWER ANTIQUE OIL.

537	1 oz. May Flower Antique Oil	3 00
538	2 oz. " " "	4 00

STICK POMADES.

540	Stick Pomatum, assorted colors, oblong	88

No.	HAIR OILS.	Doz.
548	Amber Oil, screw cap bottle, small	$4 00
549	" " " large	4 50
550	Bazin's Bear's Oil, 4 oz. oval bottles	3 50
551	" " 3 " "	2 75
552	" " 2 " "	1 88
553	" " 1 " "	1 50
555	" " shield "	1 10
556	" Rose Oil, 4 oz. "	3 50
557	" " 3 " "	2 75
558	" " 2 " "	1 88
559	" " 1 " "	1 50
564	" " shield "	1 10

BAZIN'S EAU LUSTRALE.

605	Eau Lustrale, Tonique, small	4 00

DENTIFRICES.

633	Rose Tooth Paste, in 2 oz. covered pots	3 00
637	Charcoal Tooth Paste, in oblong octagon pots	3 00
650	Kiss-Me-Quickly Tooth Powder, 1 oz. pots, ground lids	3 25

COSMETICS.

671	Amandine, for chapped hands	3 75
672	Cold Cream of Roses, in 1 oz. covered pots	3 50
684	Glycerine Cream	3 50

LIP SALVE.

686	Metal Boxes	1 25

RICE FLOUR.

691	Rice Flour, without puffs, small	1 75
696	Rose Leaf Powder	3 00
697	Violette Leaf Powder	3 00
698	Eugenie Superfine Toilet Powder, gilt boxes	3 00

LILY WHITE AND MAGNOLIA TABLETS.

699	Pearl of Beauty, photograph label	1 63
700	Lily White, round boxes, large	1 10
701	" " medium	87
702	" " small	63
705	Magnolia Tablet, oval boxes, large	1 10
706	" " " small	88
716	Cascarilla White	1 25
717	Spanish Lily White	1 38

SMELLING SALTS.

723	Bazin's Preston Salts, glass stoppers	1 88

SACHETS.

726	Frangipanni Sachets, satin	4 00
731	No. 4, Paper Sachets	1 50
734	Violette Sachets, strong perfume	1 25

VINAIGRE DE ROUGE.

758	In square bottles, large size	1 75
759	" " small	1 25
760	Lemon Rouge, large	2 00
761	" medium	1 75
762	" small	1 25

AROMATIC CRYSTALS.

763	Bazin's Aromatic Crystals, glass stoppers	2 00
764	" " " morocco cases	4 75
766	" " cut flat glass stoppers	2 75
767	Conc. ext. Aromatic Vinegar, cut glass, cap bottle, morocco cases	10 50

DEPILATORY POWDER.

770	Bazin's Depilatory Powder	2 75

Dr. JOHN BULL,

LOUISVILLE, KY.,

PROPRIETOR OF

Smith's Tonic Syrup,

BULL'S CEDRON BITTERS,

DOCKTOR KENT.

After Using. Before Using.

Smith's Tonic Syrup

BULL'S

Worm Confections

BULL'S

Sarsaparilla

You can always find a large stock of these Valuable Medicines at the Old Salamander Wholesale Drug House,

Van Schaack, Stevenson & Reid,

92 & 94 Lake St., cor. Dearborn, Chicago.

Strickland's Wholesale Prices.

We guarantee to sell you FINER and CHEAPER goods than any other house in the West.

WINE OF LIFE. ENGLISH CORDIAL GIN. AROMATIC BRITISH BRANDY.	$6.00 per case, in lots less than 10 cases, or $5.00 per case, in 10 case lots. $2.00 per gallon.

You will find these goods superior to any others you may have dealt in; we guarantee you perfect satisfaction.

	Per Case.	In Ten Case Lots.	Per Gallon.
OLD PORT WINE,	$ 7 00	$ 6 00	$ 2 50
PURE SHERRY WINE,	7 00	6 00	2 50
KENTUCKY B. WHISKY,	6 00	5 50	2 00
OLD JAMAICA RUM,	6 00	5 50	2 00
SCOTCH WHISKY,	7 00	6 00	3 00
HOLLAND GIN,	7 00	6 00	3 00
COGNAC BRANDY,	7 00	6 00	3 00
OLD IRISH WHISKY,	7 00	6 00	3 00
BLACKBERRY BRANDY,	6 00	5 50	2 00
KUMMELL,	6 00	5 50	2 00
STOMACH BITTERS,	6 00	5 50	1 75
RYE WHISKY,	6 00	5 50	2 00
NEW ENGLAND RUM,	6 00	5 50	2 00
OLD JOE WHISKY,			1 75

The above goods are **A No. 1**, and we have put the price down as low as we can offord to sell them at, as we are determined no other house shall beat us in the price and quality. We not only *say* the above goods are as good as any, but we *insist* and can *prove* they are much superior to any other, and a great deal cheaper.

OUR TERMS ARE 60 DAYS.

NOTICE.—We have all the above goods, also, in **50 cent Pocket Flasks**, which we sell to you at $3.00 per case. These flasks are of a peculiar, but very convenient shape; can be carried in the pants or any other pocket, yet they hold considerably over half a pint. You can sell a large quantity of them.

DON'T FAIL to compare Strickland's Wine of Life with any other article of the same name, then you will see what a superior article Strickland's Wine of Life is; also compare the prices; then, we think, you will want a rebate on the price you have been paying for our opponent's inferior article.

A. & W. W. STRICKLAND,

174, 176 and 178 Adams Street, Chicago.

The Trade supplied by

Van Schaack, Stevenson & Reid,

92 and 94 Lake Street, cor. Dearborn, Chicago.

JOHN WYETH & BRO.,

Manufacturers and Dispensers of

Elegant Pharmaceutical Preparations

PHILADELPHIA.

For Sale by VAN SCHAACK, STEVENSON & REID.

Elixir Cinchonæ (Calisaya).

Five pint bottles, each	$3 00
Quart " "	1 25
Per doz. pints	8 00

Ferrated Elixir Cinchonæ (Calisaya).

Five pint bottles, each	3 50
Quart " "	1 45
Per doz. pints	10 00

Elixir Pyrophosphate Iron.

Five pint bottles, each	3 50
Quart " "	1 45
Per doz. pints	10 00

Elixir Calisaya Bark, Iron and Bismuth.

Five pint bottles, each	3 50
Quart " "	1 45
Per doz. pints	10 00

Elixir Calisaya Bark, Iron and Strychnia.

Five pint bottles, each	3 50
Quart " "	1 45
Per doz. pints	10 00

Elixir Bismuth.

Five pint bottles, each	4 00
Quarts " "	1 75
Per doz. pints	11 00

Elixir Phosphate Iron, Quinine and Strychnia.

Five pint bottles, each	9 00
Quart " "	3 75
Per doz. pints	22 00

Elixir Valerianate of Ammonia.

Five pint bottles, each	4 00
Quart " "	1 75
Per doz. pints	11 00

Elixir Valerianate Ammonia and Quinine.

Five pint bottles, each	9 00
Quart " "	3 75
Per doz. pints	24 00

Elixir Valerianate of Strychnia.

Five pint bottles, each	3 00
Quart " "	1 25
Per doz. pints	8 00

Elixir Bromide Potassium.

Five pint bottles, each	5 50
Quart " "	2 50
Per doz. pints	15 00

Ferrated Cordial Elixir.

Five pint bottles, each	3 00
Quart " "	1 25
Per doz. pints	8 00

Elixir Gentian Ferrated.

Five pint bottles, each	3 00
Quart " "	1 25
Per doz. pints	8 00

Elixir Hops.

Five pint bottles, each	3 00
Quart " "	1 25
Per doz. pints	8 00

Elixir Gentian and Tr. Iron.

Per doz. pints	$8 00

Elixir Pepsin, Bismuth and Strychnia.

Five pint bottles, each	9 00
Quart " "	3 75
Per doz. pints	24 00

Compound Syrup Phosphates.

Chemical Food.

Five pint bottles, each	3 50
Quart " "	1 45
Per doz. pints	10 00

Compound Syrup of Hypophosphites.

Churchill's.

Five pint bottles, each	3 50
Quart " "	1 45
Per doz. pints	10 00

Syrup Superphosphate of Iron.

Five pint bottles, each	3 00
Quart " "	1 25
Per doz. pints	8 00

Comp. Syrup Phosphate of Manganese.

Five pint bottles, each	6 50
Quart " "	2 75
Per doz. pints	17 50

Wine of Pepsin.

Made from the stomach of the pig.

Five pint bottles, each	5 50
Quart " "	2 50
Per doz. pints	15 00

Bitter Wine of Iron.

Five pint bottles, each	3 50
Quart " "	1 45
Per doz. pints	10 00

Wine of Wild Cherry Bark.

Five pint bottles, each	3 25
Quart " "	1 35
Per doz. pints	9 00

Ferrated Wine of Wild Cherry Bark.

Five pint bottles, each	3 55
Quart " "	1 40
Per doz. pints	10 00

Wine of Ergot.

Five pint bottles, each	7 00
Quart " "	2 90
Per doz. pints	20 00

Lacto-Phosphate of Lime with Cod Liver Oil.

Per dozen	12 00

Cod Liver Oil with Phosphate of Lime.

Per dozen	8 00

Comp. Fluid Ext. Buchu and Pareira Brava.

Five pint bottles, each	7 00
Quart " "	2 90
Per doz. pints	20 00

WYETH BROS.' PREPARATIONS—continued.

Elixir Bromide Sodium.

Five pint bottles, each	$5 50
Quart " "	2 50
Per doz. pints	15 00

Syrup Chloral Hydrat.

Five pint bottles, each	5 50
Quart " "	2 50
Per doz. pints	15 00

Cod Liver Oil and Hypophosphites.

Lime and Soda.

Five pint bottles, each	4 00
Quart " "	1 75
Per doz. pints	11 00

Beef, Wine and Iron.

Five pint bottles, each	3 00
Per quart	1 25
Per doz. pints	8 00

Beef and Wine.

Five pint bottles, each	3 00
Per quart	1 25
Per doz. pints	8 00

Beef, Wine, Iron and Cinchona.

Five pint bottles, each	3 50
Per quart	1 45
Per doz. pints	10 00

Syrup Lacto-Phos. Lime.

Five pint bottles, each	5 50
Per quart	2 50
Per doz. pints	15 00

Syrup Lacto-Phos. Iron.

Five pint bottles, each	5 50
Per quart	2 50
Per doz. pints	15 00

Comp. Syrup of the Lacto-Phosphates.

Five pint bottles, each	7 00
Per quart	3 00
Per doz. pints	20 00

Elixir of the Phosphates and Calisaya.

Chemical Food with Cinchona.

Five pint bottles, each	3 00
Per quart	1 25
Per doz. pints	8 00

Elixir Calisaya Bark, Iron, Bismuth, and Strychnia.

Five pint bottles, each	4 00
Per quart	1 75
Per doz. pints	11 00

Loeflund's Concentrated Liebig's Food.

Per dozen	6 00

Loeflund's Concentrated Liebig's Ext. Malt.

Per dozen	6 00

Papoma.

Per gross	36 00

Elixir Pepsin and Bismuth.

Five pint bottles, each	8 00
Per quart	3 50
Per doz. pints	22 00

Elixir Pepsin, Bismuth, Strychnia and Iron.

Five pint bottles, each	$9 00
Per quart	3 75
Per doz. pints	24 00

Wine Calisaya.

Five pint bottles, each	3 00
Per doz. pints	8 00

Liq. Bismuth.

Five pint bottles, each	4 00
Per doz. pints	11 00

Elixir Taraxacum Comp.

Five pint bottles, each	3 00
Per doz. pints	8 00

PLASTERS.

Belladonna, Arnica.

Per doz. 1 yard rolls	8 00

Surgeons' Roller Bandages.

Per gross, assorted	5 00

LOZENGES.

Pastilles of Chlorate of Potash.

Per gross	20 00

Marvin's Cod Liver Oil.

Per doz. pints	7 00

Bishop's Granular

Effervescent Salts.

	4 oz. doz.
Granular Effervescent	
" Brom. Ammonium	$6 00
" Citrate Bismuth	8 50
" Carlsbad Salts	7 00
" Nitrate of Cerium	8 50
" Citrate Cinchona	6 00
" " " and Iron	6 50
" Bromide Iron	8 50
" Carbonate "	5 00
" Citrate "	5 00
" Iodide "	7 00
" Phosphate "	5 50
" Kissingen	5 00
" Citrate Lithia	12 00
" Magnesia Aperient	4 00
" Benzoate Potassa	8 50
" Bi Carb. "	5 00
" Citrate "	7 50
" Nitrate "	5 00
" Brom. Potassium	6 00
" Iodide "	6 00
" Pullna Salts	5 00
" Quinine	10 50
" " and Iron	11 00
" Seidlitz	5 00
" Soda Cit. Tart.	5 00
" Iodide of Sodium	6 00
" Vichy Salts	5 00
" Seltzer "	5 00
" Citrate Magnesia	5 00
" Pepsin and Bism.	9 00
" " Bis. and Stry.	10 00
Ortho.-Phos. Qui. Iron and Stry	10 00

RECOMMENDED BY THE MEDICAL FACULTY!

DR. S. B. COLLINS'

LIQUOR ANTIDOTE

Is a Perfect Cure for Strong Drink.

It destroys the Appetite for Liquors, and at the same time builds up the broken-down system.

I think that the reputation of my

OPIUM ANTIDOTE

Is sufficient guarantee of the Efficacy of this Remedy.

THE

FEMALE'S FRIEND

PREPARED BY DR. S. B. COLLINS,

A MEDICAL COMPOUND WHICH CURES PERMANENTLY

LEUCORRHŒA, PROLAPSUS, OR FALLING OF THE WOMB

Also, relieves Painful or Obstructed Menstruation, and restores the Organs to a Healthy Condition.

A LARGE STOCK

Of each of these Valuable Remedies can always be found at

VAN SCHAACK, STEVENSON & REID'S

Old Salamander Wholesale Drug Warehouse,

92 & 94 Lake Street, Cor. Dearborn, Chicago.

Message to the Suffering.

When a remedy is productive of such relief as to excel all other remedies in its curative results, it is the duty of those interested in its adoption and popularity to make it known as generally as possible Philanthropy demands such a course. The pharmaceutist who originated and who prepares it may have a pecuniary interest in its extensive sale, yet he cannot but have an honorable pride in the reflection that the medicinal agents he has introduced to the suffering bring a relief to them which cannot be measured by money, and he must derive unalloyed pleasure from the conviction that if it makes him rich, it makes those who buy it happy, which, it is to be regretted, cannot be said of all the medicines that are imposed on a people whose credulity is apt to be in proportion to their pain. These considerations have actuated the proprietor, WM. M. GILES, to bring prominently before the public his **Liniment Iodide of Ammonia.**

FOR HORSES AND OTHER ANIMALS.

Giles' Liniment Iodide of Ammonia makes the quickest and surest cures of any article yet known, and the horse may be worked during the time of cure. 50c and $1.00 sizes (yellow label.)

FOR FAMILY USE.

In Erysipelas, for Swelled Joints of the Hands or Feet, for Glandular Swellings of the Neck or Groin, Wens, Sore Throat. For Headache, brush over the temples and inhale it. **For Discolorations of the Skin,** resulting from bruises, apply the Liniment frequently to the parts. **For Blotches or Pimples, Black Heads, Moths on the Skin. For Catarrh,** inhale through the nostrils, Iodine being the greatest alterative known in medicine. **In Chronic and Inflammatory Rheumatism, Scrofulous Affections, Granular Swellings,** it has no superior. 50c and $1.00 sizes (white label.)

The Trade supplied by VAN SCHAACK, STEVENSON & REID,

Old Salamander Wholesale Drug Warehouse, Chicago.

This is not a Compound, but Common Tar, obtained from the long-leaved pine of North Carolina, the medical properties of which are found in five oils, that are *white*. These oils are disengaged from the black, impure and worthless products of the tar, by an invention of S. R. HANSCOM, of Boston, Analytical Chemist, and by this new process the "Forest Tar" has been prepared, under the advice of the *most skillful physicians*, approved and used successfully by more than one hundred of the profession during the first six months. The design is to make it a standard remedy, and as such it is offered to the Physician and the Public.

For the greater convenience of administering, and to adapt it more fully to the various forms of diseases where found beneficial, "Forest Tar" has been prepared in various forms, as follows:

Forest Tar For Consumption, Bronchitis, Catarrh, Asthma, Sore Throat, Piles, diseases of the Skin, Kidneys and Urinary Organs, and for Purifying the Blood.

Forest Tar in Solution For Inhaling, for Catarrh, Consumption, Bronchitis, Asthma, and as a Wash for diseases of the Skin.

Forest Tar Inhalers For Inhaling the Solution.

Forest Tar Troches For Bronchitis, Hoarseness, Sore Throat, and Purfying the Breath.

Forest Tar Salve For Cleansing, Purifying and Healing Indolent Sores, Bruises, Diseases of the Skin, and wherever a healing process is wanted for man or beast.

Forest Tar Soap For removing Pimples, Eruptions of the Skin, such as Salt Rheum, Chapped Hands, and all Roughness of the Skin, and for dressing Wounds—for the Toilet, cleansing the Teeth and purifying the Breath. No *Shaving* Soap leaves the face so sound and smooth. It contains no alkali, and is therefore preferable to any of the fine toilet soaps.

These preparations contain all the medicinal virtues of the Common Tar, AND NOTHING MORE.

Prepared by the FOREST TAR Co., Portland, Me., and sold by all druggists.

The Western Trade supplied by

Van Schaack, Stevenson & Reid,

92 & 94 Lake Street.

SOMETHING ENTIRELY NEW!

MILK OF MAGNESIA

A CONCENTRATED LIQUID MAGNESIA,

PREPARED BY

C. H. PHILLIPS, Manufacturing Chemist, New York.

After many investigations and chemical experiments, this ENTIRELY NEW and (as physicians who have become acquainted with it say) THE BEST PREPARATION OF MAGNESIA in use, has been brought to light, and is now rapidly coming into general use. It is a combination of PURE MAGNESIA and PURE WATER (by a new process). Being FOUR TIMES THE STRENGTH of any Fluid Magnesia before offered, and FREE FROM CARBONIC ACID, is much more agreeable, and establishes at once its claim to be the only SAFE MAGNESIA that can be used in medical practice. In cases of INDIGESTION and DYSPEPSIA, from which so many suffer, this new compound is the ONLY POSITIVE RELIEF and CURE. For excess in eating or drinking, it is most valuable, as it immediately removes the irritation and acidity of the stomach, and settles that organ in a very short time. Sir Humphrey Davy has proved the efficacy of Magnesia in ordinary cases of RHEUMATISM, GOUT and GRAVEL, for all of which this preparation has been used with the happiest results.

Home testimony! Hear a well known resident! Hon. Jno. B. Drake, for 20 years a resident of Chicago, Vice-President Ill. Loan and Trust Co., Proprietor of the Grand Pacific Hotel, Director of the Inter-State Industrial Exposition, late Proprietor of Tremont House, etc., etc.

CHICAGO, June 15th, 1874.

MESSRS. VAN SCHAACK, STEVENSON & REID:

I have used Phillips' "MILK OF MAGNESIA" in my family with the GREATEST SATISFACTION, and MOST CHEERFULLY commend it for any one troubled with acidity of the stomach, sick headache or dyspepsia, as affording positive relief and cure, and am now never without a bottle in my house.

JOHN B. DRAKE.

Directions for Use.

FOR INFANTS, "to prevent food from turning sour on the stomach," from half to a teaspoonful; as a laxative, double that quantity.

FOR ADULTS, as an antacid, a tablespoonful; for gout, etc., half wine-glass full occasionally.

In all cases it should be diluted with an equal quantity of water or milk, and its action as a laxative is greatly increased by the addition of acidulated syrup, emon juice or citric acid.

Messrs. Van Schaack, Stevenson & Reid

Are our duly authorized Agents, to whom orders should be addressed.

POND'S EXTRACT

(OF HAMAMELIS, OR WITCH HAZEL.)

People's Remedy

— FOR —

Burns, Scalds.
Bruises.
Sprains.
Wounds.
Ulcers.
Lame Back.
Frozen Limbs.
Sore Feet.
Corns.
Swelled Face.
Kidney Disease.
Broken Breast.
Ague in Breast.
Rheumatism.

People's Remedy

— FOR —

Bleeding Piles.
Blind Piles.
Toothache.
Headache.
Earache.
Sore Throat.
Sore Eyes.
Nose Bleed.
Sore Nipples.
Bleeding Lungs.
Painful Menses.
Sting of Insects.
Neuralgia.
Hemorrhages.

Prepared Only by the Pond's Extract Co.,

NEW YORK AND LONDON.

Which embraces every living person who ever made or knew HOW to make, this—the oldest, best, and only uniform preparation from the Witch Hazel.

PRICE LIST.

In lots of "six ounce" or the equivalent of assorted sizes.

Sizes.	Packed.	At Retail.	Per Doz.	Per Case.	Per Gross.
Small, - -	4 Doz.	$ 50	$ 4 00	$16 00	$ 45 00
Medium, -	2 "	1 00	8 00	16 00	90 00
Large, - -	1 "	1 75	15 00	15 00	180 00

☞ Drayage charged on any less than ten gross.

The Trade supplied by

VAN SCHAACK, STEVENSON & REID,

Old Salamander Wholesale Drug Warehouse,

92 & 94 Lake St., cor. Dearborn, Chicago.

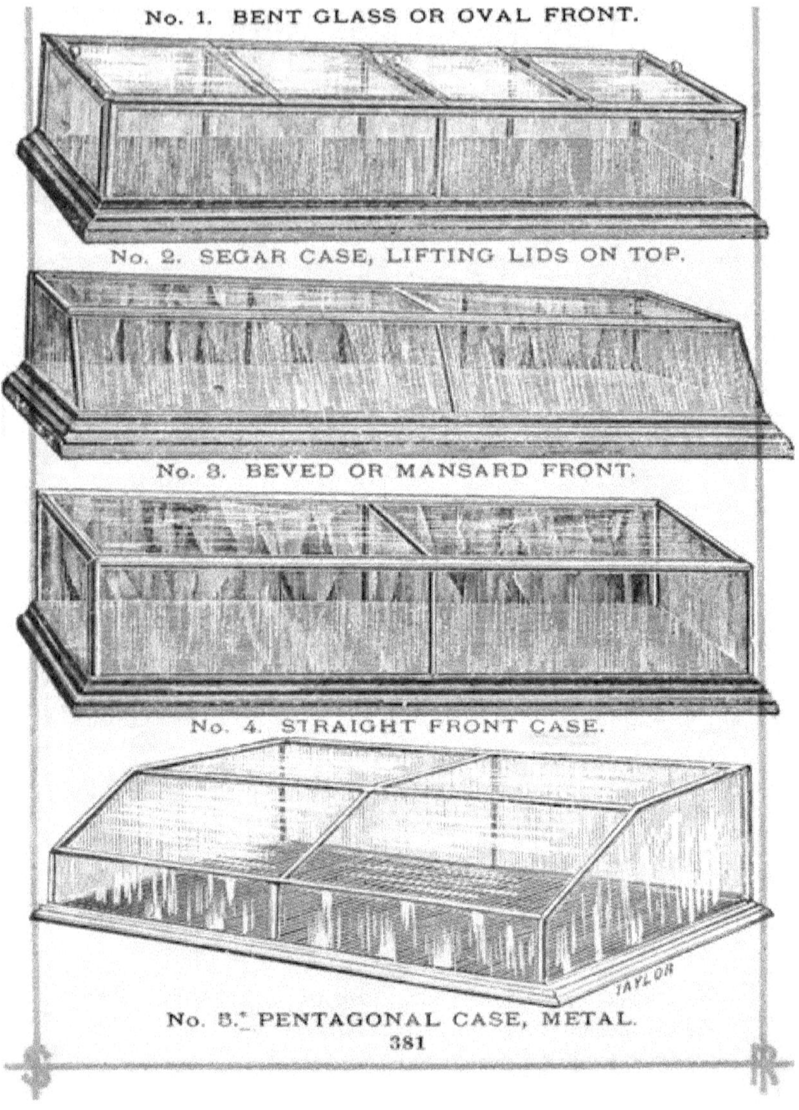

No. 1. BENT GLASS OR OVAL FRONT.

No. 2. SEGAR CASE, LIFTING LIDS ON TOP.

No. 3. BEVED OR MANSARD FRONT.

No. 4. STRAIGHT FRONT CASE.

No. 5. PENTAGONAL CASE, METAL.

WE ALWAYS CARRY A FULL STOCK OF THESE

Elegant and Reliable Goods.

Messrs. Jos. Burnett & Co. have established a fame for the *superior excellence* of all goods bearing their brand.

Unequaled by any similar preparation in the world.

Their Flavoring Extracts

Are used in all the large Hotels, Restaurants, etc.,—are put in large bottles for that trade.

BURNETT'S COCOAINE,
KALLISTON,
FLORIMEL,
COLOGNE,
ORIENTAL TOOTH WASH,

Need no recommendation from those who have used them.

Van Schaack, Stevenson & Reid,

Old Salamander Wholesale Drug Warehouse,

92 and 94 Lake St., cor. Dearborn, Chicago.

Tilden's Pharmaceutical Preparations.

ELIXIRS, SYRUPS, ETC.

Preparations of the Pharmacopœia not included in our List made to Order.

N. B. In making orders be particular to state if in pounds, 4 ounce, or bulk. When dozens are ordered, pounds will be sent if not otherwise specified. When less than a dozen is ordered, it will be charged at dozen rate.

NEW PREPARATIONS.

Aromatic Cordial, Elixir.
" " " Ferrated.
Bismuth Liquor.
Elixir Bismuth.
" " Ferrated.
" Bromide Ammonium.
" " Potassium.
" " Sodium.
" " Iron.
" Buchu and Pareira Brava.
" Calisaya.
" " Ferrated.
" " and Bismuth.
" " Columbo and Quassia.
" " Phosphate Iron and Manganese.
" " and Pyro. Iron.
" Calisaya and Pyro. Iron and Bismuth.
" Calisaya and Pyro. Iron, Bismuth and Pepsin.
" Calisaya and Pyro. Iron, and Strychnia.
" Calisaya and Pyro. Iron, Strych. and Bismuth.
" Calisaya and Protoxide Iron.
" " Strychnia and Bismuth.
" " Pepsine and Bismuth.
" Cardamom Comp.
" Chloral Hydrate.
" Cinchona, Iron and Strychnia.
" Cimicifuga.
" " Compound.
" Citrate Lithia.
" " Caffein.
" Collinsonia.
" Gentian.
" " Comp.
" " and Chloride of Iron.
" " and Pyro. of Iron.
" Helonias Compound.
" Iodide Calcium and Protox Iron
" Iodide of Lime and Protox. Iron
" Lupulin.
" Mandrake Comp.
" Matico.
" " Comp.
" Pancreatine.
Elixir Pepsine, Aromatic.
" " and Bismuth.
" " " and Iron.
" Pepsine, and Bismuth and Pancreatine.
" Pepsine and Pancreatine.
" " and Quinia.
" " and Strychnia.
" Pepsine and Strych. and Bism'h
" Pepsine and Strychnia and Bismuth, Ferrated.
" Pepsine and Strychnia and Pancreatine.
" Pepsine and Strychnia, Bismuth and Pancreatine.
" Peruvian Bark and Bismuth.
" " " and Protox. Iron.
" Phosphate of Iron.
" " and Quinia.
" " Quinia and Strych.
" Phosphate of Lime.
" Protoxide Iron.
" " with Iod. Calcium
" " with Iod. Potassium
" " and Quinia.
" Pyrophos. of Iron.
" " and Bark.
" " and Soda.
" " Quinia and Strych.
" Quinia, Bismuth and Strychnia.
" " Iron and Arsenic.
" " Iron and Strychnia.
" " Iron, Strych. and Bismuth.
" Rhubarb and Columbo.
" Rhubarb Arom. and Magnesia.
" Stillingia.
" " Compound.
" Senna Comp.
" Spigelia Comp.
" Strychnia and Bismuth.
" Taraxacum Compound.
" Valerian.
" Valerianate of Ammonia.
" Valerianate of Ammonia, Iron and Strychnia.
" Valerianate of Ammonia and Quinia.

ELIXIR, Valerianate of Ammonia and Strychnia.
" Valerianate of Iron.
" " Morphia.
" " Quinia.
" " Strychnia.
" " Zinc.

ESSENCE Jamaica Ginger.

FERRO-PHOSPHATED Elixir of Gentian.
" Elixir of Calisaya.
" Elixir of Calisaya and Bismuth.

FLUID OPIUM, Deodorized.
" " " 1 oz. vials.

SYRUP, Alterative.
" Blackberry, (Comp. Aromatic, formulæ of Surg. Gen. U.S.A.)
" Bromide of Morphia.
" " Quinia.
" " Quinia and Morphia.
" " " and Morphia and Strychnia.
" " Strychnia.
" " " and Morphia.
" Citrate of Iron.
" " and Quinia.
" " and Quinia and Strychnia.
" Citrate of Iron and Strychnia.
" Ginger.
" Hypophosphite Lime.
" " Manganese.
" " Soda.
" Hypophosphites, Comp. (Lime, Soda, Potassa and Iron.)
" Hypophosphites Iron.
" " Iron and Manganese.
" " " and Quinia.
" Hypophosphites Lime and Soda, (Churchill's).
" Iodide of Iron, U. S. P.
" " and Manganese.
" " Calcium (Lime).
" " Starch.
" " Manganese.
" Ipecac, U. S. P.
" Lacto-Phosphate of Iron.
" " Lime.
" Lacto-Phosphate of Lime, with Pepsine.

SYRUP Lacto-Phosphate of Lime, Soda, Potassa and Iron.
" Lime.
" Orange Peel (Curacoa).
" Pectoral (Jackson's).
" Phosphates, chemical food.
" Phosphate Iron.
" " and Quinia.
" " and Strychnia.
" " Quinia and Strych.
" Phosphate Lime.
" Protoxide of Iron.
" Protoxide of Iron, and Iodide Potassium.
" Protoxide of Iron, and Iodide Calcium.
" Protoxide of Iron and Quinia.
" Protox. Iron, Rhei and Columbo.
" Pyrophosphate Iron.
" Rhei.
" " Aromatic.
" " and Potassa.
" Rumex Compound. Scrofulous.
" Sarsaparilla.
" " Comp.
" " with Iodide Calcium.
" " " Potassium.
" Seneca.
" Squill.
" " Compound.
" Stillingia.
" " Compound.
" Super-Phosphate Iron.
" Tolu.
" Wild Cherry.

WINE of Beef.
" of Beef and Iron.
" " " and Cinchona.
" of Calisaya, Aromatic.
" " Ferrated.
" of Iron, bitter.
" of Pepsine.
" " 8 oz.
" of Wild Cherry.
" " Ferrated.
" " Ferrated and Iodine.
" " Phosphorated.
" " and Iodine.
" " and Iodide of Iron.
" " Iodine and Cit. Iron

Spread Eagle—zinc—gilded.

THE BUFFET.

* Bigelow's Soda Apparatus. New styles. Prices from $100 to $2,000. The old reliable Drug House of Van Schaack, Stevenson & Reid are our only local agents for the West and Northwest. Sample goods for shipment in stock. All goods guaranteed satisfactory. Copper Generators, Fountains, Tumbler Holders and Tumblers, at the lowest rates. Our goods are made for use. Measuring or Non-measuring Faucets. Block Tin Coolers for Mineral Waters. Call or send to Van Schaack, Stevenson & Reid, for a Catalogue of 1875.

THE RUSSIAN.

Style new, and contains all of our latest improvements. Block Tin Coolers, etc.

Superior in all practical points to any other make up of Counter Apparatus in use. Price, the lowest, for the amount and quality of work we give for your money.

BIGELOW MANUF'G CO.,

Springfield, Mass.

E. BIGELOW, Sup't.

VAN SCHAACK, STEVENSON & REID,

92 & 94 LAKE ST., opposite Tremont House, CHICAGO,

SOLE AGENTS FOR THE WEST.

Call and Examine. Send for Illustrated Catalogue.

BIGELOW'S ICE PLANE.

Ice Plane, Holder and Board, $5. Hints on Syrup Making sent with each Ice Plane.

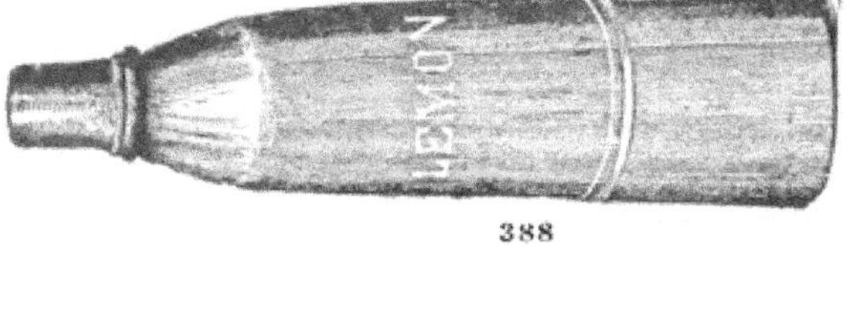

Glass Syrup Faucets.

Patented July 16th and Oct. 1st, 1872.

The above cut is an illustration of Mr. Zwietusch's Patent Crystal Glass Faucets, which are now attached to his apparatus. As can be seen, they are of a very simple construction, and made strong enough for any ordinary strain. Special attention is called to this, as they are made entirely of glass, and the syrups passing through them do not come in contact with metal, so injurious to syrups, thus insuring against all corrosion. They are self-closing, and are operated by a slider, which cuts off like a knife, with the stream, every drop of syrup, making dripping or leakage impossible. This feature cannot be overrated, and no apparatus can begin to compare with his, furnished with these Patent Glass Syrup Faucets. There are none but his make of Glass Faucets in the market.

They also greatly enhance the beauty of an apparatus; being made of Crystal Glass they set off well against the marble, and as they are transparent, show critical customers the purity of the syrups.

They can be attached to all other apparatus, if sample faucet is sent.

Send for catalogue to

OTTO ZWIETUSCH,

705 to 709 Chestnut Street,

MILWAUKEE, WIS.,

— OR TO —

Van Schaack, Stevenson & Reid,

CHICAGO.

ALL KINDS OF

Soda Water Apparatus

MADE BY

OTTO ZWEITUSCH,

705 TO 709 CHESTNUT ST.,

MILWAUKEE, WIS.

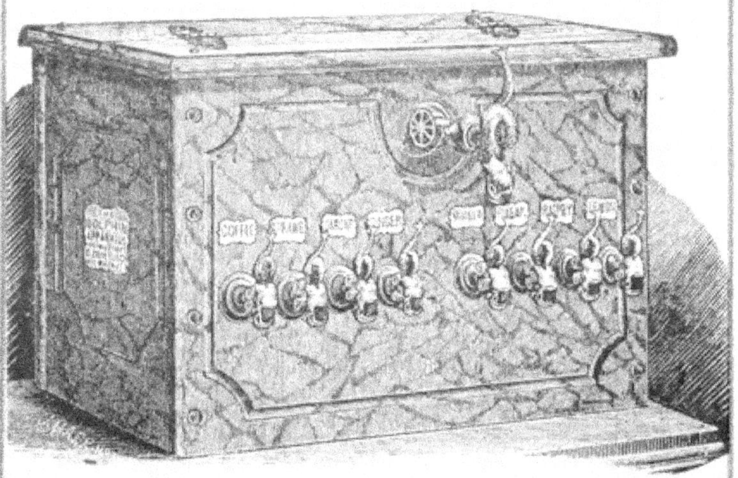

We have constantly on hand a selection of Generators, Fountains, Marble Draught Apparatus, Patent Atmospheric Apparatus, etc., etc.

All Apparatus manufactured by Mr. Zweitusch are made of the best material, and are furnished with his latest inventions and improvements. Among the most important we mention his Patent Generator, Patent Glass Syrup Faucet, Patent Double Stream Draught Tube, the Dolphin Faucet, Patent Atmospheric Apparatus, new Cooling Arrangements, Air Chamber, Improved Fountains, and the Patent Automatic Self-Regulating Generator, used mostly for preserving Beer, Cider and Wines, which, together with lower prices, render his Apparatus equal, if not superior, to Eastern manufacture, which is the testimony of all customers having them in use.

OTTO ZWIETUSCH'S
Patent Soda Water Apparatus

Patented Nov. 9th, 1869.

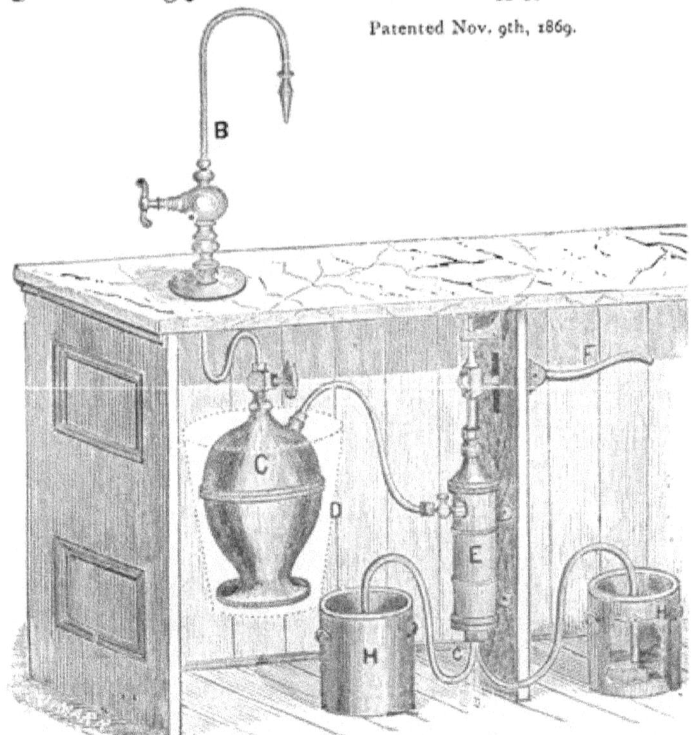

The above represents a PATENT ATMOSPHERIC SODA WATER APPARATUS, which needs but to be seen to be appreciated by all familiar with the business. It is the most economical, as well as the most durable, the simplest, and therefore the best apparatus ever constructed. By this invention, the water, including the requisite syrups, can be drawn at a cost not exceeding 1½ cents per glass, leaving a handsome margin for the enterprising proprietor. There is no three or four feet space occupied on the counter, and no cumbersome machinery to be put up. It can be placed anywhere, takes up but very little room, and never gets out of order. All parts coming in contact with acids, etc., are well protected by pure tin or silver against their corroding influences.

Bicarbonate of Soda and Acids must be used by this Apparatus to produce Soda Water, but by this Machine the result is by far better than by the old way, using common brass pumps and a single solution of Bicarbonate of Soda. No acids need be put into the Syrups, except the natural fruit extracts.

It consists of a well Tinned Pump, Copper Fountain, Tin Suction and Connection Pipes, Cooling Tub, and a heavy Silver Plated Draw Cock.

Price of Apparatus, as shown above, - - - - - - -	$ 65.00
Or, Apparatus as shown above, without Silver-Plated Counter Draw-Cock,	55.00
One Four Syrup Dolphin Draught Apparatus, - - - - -	70.00

The Western Trade supplied by

VAN SCHAACK, STEVENSON & REID,

92 and 94 Lake Street, cor. Dearborn, Chicago.

BUY THE BEST!

THE PIONEER

CALIFORNIA WINE HOUSE.

PERKINS & STERN

ESTABLISHED 1854.

108 Tremont St. BOSTON.

NEW YORK. 14 & 16 Vesey St.

Having dissolved all connection with any Wine Merchants in San Francisco, we are now receiving all our Wines and Grape Brandies direct from the Vineyards, and are thus enabled to offer an additional guarantee of their *purity*.

Our Stock comprises the largest and finest assortment ever offered in the market, and at *lower prices* than good genuine Wines can be sold for by any other house. We have also lately added to our Stock a large assortment of Fine Imported and Domestic Liquors, especially adapted to the Drug Trade. For Medicinal, Sacramental or Table use, they are unsurpassed.

For Price Current, see page 70

The Trade supplied by

VAN SCHAACK, STEVENSON & REID,

Old Salamander Wholesale Drug Warehouse,

92 & 94 Lake Street, Cor. Dearborn, Chicago.

O. P. BASSETT.

SAMUEL MITCHELL,
(Successor to MITCHELL & Co., Late MITCHELL, LAWRENCE & FORDHAM.

Bassett & Mitchell,

Nos. 12 and 14 LA SALLE STREET,

CHICAGO,

MANUFACTURERS OF

GUMMED OR UNGUMMED,

Cut for immediate use, of every description, either

Plain, Colored or Fancy Extract, Hair Oil, and Bronze and Colored Liquor Labels, Prescription Papers and Envelopes,

AND EVERY KIND OF

Druggists' Printing

— OR —

GENERAL JOB WORK,

SUCH AS

Letter, Note and Bill Heads, Cards, Shipping Tags, Circulars, Price Lists, Pamphlets and Hand Bills.

We have just issued our NEW SPECIMEN BOOK, giving New Styles and Reduced Prices, and will take pleasure in sending it to all who desire it.

☞ Orders addressed to

Messrs. VAN SCHAACK, STEVENSON & REID,

Will receive best attention.

CHICAGO

Truss and Brace Manufactory,

By OTTO KALTEICH,

116 E. MADISON STREET,

S. E. COR. CLARK STREET.

Surgical and Anatomical Mechanician, Inventor and Manufacturer of Orthopedic Instruments, such as Apparatus for

Curvature of the Spine, Wry Neck,
Anchylosis, Club Feet, Bow Legs,
Weak Ankles, Hip Diseases, etc.,

Artificial Legs and Arms,

All made to order in the best and most approved style, and fit guaranteed.

I also manufacture a variety of Trusses, among which my Truss for Radical Cure always gives satisfaction, even in the worst cases, where all other appliances have failed. In small cases of rupture my new Elastic Truss is the most comfortable ever invented.

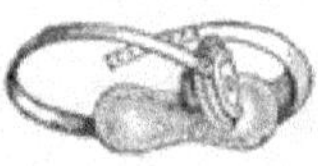

Abdominal Supporters, Shoulder Braces, Elastic Stockings, Knee Caps, Anklets, Suspensions and Bandages, of every variety & description.

☞ I wish to call particular attention to my new Bandage, called the "Ladies' Comfort," which for comfort and cleanliness during menstruation, is unsurpassed. No lady who has seen it would be without one.

OTTO KALTEICH,
116 E. Madison St., S. E. Cor. Clark St., CHICAGO, ILL.

Orders filled through

Van Schaack, Stevenson & Reid.

DR. J. DELMONICO LITTLE'S

SYRUP PECTORAL.

SURE CURE FOR

Coughs, Colds, Hoarseness, Catarrh, Bronchitis, Influenza, Croup, Raising of Blood, Whooping Cough, Asthma, and Consumption.

ACTS LIKE MAGIC ON ALL THE ABOVE DISEASES.

We can with truth assure the community that this valuable and incomparable remedy is composed exclusively of ingredients of a vegetable character, the action of which on the human system is mild and benign, and adapted to all ages, from the infant to the adult, and to every variety of temperament and constitution. Thousands are trying it daily with the happiest results, and all are unanimous in pronouncing it superior to every Cough Remedy they have ever used, and a source of indescribable consolation to those who are so unfortunate as to become victims to that fell destroyer and exterminator of the human family, PULMONARY CONSUMPTION!

CAUTION.—The genuine has the U. S. Internal Revenue Stamp engraved on the front label.

Laird's "Bloom of Youth."

This Harmless and delightful toilet preparation, for beautifying the complexion, removing tan, freckles, and all other blemishes from the skin, has been in use for 20 years; during that period over two million ladies have used it, and in almost every instance been delighted with it. The Board of Health of New York City prepared an analysis of the "Bloom of Youth," and pronounced the preparation harmless, and entirely free from any ingredient injurious to the health.

—SOLD BY—

DRUGGISTS AND FANCY GOODS DEALERS.

The Trade supplied by **VAN SCHAACK, STEVENSON & REID.**

FOR

INFANTS, DYSPEPTICS AND INVALIDS.

A Concentrated Extract, prepared according to BARON LIEBIG'S FORMULA.

☞ The acknowledged BEST FOOD for Infants, Dyspeptics and Invalids. Acknowledged at the "Foundlings' Home" to be the best food for Children.

TESTIMONIAL FROM DR. W. C. HUNT.

To Messrs. J. & W. HORLICK, 82 & 84 N. La Salle St., Chicago.

41 Clark St., Chicago, Ill., *Feb. 9th, 1875.*

Gentlemen:—After a very thorough trial of your Food for Infants, Dyspeptics and Invalids, I am satisfied that it better answers the requirements of such persons than anything of the kind of which I have any knowledge, and I can confidently recommend it to those who need and are in search of such an article.

W. C. HUNT, M. D.

Sold in large glass Bottles, $1.00 each.

Retail by all respectable Druggists, and Wholesale by

Van Schaack, Stevenson & Reid.

VAN SCHAACK, STEVENSON & REID

Are the GENERAL AGENTS for

HOFMANN'S HOP PILLS

For Fever and Ague, Dumb Ague, &c., &c. Sugar Coated.

Price, 50 *cts. per box:* $4.00 *per dozen:* $40.00 *per gross.*

McLAIN'S VERMIFUGE BON BONS,

Delicious and Effective Worm Candies. Price, 25 *cts. per box;* $1.50 *per dozen;* $15.00 *per gross.*

McLAIN'S CANDIED CASTOR OIL

[Trade Mark for the "Syrup of the Castor Oil Bean."]

A perfect substitute for Castor Oil.

Price 25 *cts. per bottle;* $1.75 *per dozen;* $18.00 *per gross.*

Thirty Years' Experience of an Old Nurse.

MRS. WINSLOW'S SOOTHING SYRUP

Is the prescription of one of the best female physicians and nurses in the United States, and has been used for thirty years with never-failing success, by millions of mothers for their children. It relieves the child from pain, cures dysentery and diarrhœa, griping in the bowels, and wind colic. By giving health to the child, it rests the mother.

To Cleanse and Whiten the TEETH,
To Remove Tartar from the TEETH,
To Sweeten the Breath and Preserve the TEETH,
To make the GUMS Hard and Healthy, USE

BROWN'S

Camphorated Saponaceous Dentifrice

MANUFACTURED BY

JOHN I. BROWN & SONS, BOSTON.

CURTIS & BROWN, Proprietors, New York.

Brown's Household Panacea,

— AND —

FAMILY LINIMENT.

Why will you Suffer ?

To all persons suffering from Rheumatism, Neuralgia, Cramps in the Limbs or Stomach, Bilious Colic, Pain in the Back, Bowels or Side, we would say,

Brown's Household Panacea

and Family Liniment is of all others the remedy you want for internal and external use. It has cured the above complaints in thousands of cases. There is no mistake about it. Try it. Sold by all Druggists.

BROWN'S VERMIFUGE COMFITS,

OR WORM LOZENGES.

Children often look pale and sick from no other cause than having worms in the stomach. BROWN'S VERMIFUGE COMFITS will destroy Worms without injury to the child, being perfectly WHITE, and free from all coloring or other injurious ingredients usually used in worm preparations.

CURTIS & BROWN, No. 215 Fulton Street, New York.

Sold by Druggists, Chemists, and Dealers in Medicines.

Always in large stock at

VAN SCHAACK, STEVENSON & REID'S,

92 and 94 Lake Street, cor. Dearborn, Chicago.

NOVELTY

PLASTER WORKS

MANUFACTURE

Mitchell's Improved India Rubber Porous Plasters,

Both in Sheets and in long Rolls.

Also, Kid Plasters in great variety, such as ***Arnica, Poor Man's, Hemlock, Burgundy Pitch, Strengthening,*** and ***Roborans***, of the following shapes:

Also make all kinds of Plaster Compounds, and sell in pound rolls, such as Emp. Belladonna, Opium, Aconite, Calefaciens, Saponis, Roborans, Adhesivum, Soap, Picis Co., Plumbi, Diachylon, Simplex, Cantharides for blistering, and any other compound desired.

Mitchell's Patent Belladonna Plaster, spread on Jeans, Swansdown and Kid. Also make it on Jeans, poroused. U.S.P. Belladonna, Opium, Aconite, Calefaciens, Saponis, Ceratum Saponis, Roborans, and Adhesivum.

Mitchell's Poor Man's Arnica Strengthening Plaster, spread on Canton Flannel. Applied by the warmth of the body. Best article known.

Surgeon's Adhesive Plaster. Patented. Applied by moisture; in one and five yard rolls.

Ready Cut Adhesive Plaster. Surgeon's Companions.

FINEST IN THE WORLD.

The Trade supplied by

Van Schaack, Stevenson & Reid,

92 and 94 Lake Street, Chicago.

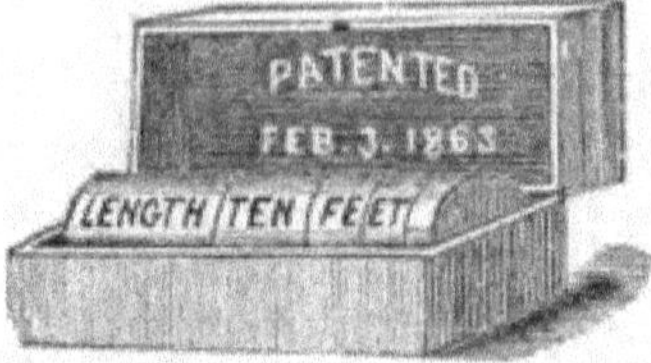

London Isinglass Plaster, a superior article. Mitchell's New Cantharides Plaster for iritating; also with more Cantharides for blistering. Mitchell's Patent yellow surface Corn and Bunion Plasters. Best stickers known. Also white surface Corn and Bunion Plasters, with a great variety of lobes. Mitchell's Tablet Court Plasters, six kinds. The nicest in the world. Novelty Plaster Works Court Plasters, twenty different kinds; very soft and pliable and very adhesive.

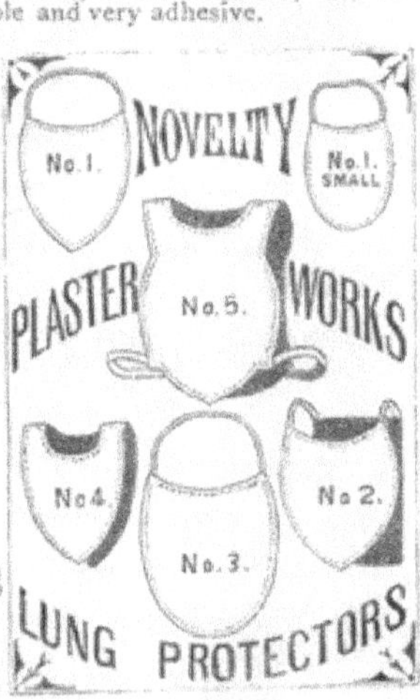

LUNG PROTECTORS

Made of Felt of the finest lamb's wool, wove and laid; the nicest and most desirable ever made.

MANUFACTURERS OF

ALL KINDS of PLASTERS

Sold on the face of the globe,

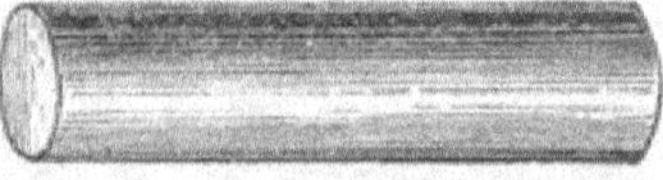

and are constantly issuing

SOMETHING NEW

and beneficial to the human race.

GEO. E. MITCHELL,

Proprietor.

SURGEON'S ADHESIVE PLASTER

The Trade supplied by

VAN SCHAACK, STEVENSON & REID,

92 and 94 Lake Street, cor. Dearborn, Chicago

BOOKS AND STATIONERY.

Conceded to be the Finest Bookstore in the United States.—*The Advance.*

FINE BOOKS A SPECIALTY.

JANSEN, McCLURG & CO.,

IMPORTERS,

AND STATIONERS,

Nos. 117 and 119 State Street, Chicago.

Keep constantly in stock, as heretofore, the largest and best selection to be found in the West, of

STANDARD LIBRARY BOOKS

In the best editions, and in all varieties of binding. ILLUSTRATED GIFT BOOKS, a preference being given to the most truly *beautiful* and *artistic* productions of the English and American presses. SPLENDID ART WORKS, those which reproduce most delicately the master-pieces of the great artists, and the most attractive scenes of nature. **The Choicest Stationery for Ladies,** including fine imported Papers and Envelopes, of all sizes and tints, Engraved Cards, Initial Monograms, Invitation and Wedding Stationery.

SCHOOL BOOKS.

We have a far larger stock than ever before. We solicit orders from the trade, and guarantee the lowest prices.

STATIONERY.

☞ Our facilities for supplying dealers at the lowest rates are equal to those of any exclusively stationery house.

☞ We are the proprietors of the superior PERI PEARL PAPERS, embracing all the different sizes and styles. These papers possess advantages over all others, at the prices. Send for samples.

☞ Buyers and dealers are invited to call upon us before making their purchases elsewhere.

☞ Correspondence solicited from Library Committees, dealers, and all persons interested in the purchase of Books. Price lists, terms, etc., furnished on application.

☞ A cordial invitation is extended to all to visit our new and beautiful store, and examine our stock at their leisure.

JANSEN, McCLURG & CO.

CINCHO-QUININE

Exerts the full therapeutic influence of Sulphate-Quinia, in the same doses, without oppressing the stomach. Does not produce cerebral distress, is nearly tasteless, and less costly than the Sulphate.

CINCHO-QUININE

AS TESTED BY

EMINENT CHEMISTS.

See Extracts.

BOSTON, 1870.

I have analyzed Cincho-Quinine, and find that it contains Cinchonine, Quinine, and other alkaloids.

J. M. MERRICK,
Prof. of Chemistry in Mass. College of Pharmacy.

BOSTON, Dec. 1, 1874.

I have made a careful examination, and have detected Cinchonidine, Quinidine, Cinchonine and Quinine.

S. P. SHARPLESS,
State Assayer of Mass.

CHICAGO, Dec. 16, 1874.

I procured a bottle of Cincho-Quinine, and have tested the same for *quinine*, and find that it gives reactions of that alkaloid very decidedly. G. A. MARINER,
Analytical Chemist, 77 S. Clark St., Chicago, Ill.

CHICAGO, Dec. 11, 1874.

This is to certify that I have submitted it to acknowledged tests for the presence of quinine; obtained reactions of that alkaloid beyond question.

JAS. R. BLANEY,
Analytical Chemist, Chicago.

University of Pa., West Phila., Dec. 21, 1874.

I have tested Cincho-Quinine by the four most characteristic tests for quinine, and have found decided quantities of this alkaloid by each of them. Yours truly,

F. A. GENTH,
Professor of Chemistry and Mineralogy.

University of Chicago, Chicago, Feb. 1, 1875.

I made a qualitative examination for Quinine, Quinidine and Cinchonia, and I hereby certify that I found the alkaloids in Cincho-Quinine.

C. G. WHEELER,
Professor of Chemistry.

BILLINGS, CLAPP & CO. call the attention of the trade to their Ether, Chloroform, various preparations of Gold, Silver, Scale Irons, Salts of Bromine, and fine chemicals used in medicine and the arts.

INDEX TO SUPPLEMENT.

www.ingramcontent.com/pod-product-compliance
Lightning Source LLC
Chambersburg PA
CBHW021348210326
41599CB00011B/796

9783744762663